Lecture Notes in Morphogenesis

Series editor

Alessandro Sarti, CAMS Center for Mathematics, CNRS-EHESS, Paris, France
e-mail: alessandro.sarti@ehess.fr

More information about this series at http://www.springer.com/series/11247

David Piotrowski

Morphogenesis of the Sign

Second Edition

 Springer

David Piotrowski
CNRS-EHESS
Paris
France

ISSN 2195-1934   ISSN 2195-1942   (electronic)
Lecture Notes in Morphogenesis
ISBN 978-3-030-07875-1   ISBN 978-3-319-89848-3   (eBook)
https://doi.org/10.1007/978-3-319-89848-3

1st edition: © Springer International Publishing AG 2017
2nd edition: © Springer International Publishing AG, part of Springer Nature 2018
Softcover re-print of the Hardcover 2nd edition 2018

Printed on acid-free paper

This Springer imprint is published by the registered company Springer International Publishing AG part of Springer Nature
The registered company address is: Gewerbestrasse 11, 6330 Cham, Switzerland

Preface

The responsibility for a text falls upon its signatory, and this work makes no exception. But if the author should assume full responsibility for any shortcomings in his work, its merits, however, deserve to be shared. Indeed, the main elements of the problematics elaborated throughout these pages borrow directly from conceptions produced by eminent thinkers. May we cite and pay tribute to Saussurean structuralism, to the phenomenologies of Husserl and of Merleau-Ponty, to Popperian epistemology, as well as to the morphodynamics of R. Thom and of J. Petitot. But it is not solely with respect to its foundations that the present study exceeds its author. Herein, one will indeed find theses and developments extending theoretical positions that are most personal and which were established during previous works. But in truth, such extensions, in their various forms (in-depth study, amendments, variations), have for source and were prompted by multiple exchanges with research companions, laboratory colleagues and friends who have woven the daily life of the reflections exposed herein. It is therefore my utmost privilege to express my vivid and heartfelt gratitude to all of them and particularly to the inexhaustible interlocutors I found in the persons of J. Lassègue, V. Rosenthal and A. Bondi. As for Y.-M. Visetti, my debt towards him is even greater, as he infused my works with new problematics, particularly as regards the dimensions pertaining to expressivity and to semiogenesis, namely through studies conducted jointly at the crossroads of Merleau-Pontian phenomenology and semiolinguistics. Several passages in the text (particularly in Chap. 7) bear his mark which the reader will easily recognize. In what concerns the text itself, it owes greatly to the remarkable translation work accomplished by L. Roussel upon material which was originally extremely dense and terminologically varied—may she be duly thanked. Finally, it is in its very existence that this work marks my indebtedness to those who contributed in making it possible. Indeed, without the friendly confidence and patient support of A. Sarti who kindly granted access to his collection, this book would not exist. May I reiterate the expression of my sincere and earnest gratitude.

Paris, France   David Piotrowski

Contents

Chapter 1
Introduction

1.1 Specific Perspective

The issues addressed in these pages, the various positions that are defended within, and, more broadly, the main theoretical perspectives which will be pursued find their unity with respect to a morphodynamic model of the Saussurean sign, which, to put it as such, constitutes their center of gravity or intersection point. Reciprocally, it is from this scope of issues that this model of semiolinguistic forms and operations—as a morphodynamic qualification of the internal articulations of the sign such as they emerge from a reading of Saussure—receives its intelligibility and its empirical value.

This morphodynamic architecture has been elaborated upon in previous works (mainly: Piotrowski 1997, 2009, 2010, 2011) but, even if its functional specifications have essentially remained unchanged, it has been progressively readdressed using new theoretical perspectives while being concomitantly placed under horizons of investigation which were at first foreign to it, in full or in part. Hence, for a same formal device, but with a reconfigured system of investigation, the empirical content and scope of the morphodynamics of the Saussurean sign, its epistemological status, its explanatory and descriptive pretensions, as well as the modalities of their validation, have undergone profound mutations.

Initially introduced as a simple model of the Saussurean sign (as a matter of fact, as a schematization, in the Kantian sense of the term), the morphodynamics of the sign was recently enhanced with a phenomenological meaning, at first of the Husserlian type, then Merleau-Pontian, all the while its empirical foundations were being established in the field of the neurosciences.

Such broadenings in the manner of approaching the question at hand and of soliciting regimes and planes of intelligibility that are inherently irreducible, could obviously not be accomplished in the form of a progressive and homogenous extension. That is, they could not be accomplished in the form of a continuous interpretational amplification which, in the end, would attribute to the components

© Springer International Publishing AG, part of Springer Nature 2018
D. Piotrowski, *Morphogenesis of the Sign*, Lecture Notes in Morphogenesis,
https://doi.org/10.1007/978-3-319-89848-3_1

of the morphodynamic model synthetic and encompassing empirical significations and gnoseological value—an interpretational amplification which, *in fine*, would bring the unlikely synthesis of existential phenomenology, transcendental phenomenology, and Saussurean structuralism under a morphodynamic format, all of this intercrossed with neurobiological experimentation. Quite to the contrary, these extensions and enhancements in the system of investigation come with stratifications and ruptures in planes of intelligibility, submitting the morphodynamic construction to fractured insights which in some cases deliver its positive empirical content whereas others reveal its gnoseological limits.

Nonetheless, if the problematic horizons retained here are in part heterogeneous, they are not, by virtue of their ins and some of their outs, without connections. It is a fact that the morphodynamics of the Saussurean sign is closely linked, albeit in various respects, with the theoretical perspectives attached to the names of Merleau-Ponty (henceforth M.-P.) and Husserl. Moreover, through the neurosciences, this model becomes subjectable to experimental testing, thereby attaching itself to a Popperian epistemology.

In this respect, we may recall the main connections which link the theoretical perspectives retained into a network of issues, one that is if not homogeneous, at least cohesive, and we should then look more specifically at how the insights they produce with respect to the morphodynamics of the sign are coherently and in part summatively distributed.

Firstly, we would recall the connection between M.-P. and Saussure: We know to what extent the former acknowledged and borrowed the latter's deep theoretical intuitions for his own purposes, namely under the concept of *diacriticity*. *Next*, we would recall the connection between Saussurean structuralism and morphodynamics, through our own propositions, but especially and more fundamentally, because as established by Petitot, morphodynamics delivers a mathematical determination of the forms of "pure structural intuition"[1,2] so as to open upon a schematization of the categories of the structural episteme and, *in fine*, so as to found its objective value. *Furthermore*, although they follow distinct gnoseological principles and postures, it is indeed the same question of the forms of semiotic manifestations which are addressed by M.-P. and Husserl and it will be necessary to take heed of their findings while keeping their specific values and scopes in mind. *Finally*, concerning the relation between Husserl and Saussure, we will show that it stems from a homology of structure between the morphodynamic architecture of the Saussurean sign and the strata of verbal consciousness recorded by Husserl's phenomenological analysis of the sign. More specifically, we will see this homology can be said to be "surjective" and that the morphodynamics of the sign delivers a system of articulations which is richer and more detailed than what is proposed by Husserl. Be it as it may, as far as this first portrait is concerned, and

---

[1] All translations of citations which are referenced in French in the bibliography are our own unless otherwise specified.

[2] Petitot (1985), p. 290.

with regard to the number of shared connections, Saussurean theory seems to prevail. In this overview of the overlappings between theoretical systems, we will opt to retain it as a "device of reference".

We therefore have theoretical layers which respond to one another and which partially overlap, but these various overlappings do not however all lie on a same plane of object determination, and, hence, do not involve the same epistemic horizons. It is as if the recourse to such or such perspective regarding the problematics had the effect of "polarizing" the conceptual device of "reference", and of "projecting" it onto specific spheres of intelligibility, which entails as a first consequence to bring to the forefront certain notions or options which other perspectives, favoring different views, may neglect.

In order to more specifically approach this conjectural configuration, we may first examine, in its main lines, the overlappings between the views of M.-P. and of Saussure, or, more specifically, Saussurean structuralism viewed through the prism of a Merleau-Pontian phenomenology.

M.-P. thus seizes certain concepts of structural analysis as introduced by Saussure, but of course without simply replicating them. As previously stated, the Merleau-Pontian options "polarize" the Saussurean structural conception, which has for main effects, on the one hand, of re-informing the descriptive content and the epistemic value of the concepts retained (in what concerns us, *difference* is promoted as the concept of *diacriticity*), and, on the other hand, of redistributing certain conceptual hierarchies. The phenomenological promotion of Saussure accomplished by M.-P. thus also constitutes a rearrangement of his structural perspective.

But this rearrangement is not without entailing major consequences inasmuch as it has the effect of bringing to the forefront a certain order of semiolinguistic factualities, namely that of *expression*, that is, the fact of a *sensible presence of meaning*. This order of factuality will also have been acknowledged by the Saussurean perspective, even from the onset, but without being recorded as such, in its paradoxical essence: It will only retain a polarized form of it (the sign as the union of a signifier and a signified), without addressing it in a satisfying manner. We know that in order to overcome the aporia of the semiotic fact, Saussure will opt to workaround the issue precisely by approaching the matter from the angle of a theory of *value*—albeit unsuccessfully since this path will eventually lead to another impasse.

The Merleau-Pontian phenomenological revisiting of the deepest of Saussure's intuitions, that of differential identity, thus goes hand in hand with the acknowledgement of the primordial character of the expressive fact which seems to trace an absolute boundary in what concerns Saussurean theory.

However, upon closer examination, it appears that the situation is more complex and richer. Because if, in his reformulation in terms of *value*, Saussure despairs to account for the fact of expression (to describe and explain it), it is because he envisions the question along a gnoseological format which is precisely that of a conceptual system which is univocal, regulated and which delivers the complete and finalized determinations of a class of empirical phenomena—a format to which he returns in the laudable view of clarifying the principles, concepts, and objects of his discipline which were at the time cruelly lacking foundations. It is therefore on another gnoseological plane that the intelligibility of the expressive fact must be sought, and, as it is, we know that existential phenomenology provides access to it, at the same time as it has the effect of marking the limits of the Saussurean device as regards semiolinguistic knowledge.

It remains that the specific characterization that Saussure provides for the notion of value, in conjunction with other fundamental concepts of his structural approach, lends itself to morphodynamic modelization. Also, this morphodynamic apparatus finds itself to be structurally paired with the articulations of verbal consciousness uncovered by Husserlian analysis, thereby endowing itself with phenomenological signification.

This is therefore to say, following another point of view, that the Husserlian polarization of Saussurean structuralism crystallizes a conceptualization of semiotic forms which lends itself to formal representation, but which, on the other hand, leaves facts of expression beyond the field of semiotic knowledge. In fact, phenomena of expression, although they may have been recognized and registered as such in Husserl's first phenomenological investigations, will have been subsequently abandoned to the benefit of a theory of verbal consciousness which partially unravels the orders of the sensible and of the intelligible.

In a certain way, therefore, the two phenomenological perspectives mobilized here operate a sort of spectral analysis of Saussurean theory: Existential phenomenology reveals its expressivist component, but without there being a conceivable conceptual representation, while transcendental phenomenology retains, on the basis of a structural homology, its morphodynamic architecture, to which the fact of expression is foreign.

After having taken heed of the partial overlaps between Saussurean structuralism and, respectively, the Husserlian and Merleau-Pontian approaches, we are quite logically and directly led to pose the question of the overlapping between these two phenomenological perspectives themselves. A question which, when addressed in itself, extends greatly beyond the scope of our argument, but which we may nevertheless approach through the narrow angle provided by what is revealed at the intersections between these two phenomenologies and Saussurean structuralism. In other words, the question of the overlapping of the two phenomenological schools mobilized herein will be restricted to the semiolinguistic field.

We could be led to think, based on the above and logically speaking, that since they retain mutually exclusive components from the Saussurean conceptual apparatus, the two phenomenological approaches hereby summoned in the arena of semiotic factualities do not overlap. But this is far from being the case, and it is something which requires to be explained.

This is because, as we will see in detail, the primary considerations of Saussure, Husserl, and M.-P. regarding the essence of the semiotic fact, in particular regarding its internal configuration as phenomenon of expression, intersect perfectly. But if these three theoretical stances share the same starting point, the horizons formed by their specific problematics differ radically in terms of format and of gnoseological requirements. And the investigations, conducted by each according to a specific set of epistemic references, will have concerned planes and orders of factuality that were retained because they were specifically adapted to the projects and methods favored by each. What may have constituted an obstacle for one approach, as it projected a certain conceptualization of the semiotic fact, could be maintained by the other and addressed in its first principles—and non-reciprocally so. In fact, we will see that Merleau-Pontian phenomenology is "overarching" since it reflects the set of facts which interest all three approaches.

Undoubtedly, on the one hand, Husserlian phenomenological analysis restitutes the fact of expression in an impoverished form, where the poles formed by the signifier and the signified, distinct albeit indissociable, are each situated on their own specific level. Also, on the other hand and also undoubtedly, Saussure, being unable to conceptualize the fact of expression in its very essence, and attempting to approach it from the angle of a theory of value, suggests a functional architecture which, as in Husserl's description, admits a polarization of the *indivisible* semiotic fact by instituting within it the two faces formed by the signifier and the signified. But what these two approaches deliver, using specific conceptual devices, is not however absent from Merleau-Pontian reflection, which recognizes its relevance with respect to its own gnoseology, to its own descriptive categories, but without delivering its internal principles and forms, and without opening the possibility for its conceptual representation.

In short, we see that the disjunction between these three viewpoints is not radical: It does not proceed from the incompatibility of the theoretical insights upon a same empirical subject matter, but from a layering of the gnoseological planes which need only to be taken into consideration in order to reinstate unity. Husserlian phenomenological analysis, which is, so to speak, the phenomenological echo to the Saussurean theory of signs, is not incompatible with the Merleau-Pontian approach which recognizes the order of facts which Husserl and Saussure will have addressed, but does so according to its own terms.

And in the end, this fabric of mutual corroborations, despite being distributed over distinct planes of intelligibility, forcefully proclaims the validity of these various analyses: The phenomenology of the sign produced by Husserl and structurally affine to the functional device of the Saussurean sign is not in conflict with Merleau-Pontian views and analyses. And if these conceptualizations do not exactly overlap, it is because each is steered towards a specific ideal of knowledge, which will have led the ones to concede in extension and the others to concede in determination.

But there is more, because if a "positive" phenomenological determination of the Saussurean device is thus conceivable, it finds, in addition, an extended sense on the plane of intelligibility which lies above it. Because it appertains just as much to the Merleau-Pontian manner of approach, the morphodynamics of the sign will find its place with respect to originary facts of expression: We will show that it partakes of a general problematics with respect to semiogenesis of which it restitutes a specific phase. More specifically, we will see that the morphodynamics of the sign constitutes a reduced form of the expressive fact as it results from a projection of the latter onto the gnoseological plane of conceptual and formal representations. In other words, it exposes this portion of the expressive fact which lets itself be grasped and put into relation with the format of a positive, explicit, and univocal determination.

It is this manner of approaching the issues at stake, at the core of which lies the morphodynamics of the Saussurean sign and from which it receives on a variety of levels its meaning and legitimacy, which we will endeavor to establish. To this end, we will begin precisely at the starting point shared by the three perspectives mobilized herein, that is, with the empirical reality of the fact of expression, conjointly with the acknowledgement of a specific ontological region.

But before doing this, it is necessary to consolidate and to amplify our perspectives regarding the issues at hand, first by broadening our scope, but most of all by introducing some measure of necessity.

1.2   General Perspective

We have stated that the morphodynamics of the sign lies at the core of our musings, and we have seen that the transcendental and existential phenomenological approaches demonstrate its empirical relevance and contribute to its intelligibility. They do so *positively*, in terms of phenomenological meaning. They also do so *negatively,* in terms of gnoseological limitation, as well as *dually,* by situating it within a global semiogenetic process.

Presented in this manner, our line of investigation and our theoretical positions, as defendable or even worthy as they may be, risk appearing contingent and limited. Their perimeter seems indeed to be circumscribed too narrowly around highly specific issues, interrogations, and concepts, thereby risking to confine eventual benefits.

In order to prevent such objections, and before getting into the heart of the matter, it will be necessary to show that our "triangular" approach—ranging between the vertices formed by the positions put forth by Saussure, Husserl, and M.-P., and with respect to which the morphodynamics of the sign occupies the central position—has a gnoseological scope which exceeds its strict perimeter, particularly as it contributes in elucidating the nature and forms of semiolinguistic knowledge. Our "triangular" approach to the matter at hand, in its most general scope, reconfigures the epistemological musings of semiolinguistics, not by introducing new components, but by establishing links between its "pre-existing" ones which determine the nature and the descriptive orientations of knowledge regarding meaning and signs. What is put forth therefore takes the form of progress rather than of radical change and, in this view, we will begin with a quick review of the main elements forming the epistemological landscape of semiolinguistic science.

1.2.1 Epistemological Landscape

1.2.1.1 Elements

The question of the nature and scope of semiolinguistic knowledge remains largely open. The most variegated positions have been defended, without any single one imposing itself, and, at the end, the resulting consensus takes the form of a pragmatic concession: Certain fundamental difficulties are to be acknowledged, but, although the foundations and horizons remain uncertain, the theoretical and descriptive work is to be pursued.

That is to say, while not proving it, that semiolinguistics acknowledges itself to be an empirical science—and everything seems to confirm this, be it the institution or the practices by virtue of which the gathering and careful observation of data nurture a descriptive finality. More specifically, the status of empirical science claimed by semiolinguistics would depend on the hypothesis of a class of factualities located both at its source, as primary data to be elucidated, and at its end, as a touchstone (instance of assessment) of the theoretical devices. These two points are those which need to be further delved into.

There is first this indubitable observation that it would be impossible to reflect upon the "objects" of semiolinguistic concern if they had not previously been delivered through the experience of speakers and of cultures.

If the "objects" of linguistics require to be "encountered", it is because they are not accessible by the sole means of thought. Linguists, for instance, do not have access to any *a priori* source from which to infer the existence of words or of classes of words, or to deduce specific regimes of functioning (passivization, pronominalization...). As argued by Auroux, "There has been no representation of the ergative for as long as an ergatively-constructed language hadn't been encountered, nor has there been a concept of penguins prior to seeing a specimen of this likable animal".[3] More broadly, "the history of linguistic discoveries [...] globally attests to the empirical character of the sciences of language",[4] and, correlatively, linguistic matter is presumed to be endowed with an autonomous reality and independent existence—at least one unlinked from the theoretical devices which take grasp of it. This position is clearly that of almost the entire community of linguists—as indicated by the following statement by R. Martin: "linguistics is above all an empirical field: it studies a preexisting object—language and forms of speech—and the linguist's primary objective is [...] to describe that which reality proposes".[5]

---

[3] Auroux (1998), pp. 210–211.
[4] Ibid., p. 210.
[5] Martin (2002), p. 16.

If this first point, that of a liminal presentation of semiolinguistic factualities, is hardly arguable, at least at this low level of precision (it would be necessary to discuss the forms of this presentation—we will return to this), the second point, concerning the evaluative function of these faculties with respect to the theoretical devices meant to describe them, is immediately and eminently problematic.

1.2.1.2 The Epistemic Circle

It is at this stage that the very fundamental question arises regarding the relations between the theoretical device and the factualities ingenuously delivered "as such" to observation. This question, which belongs to the great chapters of "classical" philosophy, comprises two dimensions. On the one hand, that of the conceptualization of empirical factualities, and on the other hand, that of the relation between qualified factualities and theoretical systems.

For the first dimension, we may simply recall this truth of a radical exteriority, of an insurmountable gap between "empirical" facts and concepts, each stemming from its own sphere of existence, its own order of necessity and of connection (causality vs. inference), and being mutually irreducible—a separation which cannot be overcome "in itself". Overcoming such a gap would thus suppose the participation of concepts to the constitution of facts. In keeping with tradition, we could refer here to Kant,[6] Frege,[7] Husserl,[8] or Popper.[9]

It is therefore appropriate to acknowledge, in general and once and for all, the "muteness" of the facts with respect to any conceptual language, and, in particular, the muteness of the facts of language with respect to any system of linguistic qualification. This observation is unanimously shared by linguists according to whom, likewise, facts do not bear their own qualifications, but receive them from the descriptive devices which are used to grasp them. Here again, we find ourselves within the realm of shared evidences, and in order to demonstrate that there is a

---

[6]"Existence is the absolute position of a thing and thereby differs from any sort of predicate, which, as such, is posited at each time merely relatively to another thing" (Kant in Walford and Meerbote 2003, p. 119), also quoted in Philonenko (1989, p. 39), or "the subtlest of concepts never abolish or produce an existing something" (Kempf, *Introduction* to Kant (1972/1763, p. 10).

[7]"The recovery of a thing by a representation would only be possible if the thing were, also, a representation" (Frege 1971, p. 172).

[8]"Natural reflection ['natural' understood here in the sense of 'naive realism'] on the relation between knowledge, its sense, and its object almost inevitably makes [mistakes]" (Husserl 1999, p. 18) because it is impossible to understand "what it could mean for a being to be known *in itself* and yet *be known in knowledge*" (Ibid., p. 23).

[9]"[I]t was [...] rightly felt that *statements can be logically justified [and refuted] only by statements*" (Popper 2005, p. 21).

consensus, it would be indeed easy to present numerous citations, ranging from Saussure[10] to our contemporary thinkers.[11]

In what concerns the second dimension, it opens onto the question of the Popperian principle of *refutation* (or *falsification*).

This principle has two roots.

On the one hand, it has a historical root since it was meant to provide an answer to the now overcome problem of the distinction between science and non-science. We will say nothing more about this. We should note, however, that although the principle of refutability entails more problems than it resolves[12] and that it has also somewhat disappeared from the debates of contemporary epistemology, its value remains intact inasmuch as it reflects a certain intellectual attitude which we can say to be constitutive of the scientific spirit[13]—in the sense that, by virtue of this principle, it is a matter of conceding to things the power of disavowing the discourses which would not admit them.

On the other hand, it has a logical root, because the connection between the always particular observations or results of experimentation and the laws submitted to experience (professed to be universal)—a connection which cannot be inductive (passage from the particular to the general)—takes the form of "deductive falsification". Such falsification consists, on the double basis of (i) a consequence stemming from a universal law and from particular empirical conditions and (ii) an observation, under these same conditions, contravening to the previous consequence, in concluding that the law in question is invalid.

From this standpoint, experience has but a refutational role, and, along the same line of thinking, the refutational dimension will be erected into a characteristic of

---

[10]"The object is not given in advance of the viewpoint: far from it. Rather, one might say that it is the viewpoint adopted which creates the object" (Saussure 2013/1916, p. 9).

[11]To cite but a few of them, beginning with illustrious figures: Hjelmslev, who almost replicates the Saussurean formulation: "Is it the object which determines and affects the theory, or is it the theory which determines and affects its object?" (Hjelmslev 1968, p. 23). Moreover: "as long as the method has not been applied, no so-called obvious facts will exist (those which some philosophers of language like to use as a starting point by appealing to naive realism, which, as we know, does not hold up to scientific examination" (Hjelmslev 1985, p. 72). Likewise, Benveniste (1966, p. 119): "[D]escription first of all necessitates specification of adequate procedures and criteria, and that, finally, the reality of the object is inseparable from the method given for its description" (Benveniste 1971, p. 101). Then, among our contemporaries: Reflecting upon what is a linguistic fact, Martin (2002, p. 22) notes that "the linguistic fact never goes beyond more or less conventional decisions."

[12]Location of the faulty component, introduction or adjustment of ad hoc protective concepts, conventional core....

[13]The falsifiability requirement, in its most general principle, is admitted as a character of intellectual probity: "most philosophers appear to now be persuaded that there exists no universal criteria of scientificity [...] though it is not uncommon to hear the same people complain that a theory [...] is not clearly testable, which presupposes that they accept the idea that if testability does not represent a necessary and sufficient condition of scientificity, it constitutes at least a desirable methodological ideal [...] testability being a virtue, and irrefutability, a vice" (Boyer 2000, p. 166).

reality. Correlatively, an empirical science will be characterized, apart from enunciating synthetic propositions, by its production of refutable statements by means of experience or observation.

This having been conceded, the previously mentioned difficulties regarding the relation between factualities "as such" and their theoretical qualifications do not fail to resurface, this being, within this particular context, in the form of a "trap" of circularity (or of self-consistency).

The problematic conjuncture, which is very clear and unanimously acknowledged, is the following: In order for observed facts to interface, in a refutational mode, with theoretical descriptions, it is indeed necessary for these facts to be established in a format that is compatible with the system of qualification of the theoretical framework being tested. Because, clearly, if such facts remain considered solely as they are "in themselves", or if they receive qualifications that are foreign to the theoretical system meant to account for them, then it is not possible to consider any connection to the propositions formulated within this system, including a refutational connection. It is therefore fitting to characterize the facts following concepts which are compatible with those of the theory destined to be experimentally tested.

Two scenarios are then possible. Either the qualifications used to account for the observations are precisely those defined by the theoretical apparatus (of which the empirical value is being evaluated), or it is a matter of qualifications which are "coterminous" with respect to the (main) theoretical framework being tested, but which are not without links to such a framework (and this will be the Popperian solution).

The first scenario entails falling into the circle of self-consistency and into vacuousness. Indeed, when the concepts meant to describe an empirical domain have a legislating power by which they fully configure the results of observation, then the latter have nothing but a corroborative effect. Such is how, by matter of force, the trap of circularity closes in: When the data are nothing but concrete counterparts to concepts, theoretical devices will necessarily see "correctly"— nothing external has the power to infirm them, since they thus subsume the legality of what can be confronted with them.

In what concerns the study of language, in the almost unanimous opinion of linguists, the linguistic sciences are elaborated accompanied by a risk of circularity. For example, we have Ducrot (1995) who recognizes that "it is not possible to distinguish the hypotheses serving for observation from those serving for explanation. To put it shortly, linguistics creates its object at the same time as it observes it", or Martin (1978, p. 5) who states while discussing the case of generative grammar, that "not without any reason whatsoever, the pitfall of tautology appears […]: the 'ideal speaker-listener' is located at the starting point but also at the end point of the model", or Culioli (1999, p. 162) who, after having established the levels of representation involved in linguistic analysis, signals the existence of level-to-level interactions, hence "the risks of circularity and the illusionary explanations which support themselves upon that which is already the product of a buried operation."

1.2.1.3 The Popperian Solution

The solution[14] proposed by Popper consists in articulating all theories into (at least) two parts that are "sufficiently" independent in terms of the concepts involved in each, but which are conjugated into a unitary theoretical apparatus. The first part, described as "auxiliary"[15]; serves as a point of observation: it entails grasping "facts" in light of a theory to which greater credibility is conferred than to the so-called "main" theoretical component which contains, for its part, the concepts which are specific to the order of facts under study. The statements produced within the auxiliary framework (called "basic statements"[16]) escape the main system's formatting, but are apt for logical confrontation (validation or refutation) with the statements of similar nature stemming from a theoretical deduction regulated following the main concepts. The auxiliary system can, from this point of view, be considered to serve as an "empirical foundation" for the main component.

Therefore, let's consider for instance an empirical factuality F and two theoretical sub-systems T_1 and T_2 conjugated into a unitary framework $T_1 \otimes T_2$. From the standpoint of T_1 (resp. T_2), F can be described as F_1 (resp. F_2). It is clear that F_2 (resp. F_1), elaborated following the T_2 regulator concepts (resp. T_1) does not have the power to disconfirm T_1 (resp. T_2). On the other hand, F_2 has the power to disconfirm $T_1 \otimes T_2$. Such is the case when $\sim F_2$ (negation of F_2) can be deduced from $T_1 \otimes T_2$. If we attribute greater credit to T_2 than to T_1, then the refutation of $T_1 \otimes T_2$ will be carried over T_1.

This solution to the problem of a (constitutive) assimilation of the theoretical forms with the empirical data has been reformulated by Milner (1989, p. 127) in the following terms:

> in order for an instance of refutation to be possible, the experimental resources [that is, following the preceding formulation, a protocol for manipulation and description that is regulated in function of the concepts of T_2] should enjoy logical independence from the propositions being tested [that is, the propositions of T_1]. This independence would of course be ensured if there existed raw observations implying no theories. […] Let's concede, however, as it seems to have been established by epistemology and even more by the history of sciences, that there is no such thing as a raw observation, that there is no observation which is not itself founded upon a theory. [Therefore,] independence in the second degree would suffice: It is only necessary for the propositions of the theory which serve as foundations for experimentation [that is, T_2] to be independent from the proposition being tested [that is, T_1].

---

[14]More precisely exposed in Piotrowski (2009, pp. 112–121).

[15]Terminology borrowed from Granger (1992, pp. 267–268).

[16]Which correspond to the "data models" of contemporary epistemology, cf. Bitbol (1998, p. 47).

1.2.1.4   An "Isolated" Science

The question is to determine whether linguistics satisfies such a coupled configuration in coordinating an observation device (T_2) that is independent from the "main" theoretical system (T_1). Regarding this point, the proposition by Milner (1989) is clear. Here is the essence of his argument:

> It is most probable that the manipulation of linguistic examples has the properties of experimental manipulation [but] these examples [and their variational manipulation] all incorporate a *minimal grammar*. It is possible [...] to treat this minimal grammar as an instrument for *observation*, [...] but doing so would be a simplification [which needs to be rectified]: a minimal grammar [...] is still a grammar, [and it therefore constitutes] an embryonic linguistic theory. The consequence of this is that *the instance of observation* [minimal grammar] cannot be made fully *independent* from the linguistic theory itself" (Ibid., p. 128). "Also, *circularity* can never be fully eliminated: Any example of language, as it enables linguistic reasoning, already supposes linguistic reasoning" (Ibid., p. 129). "In short, in linguistics, there are *experiments*, but there are no pure observations, [that is,] what is deemed an observation always includes a fragment of a linguistic theory [...], and this means exactly the following: that linguistics has no other recourse than itself for establishing the distinction between linguistic possibility and impossibility—it does not enjoy such a thing as the instance of *independent observation* provided by the structure of the *spatio-temporal* event. [...] Now the *boundary* between linguistic possibility and impossibility constitutes a *concept* in itself. Hence the *circularity* [already] described" (Ibid., p. 130). "Linguistics [is] *scientia unica*: [...] it cannot base itself on any science which is logically prior and locally independent while constructing its modalities of observation and there is no other science than itself which talks about the data that are relevant to it" (*Ibid.*, p. 131). "Never does a synthetic proposition of linguistics take into account [...] any particular proposition from biology [or from any other science]" (Ibid., p. 133).

Unable to address its phenomena otherwise than by making them comply with the principles it endows itself with, linguistics would therefore be condemned to the vacuousness of self-consistency.

It goes without saying that this thesis has been contested, but one must also acknowledge that the counter-arguments put forth are far from convincing, and that in order to enable the sciences of language to overcome this impasse, it will most probably be necessary to proceed by other means.

Let's recall for memory's sake the counter-arguments proposed by Auroux and by Lazard:

From Auroux's point of view, linguistics enjoys independent observatories, these being grammars and dictionaries inasmuch as "[if] the facts of language present themselves in a dispersed and disparate manner [...] their homogenization and collection [...] takes place in grammars and dictionaries, which *ipso facto* become observatories".[17] Moreover: "We are perfectly capable of identifying multiple linguistic observatories: writing, texts, other languages, corpora of examples, dictionaries, etc."[18]

---

[17]Auroux (1998, p. 168).

[18]Ibid., p. 215.

This point of view is fully contestable inasmuch as an observatory, in order to function as such, is required (i) to be a *qualification* system, (ii) to be *independent from the forms of linguistic analysis*, and (iii) to be *connected* with the "main" theoretical component (which comprises the hypotheses to be tested) within a global theoretical unit.

Now, "writing, texts, […] corpora of examples" are not qualification systems but sums of which the constitution, furthermore, proceeds from linguistic hypotheses.[19] Moreover, regarding the grammars and dictionaries, of which nothing ensures that their systems of qualification will satisfy the minimal principle of non-contradiction, it is unknown whether they bear any relation with the other descriptive devices to be evaluated.

Regarding this controversy, Lazard (2001) also takes position against Milner's thesis. For Lazard, *communication studies* would indeed allow to apprehend linguistic data following a secant angle of analysis, constituting the "auxiliary" system of description on the basis of which observations that are independent from any theoretical option could be conducted. For example, a confusion between phonemes could have for consequence the failure of the communication process, and, for Lazard, "such a failure is an objective phenomenon, easily observable through the reaction of the listener, directly in relation with a fact of speech, and nevertheless belonging to a field completely foreign to linguistic theory."[20]

Broadening this point of view, Lazard proposes to apprehend any sort of linguistic "distortion" in terms of communicational yield, in order to obtain thereby a descriptive basis attached to specifically linguistic phenomena (in this case the opposition between receivable/non-receivable), but to do so in accordance with a logic which is external to the linguistic system. Indeed, for Lazard, the reactions (miscomprehension, surprise) to communicational perturbations due to linguistic inconsistencies or mistakes "[…] are objectively observable and do not emanate from the [linguistic] black box, or are at least anterior to any judgment regarding acceptability."[21] Even more explicitly: "these reactions to a faulty utterance do not pertain to the field of linguistic theory, but to the connected field of communication." Which enables Lazard to conclude that "if there is a neighboring field, but which remains distinct from linguistics and which takes interest in linguistic phenomena from a different angle, it is the study of communication".[22]

Lazard's argument is indeed very convincing. But at a closer glance, a difficulty appears, because it is not certain that communicational activity proceeds from its own specific order of objectivity. Most probably, in Lazard's perspective, the consequences of a perturbation in communication are considered as *objective* data:

---

[19]Regarding this point, the consensus is obvious, as stated namely by Dalbéra (2002, p. 9): "The corpus can only be a construct and […] its construction forms an integral part of the theoretical lens through which the linguist intends to apprehend reality."

[20]Lazard (2001, p. 17).

[21]Ibid.

[22]Ibid.

"the reactions (miscomprehension, surprise) [...] are *objectively* observable..." and, furthermore, "this failure [of communication] is an *objective* phenomenon..." We will not contest that such reactions are factually observable: They are encountered or experienced through the direct occurrence of communication. But nothing ensures the availability of a well-regulated conceptual qualification satisfying the criteria of an auxiliary component and by means of which only a "raw" fact would be established as an authentic objective datum. It would also be appropriate to reflect upon the existence of specifically communicational forms of objectivity, that are independent from linguistic theorization all the while being connected to it— inasmuch as the linguistic data would pertain to such theorization on a certain level —and which would thus be capable of serving as an "auxiliary" basis for analysis.

However, communication studies define themselves as an "interdiscipline". The fact is that communicational processes are not autonomous facticities, but elaborate themselves within spaces of symbolization, of power, of social interactions, etc., each constituted following specific regimes of functioning: these are "processes which can be located in all spheres of human activity".[23] It follows that "only a scientific undertaking based on an interdisciplinary approach can claim to describe what is covered by a concept [communication] which presents such a profusion",[24] and, as a corollary, that "the study of communication bases itself on constituted sciences and discourses ranging from linguistics to semiology, from sociology to psychoanalysis..."[25]

Following this view, "communicational" objectivity is therefore not autonomous: It is elaborated on the basis of previously determined materials, including in terms of linguistics and, moreover, it would by no means be capable of constituting a framework of qualification serving to evaluate the theoretical systems upon which, precisely, it rests.

1.2.1.5   The Solution, and Other Problems

The aforementioned obstacles to the elaboration of an authentic linguistic science are very real and seem difficult to overcome. But it would be too hasty to conclude that they represent a definite impasse: There exists indeed at least two other paths for overcoming the obstructions stemming from the apparent "isolation" of the linguistic sciences.

The first, oriented towards the most recent advances in the neurosciences, is that of connecting the linguistic qualifications with the neurobiological correlates of linguistic processing—correlates of which the observation is, at least in part, independent from any linguistic hypothesis, and which should therefore provide the

---

[23]Ollivier (2000, p. 27).
[24]Ibid.
[25]Ibid., p. 20.

angle of observation which the linguistic sciences may have lacked until now—We will return to this.

In what concerns the second approach, it consists in returning to the originary forms of presentation, that is, to the constitutive forms of the phenomena under consideration—forms which therefore administer observations as "immediate" knowledge (or intuitions), and which are thus to be instituted as "auxiliary devices" in order to confer them the privileged status which is theirs within the method of the empirical sciences, because "in a last resort, it is always on the basis of phenomenological statements that theories are rejected or accepted."[26]

During the process of the construction of linguistic knowledge, the determination of the forms of the linguistic phenomenon thus represents an issue of utmost importance: Following the same relation as kinematics with respect to dynamics, they are likely to constitute the first angle of observation on the basis of which the determinations resulting from the theoretical devices are to be confronted. We will emphasize that, furthermore, the phenomenological question overlies the issue of theoretical architectures inasmuch as it proceeds from the liminalities of any empirical investigation. Indeed, it is the job of any science concerned with facts to clearly delimit the field of factualities it endeavors to study: If the ambition of the linguist is "to constitute the descriptive framework for any possible language, [it can only be accomplished] once the facts of experience chosen to define the field have been retained".[27]

But if the question of the regimes of constitution of linguistic manifestation is eminently crucial, it is apparent that it remains open—at least if we refuse, in order to maintain the essential character of linguistic phenomena, to reduce the signifier to the format of the symbol, that is, to see it as a simple concrete marking (graphical or acoustic) of which the identification proceeds from a type/occurrence relation or from a sensible perception, and which conventionally redirects consciousness towards an object of signification (Husserl's "indicative" sign, cf. 2.2).

For example, reflecting upon the sense to be given to the expression "to encounter" when it is a question of linguistic occurrences, and while discussing what an empirical proposition may be in linguistics, Milner (1989, p. 50) asserts that: "the answer is apparently clear: the fact in language X is to be encountered [or not] in time and space." But he continues by acknowledging that "we do not really know what *to encounter* means at such a time"—or, at least, that since it is a matter of linguistic data, the [expression] "to encounter" does not take exactly the same meaning as it does in the "concrete" empirical sciences. Likewise, Auroux (1998, p. 113) who, admitting that "sooner or later, the fundamental question must be asked: What exists in terms of language?", continues by noting that "[if] in the ordinary sense of existing, there only exists that which is located in time and space [then] in terms of language, the problem is to know whether this ordinary sense of existing is sufficient", following which he furthermore recognizes, without

---

[26]Boyer (2000, p. 181).

[27]Granger (1979, p. 200).

providing an answer, that "the hypothesis of insufficiency is probably the most widespread among both philosophers and linguists."

This being acknowledged—and to conclude: The phenomenological question rightly occupies a cardinal position within the landscape of issues pertaining to the linguistic sciences. The recognition of the forms of linguistic manifestation is essential inasmuch as it delivers a frame of determinations to which all theoretical devices must refer, following one mode or another (assimilation, confrontation), and this in order to serve, as it operates as an absolute reference for linguistic knowledge, as much as a touchstone for empirical evaluation than as a bedrock for intersubjective evidences.

1.2.2   Reconfiguration of the Problematics

Now, we will show that Saussurean semiolinguistics, in any case as it is restituted through the morphodynamics of the sign, delivers a mathematical determination of the form of manifestation of semiolinguistic faculties, in a manner as to thereby rightfully constitute a system of observation for phenomena of such nature. The phenomenological signification of Saussure's theory would thence enable to stay clear of epistemic circularity, which we have deemed to constitute a threat for semiolinguistic knowledge.

But the situation is now somewhat bizarre. This is because the morphodynamic device, in addition to the phenomenological signification which will be recognized for it, possesses, or claims to possess, an objective content. Also, the morphodynamics of the sign happens to make the forms of semiolinguistic objectivity and of semiolinguistic phenomenality coincide.

This point, which is very delicate, obviously requires attention. If it were a matter of addressing it face on, we would need to refer to Kant's *Third Critique*, specifically regarding the teleological principle and regarding reflective judgment, which would precisely enable to address the intrinsically meaningful character of morphologies: A morphology signifies by itself in that it delivers to intuition (i.e. as a phenomenon) the very principle of its own synthesis, that is, the concept which gives it meaning as an object.

We will choose an indirect, less philosophical approach, but one which will enable us to maintain empirical bases. Specifically, we will put aside for a few moments the objectifying pretensions of the morphodynamics of the sign in order to only retain its phenomenological signification. And it is at the term of the discussion thus engaged that we will return to the question we have just cast aside.

So let's consider the morphodynamic apparatus. We recognize that its validity requires examination. And we would of course want this validity to rest upon more tangible foundations than the "immanent evidences" to which phenomenological introspection gives access. Ideally, we would want experimental and/or empirical foundations.

It is at this point that the neurosciences intervene.

Because it is a matter of the neurobiological validation of a phenomenology of signs, we will show that, on the one hand, the circumstances surrounding the generation of the "N400" evoked EEG potential reinforces an intentionalist description of verbal consciousness, and, on the other hand, that the system of the strata of verbal consciousness given by the morphodynamic model, a system which "surjectively" qualifies the Husserlian phenomenology of signs (in that it is more finely articulated), is supported by the observations of this same evoked potential.

But at the same time as we provide ourselves with the means for empirically guaranteeing our phenomenological description, we see a singular and problematic epistemic configuration taking form.

This is because if the neurosciences are of interest for semiolinguistics, it is well beyond their usage such as considered above. Indeed, since over three decades, the neurosciences have provided a support for semiolinguistic investigation in as much as they offer an observation device of the neurocognitive processings correlated with semiolinguistic practices—a device which is in part independent from the forms of objectification as regards semiolinguistic phenomena, and which therefore likely to found, in the same capacity as phenomenological description, the possibility for an empirical confrontation of the theories which account for them, and thus to establish the empirical character of semiolinguistic knowledge.

However, this being said, semiolinguistics would seem to enjoy *two* points of observation: *two* empirical frames of reference, each calibrating, in accordance with its own order, the observables of a single science, with the one pertaining to phenomenology, and the other to the neurosciences. If these two frames of reference were mutually irreducible, the situation would be untenable simply because, in such a case, the two empirical frames of reference would support two orders of semiolinguistic objectivity.

But we are not in this situation, inasmuch as the neurobiological frame of reference administers an empirical basis for semiolinguistic knowledge as well as for its phenomenological description—this not being the case with the phenomenological frame of reference of which the foundational power is limited to semiolinguistic knowledge.

This is therefore not the hierarchical configuration we have just evoked, where the theoretical pole, situated at the vertex, would seek mutually incompatible guarantees in two distinct empirical roots. Rather, we have a more "linear" conjuncture which, following the successive relations of empirical foundation, at first aligns at the summit the theoretical pole and then the phenomenological pole. Finally, at the base, it aligns the neurobiological pole.

But such a linearization is not perfect, because the theoretical pole occupying the vertex of this column also shares a direct relationship with the neurophysiological pole at the base—a relationship which must be discussed at once.

We will first concede that the existence of neurophysiological correlates of sign-phenomena at the level of semiolinguistic functionings is by no means surprising. The operations accomplished over signs indeed suppose the moment of their presentation. It is therefore appropriate to consider the set of neurophysiological correlates to language processings to include the correlates to semiotic manifestations (signs).

The question is then to see if the neurophysiological components attached to semiolinguistic processings as such are dependent or independent from the forms of semiolinguistic phenomenality as they are manifested through the neurophysiological correlates to the perception of signs. The stakes of this question, as one may imagine, are colossal. And nothing allows to produce an answer a priori.

The alternative is the following:

Either these components are independent from those correlated to the perception of signs, and in such case, we would have the "classical" configuration of the empirical sciences, that is, on the one hand, a plane of phenomena constituted following their own order and delivering themselves as such to observation, and, on the other hand, a theoretical arrangement supposed to account for the functionings observed and to which specific neurophysiological responses are associated.

Or, conversely, we would observe that the neurophysiological components associated with the processing of signs participate in the processes through which semiolinguistic phenomenality elaborates itself, and then, we would be oriented—thereby returning to the previously suspended question—towards an *integrated conception of the orders of semiolinguistic objectivity and phenomenality*. Not, in this case, that the one would be reducible to the other, but that the one, the plane of concept and of laws—which "objectify" (describe and explain) the configurations of observable signs, their sequencings, their constraints, and their regularities—participates to the other, i.e. participates (to a degree that must be specified) to the constitution of the sign-phenomenon—and reciprocally.

In fact, and more precisely, we will show that the forms (morphodynamics) which institute the sign as an indivisible connection between a signifier and a signified, on the one hand, have phenomenological meaning, in that they regulate the manifestation of signs, and, on the other hand, they participate in the constitution of linguistic objectivity, in that they regulate the differential distribution of what is possible or impossible within language—a distribution which is correlative with an order of linguistic legality.

The forms of empirical knowledge regarding a certain class of phenomena—that is a certain conceptual device adequately qualifying such phenomena in that they account for their observable functionings—find themselves to take part, at least partially, in the very constitution of the phenomena of which they produce the objectivity. It is the Kantian break between the sensible and the intelligible which finds itself to be partially overcome here, thus making conceivable, although very partially, the edification of a science of expression ("expression" understood here as a tangible presence of meaning—cf. 2.1).

In this perspective, the practice of signs, their effective involvements in the accomplishment of acts of expression and of communication—acts which are encompassed and finalized through a certain cultural project of "producing meaning"—and of which a semiolinguistic science would thus expose the principles and modalities, incorporate the moment of their advent and conditions its form. In yet other words, the morphogenesis of the sign reveals itself to be inseparable from the normative schemas by virtue of which semiolinguistic practices regulate themselves, and this in two respects: First in that these schemas are the empirical and functional correlates of sign generation (by means of emergence and of stabilization), and then, dually, in that they constitute the frame of reference for the adjustments and innovations of which living speech is always in quest.

In any case, it is towards this conception that the neurophysiological observations will lead us, and it will then be necessary to ensure that the morphodynamics of the sign, beyond its phenomenological signification, that is, as it possesses an objective value, adequately accounts for the constitution, by means of emergence, of sign-phenomena *as they are the objects of linguistic practices that are shaped by norms*.

References

Auroux, S. (1998). *La raison, le langage et les normes*. Paris: PUF, coll. Sciences, Modernités, Philosophies.

Benveniste, E. (1966). *Problèmes de linguistique générale I*. Paris: Gallimard, coll. *Tel*.

Benveniste, E. (1971). *Problems in general linguistics* (M.E. Meek, Trans.). Coral Gables: University of Miami Press.

Bitbol, M. (1998). *L'aveuglante proximité du réel*. Paris: Flammarion, coll. Champs.

Boyer, A. (2000). Philosophie des sciences. In P. Engel (Ed.), *Précis de philosophie analytique*. Paris: PUF.

Culioli, A. (1999). *Pour une linguistique de l'énonciation; Formalisation et opérations de repérage, T. 2*. Paris: Ophrys.

Dalbéra, J.-P. (2002). Le corpus entre données, analyse et théorie. *Corpus* (1), 1–10.

de Saussure, F. (2013). *Course in general linguistics* (R. Harris, Trans.). London: Bloomsbury Publishing.

Ducrot, O. (1995). *In* Cahier "Livres" de *Libération*, 21 septembre 1995.

Frege, G. (1971). *Ecrits logiques et philosophiques*. Paris: Le Seuil, coll. L'ordre Philosophique.

Granger, G.-G. (1979). *Langage et épistémologie*. Paris: Klincksieck, coll. Horizons du langage.

Granger, G.-G. (1992). *La vérification*. Paris: O. Jacob.

Hjelmslev, L. (1968). *Prolégomènes à une théorie du langage* (p. 35). Paris: Éditions de Minuit, coll. Arguments.

Hjelmslev, L. (1985). *Nouveaux essais*. Paris: PUF, coll. Formes sémiotiques.

Husserl, E. (1999). *The idea of phenomenology* (L. Hardy, Trans.). Dordrecht: Kluwer Academic Press.

Kant, E. (1972). *Essai pour introduire en philosophie le concept de grandeur négative*. Paris: Vrin, coll. Bibliothèque des textes philosophiques.

Lazard, G. (2001). De l'objectivité en linguistique. *Bulletin de la Société de linguistique de Paris, XCVI*(1), 9–22.

Martin, R. (1978). *La notion de recevabilité en linguistique*. Paris: Klincksieck, coll. Bibliothèque française et romane.

Martin, R. (2002). *Comprendre la linguistique*. Paris: PUF, coll. Quadrige.

Milner, J.-C. (1989). *Introduction à une science du langage*. Paris: Le Seuil, coll. Des Travaux.

Ollivier, B. (2000). *Observer la communication, Naissance d'une interdiscipline*. Paris: CNRS Editions, coll. CNRS Communication.

Petitot, J. (1985). *Morphogenèse du sens: 1, Pour un schématisme de la structure*. Paris: PUF, coll. Formes Sémiotiques.

Philonenko, A. (1989). *L'oeuvre de Kant* (Vol. 1). Paris: Vrin, coll. A la Recherche de la Vérité.

Piotrowski, D. (1997). *Dynamiques et structures en langue*. Paris: CNRS Éditions, coll. Sciences du Langage.

Piotrowski, D. (2009). *Phénoménalité et Objectivité Linguistiques*. Paris: Champion, Collection Bibliothèque de Grammaire et de Linguistique.

Piotrowski, D. (2010). Morphodynamique du signe; I – L'architecture fonctionnelle. *Cahiers Ferdinand de Saussure, 63*, 185–203.

Piotrowski, D. (2011). Morphodynamique du signe; II – Retour sur quelques concepts saussuriens. *Cahiers Ferdinand de Saussure, 64*, 101–118.

Popper, K. (2005). *The logic of scientific discovery*. London & New-York: Routledge.

Walford, D., & Meerbote, R. (Eds.). (2003). *Kant, I., Theoretical Philosophy, 1755-1770*. Cambridge: Cambridge University Press.

Chapter 2
The Controversy Concerning the Nature of the Sign

2.1 The Expressive Fact

The question of the sign, of its indivisible unity and of its internal forms, is related to that of expressivity. Indeed, in its essential and most salient characters, the sign comprises an expressive feature, from the angle of which it allows itself to be somewhat approached and, moreover, to be grasped in its being as such. Although the sign configures itself with respect to expressivity, it does not reduce itself to it: in its very constitution, the sign as an expressive phenomenon finds itself to be "polarized" and even overtaken ("consumed"). Nevertheless, the phenomenon of expression, as a matrix of the sign, is naturally at the origin of semiolinguistic investigations, and for this reason, we will begin this chapter with a few reminders concerning it.

The fact of expression consists in a sensible presentation of meaning, a tangible presence of significations. The fact of expression is striking due to its *paradoxical essence*. It is *paradoxical* because "[expression] announces a 'depth' which is *concealed* and which *reveals itself directly* within it".[1] And it is paradoxical in *essence* because the contradictions which traverse it can not be lifted without annihilating the object which actually proceeds from it. That is to say that the fact of expression, when it is a matter of considering it in its fully paradoxical essence, requires to abandon the distinctions belonging to classical epistemology and to consider, with other forms of categoriality, new modes of object constitution, be they practical or intentional—an obligation to which Merleau-Ponty and Husserl will acquiesce.

In fact, expression is an inconceivable mixture of sensibility and intelligibility, of intuition and understanding, of immediate and mediate knowledge, and, *in fine*, of presence and absence. Expression, as a sensible presentation, pertains to the *immediate* and *actual* mode of knowledge, and this knowledge (intuition) is

---

[1]Rosenthal and Visetti (2008, p. 187), our emphasis.

© Springer International Publishing AG, part of Springer Nature 2018
D. Piotrowski, *Morphogenesis of the Sign*, Lecture Notes in Morphogenesis,
https://doi.org/10.1007/978-3-319-89848-3_2

complete and "perfectly true"[2]—because, in general terms, the forms of intuition which condition things as they appear (that is, phenomena) are precisely[3] the forms which deliver their appearing. As something which is intuitioned, no distance separates the expressive fact from that which it expresses: Within expression, nothing more than that which is immediately given could be sought. Consequently, as an intuitioned sense-object, expression finds itself to be closed upon itself as it is actually *present*, and by its own effect, it forbids itself from opening towards something which is *absent*. Hence, it forbids itself in a self-contradictory manner from having any signifying depth.

These difficulties concerning the expressive fact, and therefore the difficulties in clearly grasping and describing it univocally, can be avoided with the recourse to metaphorical language. For example, Taylor states that "Expression makes something *manifest* in embodying it".[4] If the term "manifest", explained as being "directly available for all to see"[5] so as to reinstate the full and immediate character of intuitive knowledge, belongs to the terminological and conceptual field of classical epistemology, that of "embodiment" is less firmly defined in conceptual terms. Nevertheless, this term still has the merit of expressing the indivisible unity of "expression and that which it expresses [...] and points towards the living, empathic presence within expression of that which is expressed."[6]

But as pertains to approaching and to thinking about the expressive facts, the difficulties are not only of a conceptual nature: They also concern the possibility of observing them. Indeed, the expressive fact, to put it as such, crumbles the moment when, by ceasing to be *practiced*, it acquires the status of an object, i.e. when it is "*thematized*". As Taylor observes: "An expression manifests something, but in an embodiment; and not any kind of manifesting-in-embodiment will do, but one that offers a physiognomic reading",[7] that is, "Expression (...) involves what we might call direct manifestation, not leaning on an inference."[8] This in contrast to readings that are "more analytical, geometric, or instrumental...".[9] In fact, from the moment we steer away from the experienced singularity and presence of the expressive fact, from the moment we lose its immediate and always effective contact, be it by retaining what it expresses so as to inscribe it in thought or, otherwise, so as to submit it to reflection, or, conversely, by retaining the expression component in order to inscribe it within an act of interpretation, each time the expressive fact finds itself to be abolished.

---

[2]Kant *in* Philonenko (1989, p. 87).

[3]"[P]henomena are of course necessarily consistent a priori with the conditions of their appearing which are the forms of intuition" (Petitot 1992, p. 61).

[4]Taylor (1985, p. 219), *our emphasis*.

[5]Ibid.

[6]Rosenthal and Visetti (2008, pp. 186–187).

[7]Taylor (1979, p. 78).

[8]Ibid., p. 73.

[9]Rosenthal and Visetti (2008, p. 187).

   As observed by M.-P., since it is a matter of linguistic expression, the thoughts which accompany texts or discourses are not coextensive to them, but occur beyond the expressive fact, at the ulterior moment of a reflexive grasp or of a thematization, in which expression is then "fulfilled"[10]: "when a text is read in front of us [...] we do not have a thought on the margins of the text itself. The words occupy our entire mind [...] The end of the speech or of the text will be the lifting of a spell. It is then that thoughts about the speech or the text will be able to arise. Previously the speech was improvised and the text was understood without a single thought; the sense was present everywhere, but nowhere was it posited for itself".[11]

   In short, the expressive fact does not suffer from the attention we place upon it, even less from our reflections concerning it. There is something of a "constitutive fragility"[12] to expression. It exists only in the moment of its encounter and in its spontaneous exchanges, "But the moment people begin to reflect upon language instead of living it, they cannot see how language can have such power".[13] Hence, the expressive fact, in that it assimilates its meaning with its manifestation, essentially signifies by weaving a world which is lived and practiced.

   The fact of expression therefore pertains neither to a logic of communication, by virtue of which predefined contents are transmitted by means of a code, nor to a dialectic between interior and exterior, through which private internal states would be made public by its means. The fact of expression, in its irreducible essence, is simply that of the actual (sensible) presence of meaning. And it is this essential character which semiolinguistic analysis will attribute to sign phenomena—at least in what concerns the analyses conducted by Husserl, M.-P., and Saussure.

2.2   A Community of Views

Saussure, M.-P., and Husserl approach semiolinguistic facts using conceptual devices and angles of intelligibility which are greatly irreducible. Nevertheless, in such a liminal moment which is required for any theoretical investigation, when it is a matter of solely delimiting the empirical field of the discipline, Saussure, M.-P., and Husserl agree on the basics.

   Because if the ambition of the linguist is "to constitute the descriptive framework of any possible language, [it can only be done] once the facts of experience chosen to define the field have been retained".[14] The epistemic project therefore requires this preliminary recognition of the specific character of the class of phenomena it considers to fall under its scope, and, regarding this specific point, Husserl, M.-P.,

---

[10]*PW*, pp. 40, 59.

[11]*PhP/L*, pp. 185–186.

[12]Rosenthal and Visetti (2008, p. 187).

[13]*PW*, p. 8.

[14]Granger (1979, p. 200).

and Saussure manifestly agree—even though, when it will be a matter of attaining its empirical truth, they will immediately solicit their own preferred categorical devices and schemas of intelligibility.

From the onset of their respective endeavors, and as if speaking through a single voice, Saussure and Husserl denounce the naïve conceptions according to which the sign would be an association between a tangible symbolic marking and a certain meaning. For both Saussure and Husserl, the sign does not rely on a distinction between sound and meaning. Such a dichotomy is fundamentally inappropriate for analyzing the semiotic fact. In *Écrits*, Saussure asserts that "it is wrong (and impracticable) to oppose *form* and *meaning*".[15] This had already been anticipated in the *Notes*: "what is opposable to the physical sound, is [...] by no means the idea",[16] or when it is question of the "obscurity and inanity of an opposition between the sign and the idea, between form and sense, or between the sign and meaning."[17] Husserl says as much: "It is usual to distinguish two things in regard to every expression: 1. The expression physically regarded (the sensible sign, the articulate sound-complex [...]); 2. A certain sequence of mental states [...] generally called the 'sense' or the 'meaning' of the expression [...]. But we shall see this notion to be mistaken".[18] Likewise, for Husserl as well as for Saussure, it is necessary to distinguish the "true" sign having an indivisible nature from the one resulting from a simple "assembly", that is, the "conventional" sign, as a correspondence between units of sound and a unit of meaning which are mutually foreign to each other with respect to their existence and to their principles of formation, and which therefore proceed from a logic of "naming-process"[19] or of "communication".[20] Husserl (*RL1/F*, p. 187) calls such signs "indicative" signs— these are the "commemorative" signs of the Stoics), and he defines them as the articulation of two moments of consciousness: There is first a certain experience of consciousness, which is the perception of the symbolic marking, and, by virtue of its constituting function, the symbol reorients consciousness towards another content which is the thing, the idea, or the state of things to be communicated: of which the listener is to be informed. The "essence of indication"[21] thus resides in the fact that "certain objects or states of affairs of *whose reality someone has actual knowledge* indicate to him *the reality of certain other objects or states of affairs, in the sense that his belief in the reality of the one is experienced [...] as motivating a belief or surmise in the reality of the other.*"[22] For Saussure, likewise, language is not organized in the manner of an index, that is, as a conventional reference of

---

[15]Saussure (2006, p. 17).

[16]N9.2 *in* Godel (1969, p. 137).

[17]*in* Godel (1969, p. 48).

[18]*RL1/F*, p. 188.

[19]*CLG/B*, p. 65.

[20]*RL1/F*, p. 189.

[21]*RL1/F*, p. 184.

[22]*RL1/F*, p. 184.

sound-units to meaning-units, each constituted within their own spheres: "The characteristic role of language with respect to thought is not to create a material phonic means for expressing ideas [i.e. Husserl's indicative sign]."[23]

Rejecting the conception of the sign as a simple assembly and in contrast to it, Saussure and Husserl defend the principle of a sign of another nature, one which is unitary and integrated. It is then necessary to distinguish, following the respective terminologies employed by each, the "meaningful sign"[24] *versus* the indicative sign for Husserl, and the sound-idea grouping *versus* the signifier/signified unit for Saussure. And for both, it is a matter of acknowledging that which constitutes the essence of the "authentic" sign, that is, a sort of reciprocal incorporation of the sign's faces which preclude soliciting the one without appealing to the other.

For Saussure, therefore "the linguistic phenomenon always has two related sides, each deriving its values from the other"[25]; "one can neither divide sound from thought nor thought from sound; the division could be accomplished only abstractedly, and the result would be either pure psychology or pure phonology."[26]

From the point of view of Husserl and in a similar manner, though already the an intentionalist inflection specific to his own system of questioning, the meaningful sign (which he also calls "expression"[27]) inscribes itself within a single moment of consciousness: The apprehension of unordered sensible data and their elaboration into a sign-phenomenon (the noetic moment) operates within a single intentional act, that is, the aim of an object of meaning. Therefore, *meaningful* signs signify in another respect than *indicative* signs do: Whereas the connection between the symbol and its meaning proceeds from an interpretative moment, which consists in redirecting the consciousness of the *actual* symbol towards the object of meaning, the connection of *expression* to meaning is *intrinsic* to it, this being its very principle of constitution: "the essence of an expression lies solely in its meaning".[28] In other words, whereas the symbol signifies in that it is "interpreted",[29] the expression signifies in the "strict sense"[30] of the term: "The essential function of expression is to signify [...]; and this signifying function, inasmuch as it is essential, exists even when the expression indicates nothing".[31]

In other words, the "true" signifier, which Husserl thus calls "expression", comprises in its phenomenal nature the orientation of consciousness towards a meaning, and it is this intentional directionality which shapes its appearance as a word-sign: "the 'meaning-intention' [...] characteristically marks off an expression

---

[23]*CLG/B*, p. 182.

[24]Or "significant sign" (*bedeutsam Zeichen*) as opposed to the "indicative sign" (*Anzeichen*).

[25]*CLG/B*, p. 8.

[26]*CLG/B*, p. 113.

[27]*RLI/F*, p. 187.

[28]Ibid., p. 199.

[29]Ibid., p. 188.

[30]*Leçons*, p. 30.

[31]Ibid.

from empty 'sound of words'"[32] and it is therefore "In virtue of [intentional] acts [that] the expression is more than a merely sounded word".[33]

These views, in what they basically assert, are largely shared by M.-P., also according to whom, and criticizing the intellectualist and empiricist approaches which place meaning outside of the word thus making it into an empty shell, *the word has a meaning*: "these two theories [empiricist and intellectualist], however, concur in the claim that the word has no signification"[34]; "Thus, we move beyond intellectualism as much as empiricism through the simple observation that *the word has a sense*".[35]

To say that "the word has a sense" is to say that meaning does not lie outside of the verbal fact, and, hence, that between the two there is no sequential link, be it of precedence (of the thought with respect to the word) or of inference (from the word to the thought). Thus, "For the speaker [...] speech does not translate a ready-made thought; rather, speech accomplishes thought. Even more so, it must be acknowledged that the person listening receives the thought from the speech itself",[36] and, moreover: "Speech is not the "sign" of thought, if by this we understand a phenomenon that announces another [...] in fact, [speech and thought] are enveloped in each other; sense is caught in speech, and speech is the external existence of sense".[37]

These considerations, which largely corroborate the positions of Husserl and of Saussure, receive an expressivist inflexion with M.-P. On the one hand, the word is approached as a sensible presence of meaning: "The word and speech [...] cease to be a manner of designating the object or the thought in order to become the presence of this thought in the sensible world, and not its clothing, but rather its emblem or its body",[38] and, on the other hand, it is the impossibility of escaping the word, therefore of moving away from it without abolishing it, which is thus emphasized: "If we push the research far enough, we find that language itself, in the end, says nothing other than itself, or that its sense is not separable from it".[39] So, as "the sense of a speech act can never in fact be delivered from its inherence in some speech",[40] it would thence be impossible to conceive it as such, in itself, without annihilating the fact of expression which it qualifies in part.

---

[32]*RLI/F*, p. 194.

[33]Ibid., p. 192.

[34]*PhP/L*, p. 182.

[35]Ibid.

[36]Ibid., pp. 183–184.

[37]Ibid., p. 187.

[38]Ibid.

[39]Ibid., p. 194.

[40]Ibid., p. 196.

2.3 Shared Difficulties

We would indeed concede that there are a few difficulties in clearly conceiving this indivisible essence, difficulties which stem from the impossibility of jointly conceiving the unity of the sign and its dual nature as a signifier/signified composite.

Indeed, if the unitary character of the sign is maintained, it will be identified with a fact of expression from which it will then inherit the paradoxes, among which the impossibility of conceptualizing its two facets without annihilating it.

Conversely, if we approach the sign while considering its two-sidedness from the onset, we will then be confronted with their unthinkable reciprocal incorporation. Certainly, by putting the emphasis on the interpenetration of the moments of sound and of meaning, it had first been a question of highlighting some of the characteristics of the essence of the "true" sign in order to distinguish it from other sorts of semiotic factualities. But this manner of approaching and of conceiving the unity of the sign cannot be maintained. Indeed, one must concede that this participation of sound to meaning, and reciprocally, either pertains to ontological teratology (in sum, to an assumed "mystery") or is stricken with inconsistency.

Because, regarding this latter point, even while choosing to acknowledge a relation of reciprocal dependency between the signifier and the signified, the necessary character of such a dependency directs against the principle of an analysis along these terms and throws into question the relevance of the poles thus identified.

As recalled by Lo Piparo (2007, p. 146 *sq.*) in his comparative study of Stoic and Saussurean semiotics, "true signs" (designated as "meaningful" by Husserl and as "indicative" by the Stoics) are dual entities of which the second term "cannot be known in an autonomous manner" (ibid.), and they pertain to an "entirely relational ontology" in that the existence of the parts which constitute them is necessarily simultaneous. For Sextus Empiricus, "indicative signs fall under the typology of the "simultaneously relative […], and here is the radical critique [he makes of them]: "The indicative sign does not exist" (ibid.), the reason being that, formulated in Saussurean terms, because the signifier and the signified reciprocally condition each other both in terms of their identities and of their existence, such a manner of dividing the sign is inconsequent and sterile. Particularly, when term A comprises term B in a constitutive manner, the inductive directionality (if A then B), at the foundation of Stoic semiotics, has no more reason to be.

Likewise for Hjelmslev: Concerning the relation of interdependence which, namely, reunites the planes of expression and of content, Jørgensen and Stjernfelt assert that it is "empty [in terms of heuristics] in the Hjelmslevian interpretation, precisely because it is relational in such a consequent manner: It welds the two terms together, to a point where they become inseparable […] and if two terms always appear together, it becomes impossible to separate them at their own level".[41]

---

[41]Jørgensen and Stjernfelt (1987, p. 90).

The difficulties to which the theory of the sign is confronted are not only of a logical or of a conceptual nature: They can also be bolstered by empirical observations—although in such a case we have trouble distinguishing whether the difficulties registered pertain "as such" to the observed fact or if they are the consequence of the system of observation by virtue of which the facts are approached. Thus, taking a more empirical stance, Tamba-Mecz identifies several facts which appear to contradict the principle of a consubstantiality between signifier and signified: "Other experience data appear to shake these first convictions: [translation, paraphrase, synonymy] demonstrate the possibility of exchanging verbal signifieds considered to be equivalent, though they may be configured by dissimilar signifiers; or [homonymy and polysemy] [...] in short, being indissolubly linked by the formulation of meaning, verbal forms and meanings may nonetheless be 'unjoined' by means of analysis".[42] Moreover, "the indefectible union of verbal signifiers and signifieds is contradicted by the exchanges between signifieds of all orders."[43]

Phenomenology is not outdone by this. The analysis of the sign proposed in Husserl's first *Logical Investigation* encounters the same difficulty as the one which was diagnosed by Sextus Empiricus when it was question of the "*simultaneously relative*": The signifier and the signified configure themselves and mutually presuppose one another to the point that the appearance of the signifier and signified fully overlap in a signitive phenomenality which is then logically indivisible.

Indeed, we have seen that, beginning with the first *Logical Investigation*, Husserl distinguishes two regimes of significance respectively at work in *indicative* and in *meaningful* signs. We have also seen that the orientation towards an object of content constitutes the essential character of the meaningful sign: Whereas the indicative sign administers a correspondence between two experiences of consciousness constituted outside of one another, the meaningful sign incorporates in its appearing the mode of a consciential aim of signification.

Now, this conception of the "meaningful" sign is unsatisfactory because if such was indeed the case, then the appearing of the sign would *fully* inscribe itself within the appearing of meaning, as an object that is a target of the meaning-intention.

Indeed, if the phenomenal identity of the sign found itself to be *integrally* configured by the sole consciential aim towards an object of signification, then the sign would never present itself otherwise than as meaning or as an integral (indissociable) part of a meaning: It would, exclusion being made of any other (manifest) character, be constitutive of a "pure" presentation of meaning "in itself"—thus obliterating the concrete dimension of the signifier. We can illustrate this conjuncture by means of an analogy with spatial perception. As the immanent adumbrating (the set of discontinuities which a spatial body projects upon the retinal surface, discontinuities which are indeed present *within* consciousness) finds itself to be spatialized and presented (*to* consciousness) as the apparent contour of a

---

[42]Tamba-Mecz (1991, p. 37).
[43]Ibid., p. 3.

three-dimensional object, so would the medium (graphical or vocal) of a signifier be semiotized into an intrinsic component of the intended object of meaning. The medium would then disappear from consciousness as regards its immanent sensible characters—as is the case with a perceptual representation where "an experienced complex of sensations gets informed by a certain act-character [...] the perceived object appears, while the sensational complex is as little perceived as is the act in which the perceived object is as such constituted".[44] But it appears that it is precisely the contrary which is revealed by phenomenological analysis: The concrete characters of the signifier persist, albeit in an altered form, in the perception of the sign. We will return to this specific point (cf 4.1 and 4.2).

Since it is a matter of approaching the reality of the sign, its contents, and its internal forms, and also due to the confrontation with numerous and considerable difficulties, the recourses to metaphors are frequent. Thus, we will encounter mentions of the "consubstantiality" of the faces of the sign, of their "fusion", of their "reciprocal assimilation", and of their mutual "incorporation." For example, Benveniste states that "there is such a close symbiosis between them [the concept and the sound image] that the concept [...] is like the soul of the sound image"[45] or: "The signifier and the signified [...] together make up the ensemble as the embodier and the embodiment [...]. This consubstantiality of the signifier and the signified [etc.]".[46] But the problem remains in full.

Saussure, on the other hand, albeit without specifically discussing the difficulties of the dual and indivisible sign, promptly abandons the conception of a "consubstantiality" of the two faces of the sign, recognizing its unintelligibility: If "neither are thoughts given material form nor are sounds transformed into mental entities",[47] it is because the sign is nothing but a "side effect." The sign is the functional consequence of a superior systemic reason (language) which operates by correlating relations of reciprocal delimitation in the substances, respectively, of expression and of content, in order to dually institute units (in these substances): "The characteristic role of language with respect to thought is not to create a material phonic means for expressing ideas but to serve as a link between thought and sound, under conditions that of necessity bring about the reciprocal delimitations of units".[48]

And, correlatively: "Language works out its units while taking shape between two shapeless masses".[49] We know that this level of elaboration of Saussurean thought (which Hjelmslev regarded with severity) is not without weaknesses, and that it is in the framework of a theory of value, developed in the *Third Course*, that he will find the key for discarding once and for all a "great illusion", inasmuch as

---

[44]*RLI/F*, p. 214.

[45]Benveniste (1971, p. 45).

[46]Ibid.

[47]*CLG/B*, p. 112.

[48]Ibid.

[49]Ibid.

"to consider a term as simply the union of a certain sound with a certain concept is grossly misleading"[50]—we will also return to this at length.

2.4   The Counter-Arguments

2.4.1   Introduction

The previously exposed difficulties regarding the problematization of the semiotic fact, and which disrupts the "community of views" about it, are not without counter-arguments.

In order to assert the indivisible essence of the sign, which is therefore in opposition with the precluded conceptions of a simple "assembly", the arguments usually take shape with a view upon the immediate experience of the states and contents of semiolinguistic consciousness—such experience being always delicate since it pertains to an evanescent object which has uncertain contours, and with respect to which a reflexive approach, if it is not methodologically regulated, risks only retaining an adventitious form or a fractured part of a global morphophonemic process of semiogenesis. In this respect, it is most likely the phenomenological perspective which offers the best assurances.

2.4.2   First Argument

There is firstly Saussure, who attributes to language a power to configure ideas. It is because without the support of signifiers, there are no distinct thoughts or circumscribed ideas: Before language takes grasp of it by means of signs, thought (an "amorphous mass of ideas") contains nothing which is defined and of which the mind could prevail itself: "Philosophers and linguists have always agreed in recognizing that without the help of signs we would be unable to make a clear-cut, consistent distinction between two ideas."[51]

2.4.3   Second Argument

The sense of a word only occurs when it is uttered. We would indeed concede that the actualization of linguistic content is fully tied with what is "uttered" through the vocable attached to it, and, correlatively, that the act of speech is never

---

[50]Ibid., p. 113.
[51]Ibid., pp. 111–112.

accomplished without being traversed by a thought which motivates it. We indeed appear to have here an "inescapable fact [...] for the semantician linguist, [that the] *being* of verbal meanings is their *utterance*, and their *existence* is to be said". [52] The roots of this certitude draw from an "immediate authenticity of that which we will call, with O. Ducrot, our linguistic experience: In the daily practice of language, signifieds and linguistic forms are so narrowly linked that they fail to be distinguished." [53]

2.4.4 Third Argument

In support of the expressivist thesis, by virtue of which meaning *inhabits* the word, M.-P. develops two sorts of arguments.

First, he confronts the logic of "thought without speech" which he interrogates with astonishment, with the observation of an effective coincidence between the moments of thought and speech and that of their factual participation: "If speech presupposed thought, or if speaking was primarily the act of connecting with the object through a knowledge intention or through a representation, then we could not understand why thought tends toward expression as if toward its completion, why the most familiar object appears indeterminate so long as we have not remembered its name". [54]

Likewise, if speech presupposed thought, we would not understand "why the thinking subject himself is in a sort of ignorance of his thoughts so long as he has not formulated them for himself, or even spoken or written them". [55] Finally, why "would thought seek to double itself or to clothe itself in a series of vocalizations, if the latter do not carry and do not contain their sense in themselves?" [56]

Then, and more fundamentally, because supported by a partial phenomenological analysis from which he draws all the consequences, M.-P. introduces and discusses the idea of an "algorithmic" language, that is, a language conceding to the principles of the theory of models in that its symbolic and significant sides (namely the syntactic and semantic components) are mutually external, and, correlatively, in that their respective units, of a distinct nature, are linked following the mode of conventional correspondence.

This conception of an "algorithmic" language is not without phenomenological support, since it must indeed be acknowledged that among the powers of language, and even as a superior form of its accomplishment, there is that of referring to universes of meanings or of things, endowed with their own respective orders and

---

[52]Tamba-Mecz (1991, p. 44).

[53]Ibid., pp. 36–37.

[54]*PhP/L*, p. 182.

[55]Ibid., p. 183.

[56]Ibid., p. 187.

situated beyond language. Indeed, it is a fact that, in their actual usage, "signs are immediately forgotten; all that remains is the meaning [and] the perfection of language lies in its capacity to pass unnoticed. *But therein lies the virtue of language:* it is language which propels us toward the things it signifies. In the way it works, language hides itself from us. Its triumph is to efface itself and to take us beyond the words to the author's very thoughts".[57] Furthermore, the expression "is most complete when it points unequivocally to events, to states of objects, to ideas or relations, for, in these instances, expression leaves nothing more to be desired, contains nothing which it does not reveal, and thus sweeps us toward the object which it designates".[58]

If, from a phenomenological description of speech in action, we retain nothing but the projection of speech towards an exterior made up of thoughts or of objects, then the order of language must be conceived with regard to such dispositions, hence as an order which, in view of delivering meaning by detaching it from signs, reduces them to the role of transition towards meaning. Thus, "expression involves nothing more than replacing a perception or an idea with a conventional sign that announces, evokes, or abridges it."[59]

It is also in this direction that we are steered by the illusion of a "constituted language", that is, by the illusion of a common repository of signs related to established meanings and which seem available to the latter for their own benefit (cf. 7.6 and 7.7).

But if this were the case, any speech would find itself, by construction, enclosed within a universe of primeval ideas, of which it would enable original compositions, but without ever finding a way to unhitch itself from them. Speech would then have no other content than that which it takes from a repository of possible meanings, without ever broadening its perimeter or modulating its elements. And its innovative power would therefore be illusionary: "Of course, language [...] can refer to what has never yet been seen. But how could language achieve this if what is new were not composed of old elements already experienced."[60]

But in keeping with this "myth" of *universal* language, "which contains in advance everything it may have to express, because its words and syntax reflect the fundamental possibles and their articulations"[61] nothing novel can be said nor transmitted. Because if language "is coding [the] thought, [as it replaces it] with a visible or sonorous pattern [...] [then] we never find among other people's words any that we have not put there ourselves. Communication is an appearance; it never brings us anything truly new [...] since the signs communication employs could never tell us anything unless we already grasped the signification."[62]

---

[57]*PW*, p. 10.

[58]Ibid., p. 3.

[59]Ibid.

[60]Ibid.

[61]Ibid., p. 7.

[62]*PW*, p. 7.

In short, and relying on a phenomenological reading which is consistent but faulty because incomplete, we are logically led to believe that "speech that is heard can bring [...] nothing: [because it is the listener who] gives the words and the phrases their sense".[63] It remains that "we have the power to understand beyond what we could have spontaneously thought. [...] Here, then, the sense of words must ultimately be induced by the words themselves [...] There is then [...] a *thought in the speech*".[64]

2.4.5 Fourth Argument

We know that, from a phenomenological point of view, Husserl establishes a distinction between meaningful and indicative signs, a distinction which the first *Logical Investigation* seeks to firmly establish from the start. Whereas the indicative sign links two contents of consciousness, the meaningful sign (or expression) belongs to a single and same moment of consciousness administered by an intent to signify—an intent which constitutes the phenomenological characteristic of expression (cf. *RL1/F*, p. 194): The apprehension of unordered sensible data and its elaboration into a sign-phenomenon (the noetic moment) operates within a single intentional act which is the aim of a signifying object (for a presentation of intentionality, of the problematics and concepts surrounding it—cf. 3.2).

The empirical truth of the meaningful sign should therefore be sought by means of a phenomenological examination which could reveal the intentional tension configuring its appearing. This examination will be of a variational nature.

It is in this view, as a matter of fact, Husserl will examine the phenomenal mutations induced by intentional variations. He thus discusses the case of an arabesque,[65] first perceived as an ornament than recognized as a sign, and, following other intentional modalities, he discusses[66] the encounter of a human form first apprehended as a living being, and then reconsidered to be a mannequin: an inert object. In each case, the object which is firstly constituted in consciousness undergoes a radical phenomenal modification. Whereas the "medium" remains the same, the noetic act's substitution, which firstly aims at a simple outline and then at a semiotic object, translates as a modification of the characteristics of appearing. In other words, the same matter involved in different intentional orientations produces in consciousness phenomenal identities of distinct natures: "In a certain way, [...] the 'foundations' remain the same. But phenomenologically speaking, it is not at all the same thing which appears: on the one hand, we have intuition, and on the other

---

[63]*PhP/L*, p. 184.
[64]Ibid., pp. 184–185.
[65]*RL5/F*, p. 105.
[66]Ibid., pp. 137–138.

[expression]. [...] The modification [...] refers to the differences between two *aims*, which differ in the manner whereby they invest a same 'content'".[67]

Correlatively, the substitution of the semiolinguistic intentional aim with, we might say, an aim towards an object of the physical world has for consequence the modification of the appearance of the apprehended graphical marking or sound: "A word only ceases to be a word when our interest stops at its sensory contour, when it becomes a mere sound-pattern".[68] And, conversely, "the intuitive presentation, in which the physical appearance of the word is constituted, undergoes an essential phenomenal modification when its object begins to count as an *expression*".[69]

Thus, the word which is printed on the sheet of paper, beneath my eyes, has a way of being there, of offering itself to be seen, which is essentially different whether it is targeted in its nature as an expression, that is, whether it is apprehended following a noetic act which "infuses" the printed word with a meaning, or whether it is grasped through an act which has a view on the word "qua external singular",[70] as a concrete object of perception.

This modification of the word's way of being present when its power of signification finds itself neutralized has also been observed by M.-P.: "when it loses its sense, the word alters even in its perceptible appearance, it *becomes empty* [...] just as names are for us when we have repeated them for too long [...] it has altered itself, like an inanimate body".[71]

2.5   Transition

With respect to the nature of the sign and of the fact of expression, the "community of views" thus finds itself to be confronted with "shared difficulties", although these do not remain without answers ("counter-arguments"). Nevertheless, in the end, the resulting portrait is not free of confusion, because despite the very strong arguments in favor of an essentially indivisible sign, in which its form and meaning are merged, the explanations or descriptions of such a sign do not dispel the objections against it. We therefore retain the right to question the empirical authenticity of a phenomenon of this nature.

It will therefore now be important to take heed of the manners of approach and of the conceptual devices established for recognizing the sign in its paradoxical being and to overcome the difficulties mentioned, precisely by situating them within an appropriate perspective which give them meaning and which neutralize them.

---

[67]Benoist (2001a, p. 33).

[68]*RL1/F*, p. 190.

[69]Ibid., pp. 193–194.

[70]Ibid., p. 192.

[71]*PhP/L*, p. 199.

We will therefore devote the next section to presenting the theoretical devices which Husserl and M.-P. developed to address the matters of the sign and of expression. Adopting here a resolutely utilitarian approach, we will limit our expositions of transcendental and existential phenomenologies to the sole elements necessary to our demonstrative progression. But as our investigation involves the epistemological dimension—precisely, we should recall, in that it is a matter for us to base the morphodynamics of the sign on experience data, and, more generally, to circumscribe the legitimate field of empirical knowledge concerning signs and meaning—we will introduce the Husserlian and Merleau-Pontian perspectives by beginning with the more general questions, those concerning the conditions of possibility of knowledge. Though initiated while adopting this very wide angle, our discussion will promptly be narrowed down by the finality of our study, so as to bring the theme of intentionality to the forefront and to address the various related issues (noesis/noema, nature of the intentional object, fulfillment, absence/presence duality…) which delimitate the problematics relating to semiolinguistic objectivity.

References

Benoist, J. (2001). *Intentionalité et langage dans les "Recherches logiques" de Husserl*. Paris: PUF, coll. Epiméthée.

Benveniste, E. (1971). *Problems in general linguistics* (M. E. Meek, Trans.). Coral Gables: University of Miami Press.

de Saussure, F. (2006). In S. Bouquet, R. Engler, C. Sanders, & M. Pires (Eds.), *Writings in general linguistics*. Oxford: Oxford University Press.

Godel, R. (1969). *Les sources manuscrites du Cours de Linguistique Générale de F. de Saussure* (Vol. 61). Genève: Droz, coll, Publications Romanes et Françaises.

Granger, G.-G. (1979). *Langage et épistémologie*. Paris: Klincksieck, coll, Horizons du langage.

Jørgensen, H., & Stjernfelt, F. (1987). Substance, substrat, structure. *Langages, 86,* 79–94.

Lo Piparo, F. (2007). Saussure et les Grecs. *Cahiers Ferdinand de Saussure, 60,* 139–162.

Petitot, J. (1992). *Physique du sens: de la théorie des singularités aux structures sémio-narratives*. Paris: Editions du CNRS.

Philonenko, A. (1989). *L'oeuvre de Kant* (Vol. 1). Paris: Vrin, coll, A la Recherche de la Vérité.

Rosenthal, V., & Visetti, Y.-M. (2008). Modèles et pensées de l'expression: perspectives microgénétiques. *Intellectica, 3*(50), 177–252.

Tamba-Mecz, I. (1991). *La sémantique* (Vol. 655). Paris: PUF, coll, Que sais-je ?.

Taylor, C. (1979). Act as expression. In G. E. M. Anscombe, C. Diamond, & J. Teichman (Eds.), *Intention and intentionality: Essays in honour of GEM Anscombe*. Ithaca: Cornell University Press.

Taylor, C. (1985). *Philosophical papers: Human agency and language* (Vol. 1). Cambridge: Cambridge University Press.

Chapter 3
Theoretical Elements

3.1 Introduction

The concept of intentionality runs across and connects all sectors of our field of investigation. We have already referred to it (cf. 3.2) when we addressed the aporia of the expressive fact—we shall now examine it in greater detail, without overlooking its general scope. This concept will obviously be necessary for introducing and discussing Husserl's phenomenological analysis of the sign, as well as for uncovering the phenomenological content of Saussurean semiotics—something which will be done by means of morphodynamic characterization. Moreover, intentionality will serve as an articulation with the Merleau-Pontian perspective. Finally, the concept of intentionality interests us on an empirical level in that we will be led to propose an intentionalist interpretation of the "N400" EEG component.

3.2 Intentionality: Introduction and Generalities

It is with respect to the epistemological and critical issue of the possibility for empirical knowledge that the concept of intentionality finds its greatest *gnoseological* scope.

In its more abrupt form, one in which the existence of a self-given world is presupposed and being endowed with its own order so as to be independent from the thinking subject, the problem of knowledge reveals itself to be an "abyss for the mind". Because if knowing means to grasp in full consciousness, to thus establish "within oneself" an order of things "in themselves", or to produce in the format of thought the same schemes of necessity which regulate the world in its objective

© Springer International Publishing AG, part of Springer Nature 2018   37
D. Piotrowski, *Morphogenesis of the Sign*, Lecture Notes in Morphogenesis,
https://doi.org/10.1007/978-3-319-89848-3_3

being—"objective" being understood here in the naive sense of being independent from the subject—then of course, no knowledge is possible.

Indeed, "How is one to acquire knowledge regarding an object one did not posit? How is one to know that which, by its very *existence*, does not depend upon oneself?"[1] Because thought, be it a labile psychological flux or a categorical device, remains forever confined to its fabric of subjective processes or to its own system of forms: "How are we to understand the fact that the intrinsic being of objectivity becomes 'presented,' 'apprehended' in knowledge, and so ends up by becoming subjective?"[2]

We know that the Kantian solution consists in overcoming this opposition between subject and object, firstly by relating the theater of the world to the forms of intuition pertaining to physical knowledge (specifically mechanistic knowledge), that is, spatial extension and temporal succession. The world, such as it presents itself through immediate experience and in view of scientific knowledge, is not a world of *objects "in themselves"* that are encountered as such but a world of *phenomena* conditioned by forms of intuition that are internal (time) or external (space)—forms which, following the conceptions of the School of Marburg,[3] are the methods of an intrinsic determination of phenomena and the foundations of a mathematization of nature (cf. the *transcendental exposition*[4]).

The problem of the correlation between a subject and the world then becomes that of the unity of the *a priori* forms of the sensible data with the *a priori* forms of intelligibility—a unity which appears to be compromised, since the forms of intuition and those of understanding are mutually irreducible. Overcoming this obstruction is achieved through transcendental deduction, by means of which the common source of understanding and of intuition is uncovered: It is indeed under the pure unity of an "I think" and in its power of synthesis distinctively accomplished by the various categories of understanding regarding the manifold of intuition that the dimension of time is constituted, a dimension which, by encompassing that of space, traverses any possible experience.

Categories, which are the general forms of thought regarding objects, therefore essentially carry, as regimes of synthesis, the order of time which, reciprocally, conditions its applications. Correlatively, the "method of [constitution] of the internal sense is at the same time the method of determination of phenomena, and consequently it founds the object [the phenomenon determined under a categorical system] and the empirical subject (the internal sense)."[5] Thus, the empirical subject and the empirical world, in their relationship of mutual externality, appear as poles constituted under the unity of a "pure, originary and immutable consciousness"[6]: *transcendental apperception.*

---

[1]Philonenko (1993, pp. 16–17).

[2]*RL1/F*, p. 169.

[3]Of which the main representatives are Hermann Cohen, Paul Natorp, and Ernst Cassirer.

[4]Cf. Kant, *Critique of Pure Reason*, Part. 1: Transcendental aesthetic.

[5]Philonenko (1989, p. 163).

[6]Ibid., p. 160.

But if Kant overcomes the gap between intuition and understanding, between sensibility and intelligibility in order to establish both the objective power of thought and to circumscribe its legitimate scope, the difficulty which he surmounts remains in a way embedded within the plane of intuition, that is, within the plane of immediate knowledge.

Indeed, we may recall,[7] if necessary, that the phenomenon is the "undetermined object of an empirical intuition" and that the "manner in which [knowledge] immediately relates to [objects] is by means of an intuition".[8] Whereas knowledge approaches empirical reality by means of interposed concepts, that is to say, mediated by general and abstract forms of determination, intuition involves a direct relationship with its objects in that it grasps them as they present themselves. The object of an intuition therefore comprises the properties of its *gnoseological* immediacy, that is, its current presentation and its singular individuality. But if intuition "is directed towards that which is concrete as it immediately presents itself",[9] as a mode of knowledge, it no less presumes a gap between the knowing subject and the object. Intuitive knowledge remains knowledge in that it institutes itself with respect to an object, even when its content adheres to it, which is not without being paradoxical.

Thus, with intuition, the problem regarding the possibility of knowledge is posed in reverse: It is not a matter of finding the principle of a necessary unity between the distinct forms of understanding and of sensibility, hence between two mutually exclusive regimes of knowledge, but of uncovering the internal forms of a regime of immediate knowledge, that is, of a regime which distributes the facing poles of object and of subject, but without introducing alterity, and especially without breaking the directness of their connection.

Husserl will address this question regarding the forms of intuitive knowledge by returning to the Kantian issue of the possibility of objective knowledge—a problem he will instruct following the terms of the conjuncture and of the scientific debates of his time (namely against psychologism and empiricism), that is, along the terms of an opposition between the thinking subject and the world, between mental states correlated to acts of thought and the objects of knowledge presumed to be external. Thus: "In all of its manifestations, knowledge is a mental experience: knowledge belongs to a knowing subject. The known objects stand over against it. How, then, can knowledge be sure of its agreement with the known objects? How can knowledge go beyond itself and reach its objects reliably? What appears to natural thinking as the matter-of-fact givenness of known objects within knowledge becomes a riddle."[10]

The *Husserlian* solution, for overcoming the gap between the objective and the subjective, between the "things in themselves" and the "solitary mental life",

---

[7]Cf. Kant, *Critique of Pure Reason*, Part. 1: Transcendental aesthetic, §1.

[8]Kant (2003, p. 21).

[9]Ladrière (1996, p. 723).

[10]Husserl (1999, p. 17).

is comprised within the Brentanian observation that "all consciousness is consciousness of something other than itself." That is to say that consciousness is intrinsically an opening towards something beyond itself. In other words, it instructs within itself a form of exteriority: "Consciousness is open towards something other than itself and it becomes itself by its being penetrated by this other."[11]

This essential property of consciousness bears the name of *intentionality*: "The word intentionality signifies nothing else than this universal fundamental property of consciousness: to be consciousness *of* something"[12] or, furthermore: "Intentionality is the way for thought to *contain ideally something other than itself*."[13]

3.2.1 *Appearance/Appearing*

Intentionality confers to consciousness the power of extending beyond its own actual and psychological self, and of producing an outside object that is other than itself. Thus, a distinction is made "between the *components proper* of intentive mental processes and their intentional correlates",[14] that is, between that which consciousness effectively encompasses, i.e. what is immanent to it, and what constitutes as such the object of consciousness, that which appears to it.

It is thus necessary to clearly distinguish between what is effectively present *in* consciousness from what is present *to* consciousness. Thus, while perceiving something of a physical nature, "the appearing of the thing (the experience) is not the thing which appears (that seems to stand before us *in propria persona*)."[15] It is necessary to recognize, *on the one hand*, that which pertains to *appearing* as such, that is, a certain sensation, configured following the mode of an intentional orientation, and which finds itself actually present *within* consciousness, "as belonging in a conscious connection."[16] This must, *on the other hand*, be distinguished from the object as "*the thing which appears*", that is, the object such as it is made visible by the natural activity of consciousness, and which "is not conscious in the 'real' sense of the term, that is, in the sense of a real inclusion within consciousness, as one of its constituting moments."[17]

Thus, *on the one hand*, there are truly *immanent* mental contents which are not the referents to which consciousness attaches itself in its "natural" manner of being,

---

[11]Depraz (2012, p. 8).
[12]Husserl (1960, p. 33).
[13]Levinas (1998, p. 59).
[14]Husserl (1983, p. 213).
[15]*RL5/F*, p. 83.
[16]Ibid.
[17]Benoist (2001, p. 34).

and, *on the other hand*, there is the "thing" which is the objectal correlate of actual experiences that are intentionally linked to it.

3.2.2 Directedness

Due to its intentional structure, consciousness operates in the mode of "directedness", that is, in the mode of a directional activity transcending current mental contents in order to produce an *effect of reference* to objects of various natures: "It is the act-character [the intentional experience] which as it were ensouls sense, and is in essence such as to make us perceive this or that object."[18] Intentionality establishes consciousness in *relation* to objects. By means of intentionality, current mental contents will be instituted *as referring* to certain objects: "Intentional experiences have the peculiarity of directing themselves in varying fashion to presented objects, but they do so in an *intentional* sense. An object is 'referred to' or 'aimed at' in them."[19]

In the case of an empirical intuition, for example, one would distinguish, on the one hand, matter of sensation: the (highly hypothetical, cf. below) manifold of sensation being linked to our receptive physiological faculties, on the other hand, such matter as it is animated in consciousness in an intentional mode, and finally, one would distinguish the object towards which this effective experience directs consciousness.

We should emphasize that "the data of sensations", or "sensorial experiences", are by no means a "pure" consciousness of sensation, that is, a consciousness of the manner in which we experience the excitation "as such" of our receptive apparatus when it is stimulated. As has been firmly established by Merleau-Ponty,[20] the notion of pure sensation is but a theoretical abstraction devoid of phenomenal relevance. Likewise for Husserl: "*Sensations*, and the acts 'interpreting' them or apperceiving them, are likely experienced, *but they do not appear as objects:* They are not seen, heard or *perceived* by any sense. *Objects* on the other hand, appear and are perceived, but they are not *experienced.*"[21] The "data of sensations" of which it is question here are immanent contents of consciousness which "intentionally" refer to a quality or character of the object as it appears. Particularly, some data perform a function of "adumbration" (or profile, aspect, touch…)[22] when, as part of a logic of "fragmented and progressive revelation"[23], they refer consciousness to a single thing through their incessant variations. Such is the case, for example, with

---

[18]*RL5/F*, p. 105.

[19]Ibid., p. 98.

[20]*PhP/L*, Introduction, Chap. 1.

[21]*RL5/F*, p. 105.

[22]Ricœur, Note 1 *in* Husserl (1983, p. 132).

[23]Ibid.

color-data: "*The same* color appears 'in' continuous multiplicities of color *adumbrations.*"[24] In this case, "the physical thing is the intentional unity, the physical thing intended to as identical and unitary in the continuously regular flow of perceptual multiplicities."[25] Spatial perception provides a good illustration of this. Although the spatial object presents itself *effectively* to consciousness (as immanent content) in the form of a flux of apparent contours, it is indeed a three-dimensional form, a unit to which each apparent contour has the power to refer, which is made present *for* consciousness.

To complete this point, we should emphasize that the concrete contents of consciousness, be they intentionally "animated", should not be conflated with the aspects or qualities of the thing-as-perceived towards which they direct consciousness. Thus, care should be give to not amalgamate

"[the] 'sensation-contents' such as color-Data, touch-Data and tone-Data, [...] with appearing moments of physical things—coloredness, roughness, etc.—which 'present themselves' to mental processes [...] by means of those 'contents.'"[26] More generally, as Husserl stresses, "The adumbration, though called by the same name, of essential necessity is not of the same genus as the [object] to which the adumbrated belongs",[27] and particularly in the case of spatial objects which adumbrate into apparent contours: "The adumbrating is a mental process. But a mental process is possible only as a mental process, and not as something spatial."[28]

3.2.3 Noesis/Noema

Precisely, it is by means of a "noetic" act, through which the relation to the object is elaborated, that matter is informed and instituted as intentional experience: In the noetic act, "Sensuous Data present themselves as stuffs for intentive formings"[29] and in the resulting immanent mental processes, consciousness finds itself to be in a state of directedness towards something. As a counterpart to *noesis*, as an act of apprehension configuring and intentionally animating sensible matter, we have the unitary moment of the *noema*, both as a convergence point for the intentional aims, as the principle regulating the flux of effective appearings and operating a synthesis so as to produce an object of consciousness. In other words, as an abstract structure instituting the intentional character of mental contents (of representations) "a noema *normatively* anticipates on the objects of a regional ontology."[30] And in its

---

[24]Husserl (1983, p. 87).

[25]Ibid., p. 88.

[26]Husserl (1983, p. 203).

[27]Ibid., p. 88.

[28]Ibid.

[29]Ibid., p. 204.

[30]Petitot (1992, p. 84).

transcendental sense, "the noema as object = X [is] the pole of identity and of unity of the synthetic rules and connections of appearances."[31]

The noesis/noema correlation thus enables to overcome the alterity between the thinking subject and the surrounding world: "intentional directedness is what nullifies the very idea of an opposition between the subject and object, where these two poles would be exterior from one another and would exist independently from one another."[32] Thus, through intentional directedness, an object is presented to consciousness, and in the fabric of intentional experiences of which this object is the focal point, it indeed finds itself to reside within consciousness.

It is also the immediate character of consciousness with respect to the object, a character specific to intuitive knowledge, which is reconstructed all the while a separation is maintained between the object and the empirical subject, not an originary separation which would then need to be overcome, but a separation instituted by way of the intentional act which establishes, at its poles, the empirical subject bearer of mental states and the object as the target of consciousness.

3.2.4 "Paradoxical" Presence

Even if the targeted object is by essence in an immediate relation to consciousness, it is not necessarily present in perception as such with the fullness of an actual empirical phenomenon.

Because if, "as seen with the traditional interpretation[,] any intentionality comprises a part of transcendence",[33] the object which is *simply* aimed at is nothing more than a reckoned object, *pending* to be "true": Its forms are established by its directionality which does not, however, deliver it as a full presence. This is because the intentional object integrates in its own order the modality of *absence*: Its apprehension is essentially incomplete, and even though its absent components are fully accounted for in its elaboration as phenomenality within consciousness, *i.e.* as an "object for oneself", it continues to be considered as an "empty locus for the object which could be 'real'."[34]

This point is important: The intentional object is not a real object in the strict sense of the term, but an object which is simply presumed by consciousness, in full or in part: "[The] thing which is targeted (perceived) by consciousness only acquires existence under the gaze of consciousness",[35] rather than existing in itself.

Indeed, the intentional object, though aimed at by consciousness, does not reside within consciousness as such—its "inclusion" within consciousness is "not real":

---

[31]Petitot (2002, p. 429).

[32]Depraz (2012, p. 7).

[33]Benoist (2001, p. 129).

[34]Ibid., p. 59.

[35]Depraz (2012, p. 8).

"We [...] can speak with Husserl of an inclusion of the world within consciousness, because consciousness does not only consist in the 'I' pole (noesis) of intentionality, but also consists in the 'that' pole (noema). However, it is necessary to specify that this inclusion is not *real* (the smoking pipe is in my room), but intentional (the pipe phenomenon appears to my consciousness). It is because the inclusion is intentional that it is possible to found the transcendental *within* the immanent without degrading it."[36]

Benoist rightfully insists on the importance of the epithet *simple* which modalizes the status of the intentional object: "When the object is not present, the directedness is relegated to a rank of *simple directedness*."[37] More specifically, for Husserl, "when an object is said to be 'simply intentional,' this obviously does not mean *to say* '*It exists*,' albeit simply in *intention* [...], but rather that what exists is the intention, the 'directedness' towards such an object, but *not* the object. If, on the other hand, the intentional object exists, it is not only the intention, that is, the act of directedness which exists, but *also* that which is aimed at."[38] In other words, Husserl refuses "to grant the slightest ontological status to the object as the locus of a 'there is' which would differ from and in a way be preliminary to what occurs in reality."[39] When the object is simply intended, "it is then merely entertained in thought, and is nothing in reality."

The intentional object, though it is in a way anchored to current mental processes, does not possess the attributes of reality: By means of intentionality, the real components of consciousness "refer to what is not really inherent, namely by means of the heading of sense"[40], and intentionality therefore operates "a 'non-real' inclusion of the correlate [*i.e.* the target object] within consciousness."[41]

As it "refers" to "non-real" components, intentionality sets forth and articulates the modalities of *presence* and of *absence*, of the *effectively given* and of regulated *expectation*, of that which is perceived as "truly there" and of what is "presumed" to pertain to the object: "[The] 'intended content' [is] precisely that which is not actually given and that to which we nevertheless refer, in a mode which is adapted to this non-presentation."[42] Furthermore, "[intentional] means as much as: to tend, by means of any contents given to consciousness, towards other contents which are not given."[43]

It is this dialectic of presence and absence, this paradoxical form of the presentation of something which is nevertheless not there, which opens a field of specific objectivity, that of the phenomenal world. As highly stressed by M.-P.,

---

[36]Lyotard (1982, p. 29).

[37]Benoist (2001, p. 57).

[38]Husserl in Benoist (2001, p. 130).

[39]Benoist (2001, p. 131).

[40]Husserl (1983, p. 214).

[41]Ricœur, Note 1 *in* Husserl (1983, p. 302).

[42]Benoist (2001, pp. 26–27).

[43]Ibid., p. 28.

"how should we describe the existence of these absent objects or the nonvisible parts of present objects?"[44] They cannot be said to be *represented* or *imagined*, because that "[would] imply that they are not grasped as actually existing."[45] Neither is it possible to say that they are deduced according to the laws of the object, because "under this hypothesis I would know the unseen side as the necessary consequence of a certain law",[46] whereas it is not a matter here of *knowledge*, but of presence: "The hidden side is present in its own way. It is in my vicinity."[47]

It is at this point that the rupture with the world of objective reality, subject to determined and actual representations, is completed—because for the "categories of the objective world [...] there is no middle ground between presence and absence."[48]

3.2.5 Fulfillment

Also, the act of directedness, which refers thought to an "object", calls for an act that will bring it forth in the mode of effective presence, this being the *intention of fulfillment*[49] by means of which the anticipated components of the intended object find their actuality in terms of real components. As a structure of directedness, "[intentionality] constitutively [opens] a space of fulfillment (may it have been called to remain empty)"[50]—a fulfillment which is "nothing more than the

---

[44]Merleau-Ponty (1964, p. 13).

[45]Ibid.

[46]Ibid., p. 14.

[47]Ibid.

[48]*PhP/L*, p. 82.

[49]Concerning the act of fulfillment, it should be mentioned that it is in the context of the study of signifying intentionality that fulfilling intention is introduced. Such is the case starting with the first of the *Logical Investigations* where a theory of meaning is elaborated. Likewise in the sixth of the *Logical Investigations* where the concept of fulfillment is subject to a thorough examination, it is a correlate of meaning-intentions that it is at first presented. Specifically, in §8: "[when] an expression [...] is accompanied by a 'more or less' corresponding intuition [...] we experience a descriptively peculiar *consciousness of fulfilment*: the act of pure meaning, like a goal-seeking intention, finds its fulfilment in the act which renders the matter intuitive. In this transitional experience, the *mutual belongingness* of the two acts, the act of meaning, on the one hand, and the intuition which more or less corresponds to it, on the other, reveals its phenomenological roots" (*RL6/F*, p. 206). "[Where an expression] [...] is accompanied by a 'more or less' corresponding intuition [...] we experience a [...] *consciousness of fulfilment*: [...] the act of pure meaning, like a goal-seeking intention, finds its fulfilment in the act which renders the matter intuitive. In this transitional experience, the *mutual belongingness* of the two acts, the act of meaning, on the one hand, and the intuition which more or less corresponds to it, on the other, reveals its phenomenological roots." (Husserl 2013, p. 206).

[50]Benoist (2001, p. 31).

resorption of any 'intention' (as a directedness towards that which is not there) in the effective presence of that which was aimed at and which is now given."[51]

Fulfillment is therefore the act by which the prefigured attributes of the intentional object are made to coincide with object determinations (perceptive, imaginative…) produced following complementary intentional experiences, and it is the act which institutes the simply targeted object at superior levels of presence and of effectivity: "All intentions have corresponding possibilities of fulfilment (or of opposed frustrations) […] characterizable as acts, which permit each act to 'reach its goal' in an act specially correlated with it. These latter acts inasmuch as they fulfil intentions, may be called 'fulfilling acts' […] on account […] of self-fulfilment."[52]

Fulfillment therefore joins the trial of concordance with the process of saturation. The data invested in terms of actual correlates of the intentional object are appreciated, in what concerns their compatibility with the forms "to fulfill", in terms of failure or success of the act of fulfillment undertaken. Thus, when "[the fulfilling] intuition may not accord with a significant intention, [it] may 'quarrel' with it."[53] And to the intention which is then "frustrated in conflicts"[54] corresponds a specific consciousness of "frustration". Conversely, when there is assimilation, due to their adequation, between the intended forms and the presented forms, intention "[is] fulfilled in identifications."[55] In such case, "instead of the emptiness of absence (of the "emptiness" of the simple directedness), we now find the fullness of real presence",[56] which gives way to a "consciousness of "saturation", of fulfillment of that which had previously only been intention."[57]

But if intentionality maintains an essential relation with its fulfillment, it is not, however, a relation of necessity. This is because the tension towards fulfillment which inhabits the intentionally constituted object in no way presupposes the possibility of fulfillment. Intentionality is simply a *preparation* with respect to the object, in the sense where the forms of the object it outlines (in a more or less determinate manner) are those which open towards the *encounter* of an object, but without presuming the possibility of such a coincidence.

In this theoretical apparatus, *intuition* thus finds itself to be grasped anew as an act of fulfillment: Within and by means of intuition, the intended object acquires a higher degree of actuality. For example, an imaginative act, which delivers the mental representation of an object or of a scene, fulfills and presentifies the signification of an "unfulfilled" meaning-intention carried by any semiolinguistic item. We will speak of *perception* when the fulfilling act delivers the object "in itself":

---

[51]Ibid., p. 29.

[52]*RL6/F*, p. 216.

[53]Ibid., p. 212.

[54]Ibid.

[55]Ibid.

[56]Benoist (2001, p. 29).

[57]Ibid.

"Any perceiving consciousness has the peculiarity of being a consciousness of the own presence 'in person' of an individual Object."[58] Perception is therefore *originary presentive intuition*, and, as it makes present the being-in-itself, it occupies a fundamental position in the gnoseological apparatus of transcendental phenomenology—hence the "principle of all principles" by virtue of which "every originary presentive intuition is a legitimizing source of cognition."[59]

3.2.6 Meaning-Intention

It is of no surprise that the concept of intentionality, as it configures immanent experiences in order to refer consciousness, through them, to its objects, is first approached and explored in the facts of meaning (*Philosophy of Arithmetic* and the first of the *Logical Investigations*). The similarity between intentional directionality and the reference of the signifier to the signified is indubitable and signifying phenomenality thus seems indeed to be the terrain where the mode of intentionality is exhibited in the most straightforward manner. As Benoist points out: "The birthplace of the Husserlian phenomenological theory of intentionality is clearly the theory of signification" and "the meaning-intention [...] serves at a certain level as a general model for intentionality."

Indeed, the reference to the object which functions in "significant" intentional acts is much more than the simple passage from a symbol to a concept, in short, of a perceptive identity to an intellective identity of which, following the classical definition of the sign, the first would "be in place of." We should recall (cf. 2.2) that Husserl insistently and steadily distinguishes (even in the *Lectures on the Theory of Meaning*) the *meaningful* sign (the "true" sign which he calls *expression*) from the *indicative sign*. Whereas the second correlates two objects of consciousness that are of distinct natures and which are constituted independently from one another, the first proceeds from a specific intentional tension called "significative", which "animates" perceptual matter so as to establish it as an expression, as far as a signification is aimed at through it.

3.2.7 The Fulfillment of Signification

The act of meaning—like any intentionality—thus comprises in its nature and in its internal logic a kind of "ontological" incompleteness, the resolution of which is incumbent upon the act of fulfillment. The intentional act, which confers signification and by which a name finds itself endowed with meaning, is logically

---

[58]Husserl (1983, p. 83).
[59]Husserl (1983, p. 44).

accompanied by an additional act by which the intended object finds itself to be "presentified". Complementarily, therefore, the significatory aim awaits an act of fulfillment by which consciousness acquires from its previously *simply* intended object a new determination that will carry it to a greater degree of "*actuality*" and of "*effectivity.*"

Three types of contents may then be distinguished: "[i] the content as intending sense, or as sense, *meaning simpliciter*; [ii] the content as fulfilling sense; and; [iii] the content as object."[60] Whereas the "intending sense" is that of the object as it is *simply* intended following the mode of meaning-intention, the "fulfilling sense" designates the actual contents coinciding precisely with the intended object of meaning—in other words, it is "*the idea of the object such as could be given in accordance with what is said about it*"[61]—and it is through the fulfilling sense that the relation between the designation and the thing designated is likely to be achieved. Signification as fulfillment thus establishes a bridge between the expression and the designated thing (when such thing exists), and therefore plays the role of 'sense,' taken in the very classical acceptation following Frege as a means of access to the object, and reciprocally, as its mode of presentation—that is, as the way in which this object finds itself presented by the word which denotes it. In short, given that "[…] the originally empty meaning-intention is now fulfilled, the relation to an object is realized, the naming becomes an actual conscious relation between name and object named."[62]

But if it opens a path towards objects, expression in its essential principle is in no way suspended to them, specifically in two respects: *Firstly*, the act of fulfillment does not necessarily constitute mediation towards the object. *Secondly*, as we have glimpsed, the fulfillment bears no necessary relation to the meaning-intention.

Indeed, *on the one part*, the fulfilling sense is not necessarily the act which brings the meaning-intention into coincidence with a referent in the empirical world; it can very well be matter of conceptual determinations or of products of the imagination: "Expressions and their meaning-intentions do not take their measure […] from mere intuition—I mean phenomena of external or internal sensibility— but from the varying intellectual forms through which intuited objects first become intelligibly determined, mutually related objects."[63] *On the other hand*, the acts fulfilling meaning, if they are effectively in a fundamental logical relationship with expressions, "[are] not essential to the expression as such."[64] Thus, "beyond [any] presentation the *directedness* subsists. And this directedness is indeed, in a way, directedness towards an object."[65] *Moreover*, the phenomenological examination of

---

[60]*RL1/F*, p. 200.

[61]Benoist (2001, p. 60), or: "This is what would be perceived inasmuch as we would *see what we say.*"

[62]*RL1/F*, p. 192.

[63]Ibid., p. 199.

[64]Ibid., p. 192.

[65]Benoist (2001, p. 55).

the experiences of meaning indeed indicates the possibility for receiving expressions in their significatory dimension without however having recourse to any act of fulfillment: "Verbal expressions are no doubt often accompanied by images [...] but to treat such accompaniments as necessary conditions for understanding runs counter to the plainest facts [...] true illustrations which genuinely carry out or confirm the meaning-intention of our expression, can often only be evoked with difficulty or not at all."[66] *Furthermore*, the fact of legitimate expressions (from the standpoint of the language system) which do not have any possible fulfillment cannot be ignored, for instance, "round squares" or other oxymorons which exclude any possibility for intuitive illustration: "The sense of an absurd expression is such as to refer to what cannot be objectively put together."[67] In such case, "we apprehend the real impossibility of meaning-fulfillment through an experience of the incompatibility of the partial meanings in the intended unity of fulfilment."[68]

The existence of meaning intentions devoid of any fulfillment whatsoever would then attest to the (previously assessed) power of systems of signification to produce their own objectivity. In sum, it would be the proof "of a capacity of linguistic intentionality for having objects which would have no other consistency than what intentionality had provided them with."[69] Indeed, "round squares and other objects of this type—objects which can only be signified, but not perceived nor even intuitioned or 'illustrated' in the broad sense of the term—[seem to attest to] the autonomy of the significatory modality of intentionality, [and] to its capacity to comprise objects itself."[70] But the strictly transcendental conception of meaning-intention, the thesis of a linguistic idealism without nuance is, as we know, by no means defended by Husserl, for whom "*the fact of being intended will never suffice, in the proper sense of the term, to produce an object.*"[71]

It is that intentionality, due to its functional and categorical signification as pure directedness, integrates within the forms of its object the dimension of absence. Thus, as a tension calling for complementary acts of fulfillment by which it would fill the incompleteness of what it elaborates, intentionality "it *nothing more that a path towards the object.*"[72]

---

[66]*RLI/F*, p. 206.

[67]Ibid., p. 209.

[68]Ibid., p. 202.

[69]Benoist (2001, p. 54).

[70]Ibid., p. 53.

[71]Ibid., p. 130.

[72]Ibid., p. 46.

3.2.8 Nonsense and Absurdity

The uncoupling and the relation between intentional acts and fulfilling acts have a direct translation with respect to linguistic dysfunctions.

Considering complex expressions, Husserl distinguishes[73] between the "sense-less" ("or nonsensical"[74]) and the "absurd" ("or counter-sensical"[75]): It is necessary to not confuse "the true meaningless [...] with another quite different meaning-lessness, *i.e. the* a priori *impossibility of a fulfilling sense*."[76]

Whereas the first (nonsense) concerns the true forms of linguistic objectivity, that is, the regime of meaningful intentions and the laws of their complexions which condition the very existence of meanings, the second (absurdity) concerns the intuitive or imaginational correlates by which the intended meaning takes the form of an actual representation within consciousness.

In the first case, what is in question is the very *existence* of an object of meaning. Thus, confronted with a random assortment of words[77], "it is apodictically clear that no such meaning can exist, that significant parts of these sorts, thus combined, cannot consist with each other in a unified meaning",[78] and the apodictical con-sciousness of the impossibility of such an assortment demonstrates the existence of *essential laws of meaning*, in other words, "[of] laws governing the existence or non-existence of meanings in the semantic sphere."[79] Also, the expression which would be "nonsensical", inasmuch as it contravenes to linguistic regimes of legality, would be devoid of any intentional capacity, and, therefore, would be devoid of linguistic existence: "nonsense" is the *annihilation* of any form of lin-guistic object.

In the second case, the expression complex presents an absurd character which does not put into question the existence of a meaning, but expresses the impossi-bility of fulfillment, for example, by an illustrative explicitation: The absurdity, or counter-sense, is the impossibility of conferring to an *existing* meaning a "mental image"[80] which actualizes it. Thus, in any oxymoron, we will recognize the exis-tence of meaning, however without this meaning being able to correspond to an object of the world or to an imaginational illustration: The absurd expression *round square* "really yields a unified meaning, having its mode of 'existence' or being in

---

[73]*RL1*, pp. 61–65 & pp. 76–77, and *RL3*, §12 & §14.

[74]*RL4/F*, p. 67.

[75]Ibid.

[76]*RL1/F*, p. 202.

[77] "[I]f we say 'a round or', 'a man and is' etc., there exist no meanings which correspond to such verbal combinations [...] The coordinated words give us the indirect idea of some unitary meaning they express, but it is apodictically clear that no such meaning can exist, that significant parts of these sorts, thus combined, cannot consist with each other in a unified meaning" (*RL4/F*, p. 67).

[78]*RL4/F*, p. 67.

[79]Ibid., p. 68.

[80]*RL1/F*, p. 207.

the *realm of ideal meanings, but it is apodictically evident* that no existent object can correspond to such an existent meaning"[81]: "the sense of an absurd expression is such as to refer to what cannot be objectively put together."[82]

3.3 The Prism of an Existential Phenomenology

3.3.1 Transition and Introduction

We have therefore seen that the intentional object is the focal point, the synthetic unity of a cluster of mental states which converge towards it as they seemingly carry consciousness towards its object. In the lengthy quote which follows, in which we find problematic terms closely resembling those of Husserlian phenomenology, it seems that M.-P. concurs, although by means of different routes, with this conception of intentionality and of consciousness as the faculty of animating and ordering the manifold of immanent experiences under the unity of a specific objectivity.

Thus, considering this power "that consists in treating the sensory givens as representatives of each other and, when taken together, as representatives of an 'eidos,' [that] consists in giving them a sense, in animating them from within, in organizing them into a system, in centering a plurality of experiences upon a single intelligible core, and in making an identifiable unity appear in them under different perspectives. In short, [that] consists in arranging behind the flux of impressions an invariant that gives the flux its reason and in articulating the material of experience."[83] So considering this power, M.-P. writes that "it cannot be said that consciousness has this power; rather, it *is* this power itself."[84] And, as if in an echo to Husserl's cardinal considerations, he continues: "From the moment there is consciousness, and in order for consciousness to exist, there must be something of which it is conscious, an intentional object, and it can only bear upon this object insofar as it 'irrealizes' itself and throws itself into the object, insofar as it is entirely within this reference to something, and insofar as it is a pure act of signification."[85]

But even if M.-P. and Husserl agree in what concerns words and the surface of the operations of consciousness, the forms and content of the regimes of intentionality they see at work differ completely.

In order to approach the Merleau-Pontian conception, it suffices to place the preceding quote within *PhP's* progression. This citation, apparently so close to the Husserlian spirit, comes from the discussion regarding *concrete* and *abstract*

---

[81]*RL4/F*, p. 67.
[82]*RL1/F*, p. 209.
[83]*PhP/L*, p. 123.
[84]Ibid.
[85]Ibid.

movements. In the lines preceding the passage in question, we can read that this power "that consists in [animating from within] the sensory givens, [etc.]",[86] and which is therefore the signature of a consciousness, is eminently locatable, but without being exclusive, in a class of bodily movements qualified as abstract: "[abstract movement] is thus inhabited by a power of objectification, by a 'symbolic function,' a 'representation function,' or a power of 'projection' [...] that consists in treating... [*we return here to the previous citation*]."[87] It therefore appears that the source of intentionality does not reside with a constitutive consciousness with a noematic aim, but within a bodily and action-oriented power which installs a universe of objects within its field of action.

At this point, it is fitting to return somewhat and very rudimentarily to the presuppositions and foremost considerations of Merleau-Pontian phenomenology. In doing so, it will essentially be question of introducing the elements necessary for the recognition of facts of expression and of the concrete/abstract opposition to which we will return later when the activity of speech will be approached as forming a gestural practice. But also, and foremost, it will be question of preparing the later introduction (cf. 7.16–7.18) of the concept of *diacriticity* establishing a link between speech and perception.

3.3.2   Positing the Issue, and as an Introduction

"All begins", to put it as such, with an interested and interrogative meeting between a bodily schema and an environment of solicitations, one which directs towards a constitution made of the crossings of body and world, and having, from the onset, a value as co-expression. Thus, M.-P. emphasizes motor projects, the rhythms of existence, the solidary differentiation of sensible things and sensorial modalities, to posit the body as the central actor of an "expressive saga", inasmuch as it outlines through each of its gestures a world of signifying presences.

A first modality of Merleau-Pontian being-in-the-world would therefore correspond to a desiring interrogation sparked by an environing halo within which meetings are sought. And the local figure of such exploratory experience is the gesture, being already in itself a response to a prior solicitation, an interrogation directed towards its source and an attempt at obtaining acknowledgement, at finding some issue within, or some motive.

It is therefore necessary, from the onset, to posit the distinction between *world* and *universe*. The first being the set of things as they find themselves *for* me, around me, as they are addressed to me and as I respond to them through reciprocated interactions, and the second being the set of objects amongst which I find myself to

---

[86]Ibid.
[87]Ibid.

be immersed, organized collections of mute objects installed for their own sake within an immutable space and which would remain as such without me.

Two orders of intelligibility are associated with these two orders of things: That of existence and that of reason. The *universe* does not have an essential centre. It arranges all things in function of specific interactions: It is "a completed and explicit totality where relations [are] reciprocally determined."[88] The *universe* does not stem from one point of view; rather, it integrates them all.

Conversely, the World is my world inasmuch as it responds to my power of action—it is not a flat world that can be encompassed from afar in its entirety under a system of representation. It is, to the contrary, the progressive world of an open and indefinite multiplicity which I embrace in the exploration my body engages within it, and which correlatively constitutes itself before me as a distribution of forms and qualities.

Since it is the *World*, and since it is with respect to my body and to its vital power that it takes shape as the theater of my activities, so since it is not a set of things, of qualities and of relations that all exist independently from me, that precede me, and that are simply delivered to me, things which I would visit and, if required, which I would seek to reach in their hypothetical objective essence while in search for knowledge, this world is configured in its sensible qualities, in its forms of appearings, inasmuch as it responds to the style of existence which I carry and which attempts to approach, to grasp, and to domesticate it.

It is therefore necessary to look prior to one's relation as an instituted individual with one's surroundings as a defined set of objects and of qualities. And prior to such relation, there remains but one's body as a carrier and performer of a certain life force and a hazy environment which "vaguely solicits", a sort of "poorly formulated question.[89] "Without the exploration of my gaze or my hand, and prior to my body synchronizing with it, the sensible is nothing but a vague solicitation"[90]—with which I will attempt to syntonize and the effect of which will flourish into sensible qualities.

3.3.3   The World as Expression

This is therefore to say that "vision is [...] inhabited by a sense that gives it a function in the spectacle of the world and in our existence."[91] *A contrario,* "The pure quale would only be given to us if the world were a spectacle and one's own body a mechanism."[92] Furthermore, "sensing [...] invests the quality with a living

---

[88]Ibid., p. 73.

[89]Ibid., p. 222.

[90]Ibid.

[91]Ibid., p. 52.

[92]Ibid.

value, grasps it first in its signification for us, for this weighty mass that is our body"[93] and "sensing is this living communication with the world that makes it present to us as the familiar place of our life."[94]

Thus, the world in which my life unfolds is a world which is *significant* with respect to this very life, precisely because it is my living power which establishes it in its actual forms. The world of experience is a world of meaning. Each sensible character of a thing expresses the sense of my investment in its direction as a response to its address.

Also, in the most general way, there is the co-constitution of an environment made of meanings which we can describe (for the moment) as being experienced and of a praxis: a bodily configuration which manifests through familiar gestures, through reiterable gesticulations. This structure, which is almost dialogical, of the gesture and its vital horizon, first requires to be characterized as a bodily movement, relatively to its expressive ranges and to the modalities of the engagements which traverse it.

Finally, movement, understood as a modality of corporeal action, will also shape the aspects by which an environment actualizes itself before me: The objects, qualities, geometries, horizons which take shape and form my actual empirical horizon all represent significations instituted with respect to my latitudes of action and logic of approach. If we call "background" (of movement) the geometry of meanings that movement institutes as an environment—the background of movement is not "a stock of sensible qualities, but by a certain manner of articulating or of structuring the surroundings[95]—we understand that movement "adheres to its background" or, correlatively, that the background fully traces the lines of movement ("movement and its background form a unique totality"), and the background of movement "is not a representation associated to movement, it is immanent to it: It animates and carries it at each instant." Such is the case at least for concrete movement, the "movement necessary for life" by which an ecological framework takes shape. For this type of movement, movement and background mirror one another, are contiguous to one another: It is "[from the affective situation of the whole that] the movement flows"[96] and it is movement, as appropriation of the world, which institutes the characters of the situation where it unfolds.

---

[93]Ibid.

[94]Ibid., p. 53.

[95]Ibid., p. 117.

[96]Ibid., p. 107.

3.3.4 Abstract/Concrete

It is at this point that M.-P., in one of the first chapters of *PhP* called "The Spatiality of One's Own Body and Motility", introduces the distinction between "abstract" and "concrete" movements.

If bodily space is what circumscribes action, this only holds for concrete movements—in general, an object or a tool "are presented to the subject as poles of action; they define, through their combined value, a particular situation that remains open, that calls for a certain mode of resolution, a certain labor."[97] Concrete movement is the direct and appropriate response to this solicitation for action, which "obtains the necessary movements from [the subject] just as [...] the customs of our milieu [...] obtains from us the words, attitudes, and tone that fits with them."[98]

Concrete movement, in a radicalized form induced by pathological states, rather than being a flexible and adaptable response to the solicitations of action, appears to be but a forced response, as if the environment imperiously commanded and got the subject to perform the required movements without any choice. In such cases, the affected person acts without the ability to resist in order to produce imposed, oriented gestures: "[the patient] experiences movements as a result of the situation, [he and his] movements are, so to speak, merely a link in the unfolding of the whole."[99]

Then, the gesture and its background form a whole: The movement and the situation become one, the gesture institutes an environment and a geometry of objects as signifying presences and these, in return, canalize the action of which they express the unfolding—thus, the affected person only succeeds in performing the movements "on command"[100] "[only] on condition of placing himself into the spirit of the actual situation."[101]

But if the concrete movement only concerns, in its imperious form, the register of vital gestures, it also pertains to "habits": The "assimilated" gesture is indeed what responds to an environment perceived as the fitting and guiding receptacle for its accomplishment.

This sort of movement may be considered as a pathological or rigidified version of other movements, be they vital or assimilated: the situational context opening naturally onto unquestioned movements which deliver its sense, all the while leaving the subject with the liberty to organize the situation (such as in the case of a perfectly healthy individual who would casually stretch his or her hand towards the fruit basket).

---

[97]Ibid., pp. 108–109.

[98]Ibid., p. 109.

[99]Ibid., p. 107.

[100]Ibid.

[101]Ibid.

On the other hand, *abstract* movements are free from conditioning by any more or less assimilated situations. The abstract movement is "on command" and "[is] not directed towards any actual situation."[102]

To accomplish an abstract movement is in a way "to possess my body independently of all urgent tasks, in order to make use of it in my imagination."[103] In the pathological situations described by M.-P., these movements are very difficult to accomplish for some patients who may lack this aptitude of relaying a definite program in "abstract" form to their motor projects, be it on their own accord or in response to directives.

Dually, the abstract movement projects its fabric of meanings so as to set up the scene of a new situation. Being capable of abstract movements entails liberating oneself from the situation's conditioning through them. It means to cause things to recede and to introduce a plane of novel meanings between them and the acting subject: The motor project of abstract movement "aims at my forearm, my arm, my fingers, and it aims at them insofar as they are capable of breaking with their insertion in the given world and of sketching out around me a fictional situation"[104]—the abstract movement "carves out within that plenum of the world in which concrete movement took place a zone of reflection and subjectivity; it superimposes upon physical space a virtual or human space"—in other words, if "[the concrete movement] adheres to a given background, the [abstract movement] itself sets up its own background."[105] With abstract movement, we "invert the natural relation between my body and the surroundings."[106]

Furthermore, with abstract movements, subjects "polarize the world, causing a thousand signs to appear there, as if by magic, that guide action, as signs in a museum guide the visitor"[107]; they are capable of "marking out borders and directions in the given world, of establishing lines of force, of arranging perspectives, of organizing the given world according to the projects of the moment, and of constructing upon the geographical surroundings a milieu of behavior and a system of significations that express, on the outside, the internal activity of the subject."[108] Abstract movement therefore is "a voluntary movement [which] takes place in a milieu, against a background determined by the movement itself"[109]—meaning that the movement projects its background.

Concrete and abstract movements are penetrated with two sorts of consciousnesses of the surrounding world: Concrete movement occurs in a world that is perceived and experienced as being built according to a guiding schema, whereas

---

[102]Ibid., p. 105.

[103]Ibid., p. 115.

[104]Ibid., p. 114.

[105]Ibid.

[106]Ibid., p. 115.

[107]Ibid.

[108]Ibid.

[109]Ibid., p. 139.

abstract movement operates within a world recognized as an "objective environment", a world of objects liberated from one's own body's power of action, instituted in themselves, and which, in return, require nothing from it.

Abstract movement, in its power to produce an environment of novel values, would thus present two concomitant facets: The residual side of a world devoid of its former "concrete" meanings, in sum, the "objective milieu" of a transcendental consciousness, and a projected world of which the "objective milieu", passive and mute, constitutes a possible receptacle.

But not only does this process of objectifying refiguration concern objects, it is also the *gesture* which intervenes and which finds itself to be precipitated into an in-between within which, without necessarily losing its spontaneity, it might be *traced*, promoted as a figure, inscribed within an external spatiality, and already almost articulated into a diagram. Thenceforth likely to be reflected upon and broken down so as to be addressed in the form of distinct parts, it is like an elucidated component of the bodily praxis which had until then been tacit.

3.3.5   *"Paradoxical" Presence*

Let's examine how the dialectic of *absent presence* presents itself in existential phenomenology.

There is, from the onset, a set of diffuse solicitations and the body conceived as a muffled resource for action, with its life force steered towards surroundings which interrogate it.

Perception will primitively and fundamentally be this aptitude of receiving solicitations and, dually, of syntonizing with them so as to establish them within a world of objects and of qualities which are the expression, the living meaning, of this successful coordination: "the subject of sensation is a power that is born together with a certain existential milieu or that is synchronized with it."[110]

To perceive is therefore to already be capable of being drawn into a certain fabric of solicitations, namely as it indistinctively resides in the laterality of our perceptual fields. Thus, to see an object is "either to have it in the margins of the visual field and to be able to focus on it, or actually to respond to this solicitation by focusing on it."[111]

In its simple attentional aspect, the perceptual act thus consists in promoting to a higher level of determination a certain marginal and uncertain component of the field, for example through an emphasis which thematizes in function of the figure/ ground articulation: "To pay attention is not merely to further clarify some preexisting givens; rather, it is to realize in them a new articulation by taking them

---

[110]Ibid., p. 219.

[111]Ibid., pp. 69–70.

as figures."[112] Thus, perception, in its attentional moment is a "passage from the indeterminate to the determinate."[113]

But this marginal and uncertain portion of the perceptual field, its power of attraction, stems from its being full of announced presences, from its promise of a universe of things to be met—things that "are only pre-formed as *horizons*."[114] In other words, "attention [...] is the active constitution of a new object that develops and thematizes what was until then only offered as an indeterminate horizon."[115]

But to promote an object as a figure means to correlatively put back that which accompanies it into an indistinct background. The logic of vision therefore consists in the following: "I close off the landscape and open up the object"[116]—it is a necessity of perception that "to suspend the surroundings in order to see the object better, and to lose in the background what is gained in the figure."[117]

But if to focus on an object is "to plunge into it"[118] and "[to] continue within one object the same exploration that, just a moment ago, surveyed all of them",[119] the fact remains that the objects relegated to the background "do not cease to be there."[120] In other words, "the inner horizon of an object cannot become an object without the surrounding objects becoming a horizon."[121]

Thus, just as the object, taken in isolation, only reveals itself to me through one of its faces, hence, is actualized according to a synthesis of the presence and of the absence of its parts, one which is seemingly woven into its inner horizon, likewise, an object always appears in a field making way for other objects, meaning that it is accompanied by other objects that are present only in a lesser manner: This is the outer horizon.

Through the horizons, one perceives an object in company of other objects of which the presence is more or less actualized and which, together, "form a system or a world, [thus] insofar as each of them arranges the others around itself like spectators of its hidden aspects."[122] For example, "the back of my lamp is merely the face that it 'shows' to the fireplace."[123]

And this is the case despite that "my human gaze never posits more than one side of the object, even if by means of horizons it intends all the others"[124] and "with

---

[112]Ibid., p. 32.

[113]Ibid., p. 33.

[114]Ibid., p. 32.

[115]Ibid., p. 33.

[116]Ibid.

[117]Ibid., p. 70.

[118]Ibid., p. 70.

[119]Ibid.

[120]Ibid.

[121]Ibid.

[122]Ibid.

[123]Ibid., p. 71.

[124]Ibid., p. 72.

these other objects, I also have their horizons at my disposal, and the object I am currently focusing on—seen peripherally—is implied in these other horizons."[125] Thus, "each object [in itself] just is all that the others 'see' of it [and as it] is seen from everywhere."[126]

In this perspective, the absent faces of an object are not given through an act of directedness, and thus being simply presumed, but as they are accessible to my body, as a potential target of (visual) action. The horizon replaces the noema, in a manner of speaking. It is the correlative of the active power of my gaze rather than being the unity of the flux of mental processes. The horizon "then, is what assures the identity of the object throughout the exploration."

References

Benoist, J. (2001). *Intentionalité et langage dans les "Recherches logiques" de Husserl*. Paris: PUF, coll. Epiméthée.

Depraz, N. (2012). *Introduction* à Husserl, E., *La crise de l'humanité européenne et la philosophie*. La Gaya Scienza.

Husserl, E. (1960). *Cartesian meditations* (D. Cairns, Trans.). Dordrecht: Springer-Science.

Husserl, E. (1983). *Ideas pertaining to a pre phenomenology and to a phenomenological philosophy* (F. Kersten, Trans.). The Hague: Martinus Nijhoff.

Husserl, E. (1999). *The idea of phenomenology* (L. Hardy, Trans.). Dordrecht: Kluwer Academic Press.

Husserl, E. (2013). *Logical investigations, Volume 2* (D. Moran, Trans.). London & New-York: Routledge.

Kant, E. (2003). *Critique of Pure Reason* (J. M. D. Meiklejohn, Trans.). Mineola: Dover Publications.

Ladrière, J. (1996). *Sciences*. Paris: Encyclopaedia Universalis.

Levinas, E. (1998). *Discovering existence with Husserl* (R. A. Cohen & M. B. Smith, Trans.). Evanston: Northwestern University Press.

Lyotard, J.-F. (1982). *La phénoménologie*. Paris: PUF, coll. *Que sais-je?*, 625.

Merleau-Ponty, M. (1964). *The primacy of perception and its philosophical consequences* (J. M. Edie, Trans.). Evanston: North Western University Press.

Petitot, J. (1992). *Physique du sens: de la théorie des singularités aux structures sémio-narratives*. Paris: Editions du CNRS.

Petitot, J. (2002). Eidétique mophologique de la perception. In J. Petitot et al. (Eds.), *Naturaliser la phénoménologie: essais sur la phénoménologie contemporaine et les sciences cognitives*. Paris: CNRS Éditions, coll. CNRS Communication.

Philonenko, A. (1989). *L'oeuvre de Kant* (Vol. 1). Paris: Vrin, coll. A la Recherche de la Vérité.

Philonenko, A. (1993). *Introduction à: Kant, Critique de la Faculté de Juger*. Paris: Vrin, Bibliothèque des Textes Philosophiques.

---

[125]Ibid., p. 70.
[126]Ibid., p. 71.

Chapter 4
The Husserlian Perspective

Lengthy research on the constitution of the sign conducted by both Saussure and Husserl propose, through specific characterizations and within their own theoretical territory (structuralism and phenomenology, respectively), a solution for overcoming the "difficulties" of the sign. We will devote this chapter to a presentation of the phenomenological analysis of the sign developed by Husserl, starting with the first *Logical Investigations* and progressing towards the *Lectures*. We will focus on the series of problems Husserl was required to resolve—starting with the initial problems we evoked earlier (cf. 2.3) and followed by those which arose at various stages of his work.

4.1 A First Approach

The constitutive characters of the semiolinguistic phenomenon, as they are exposed in the first stages of the first *Logical Investigation*, are not devoid of ambiguity nor of difficulty.

Firstly, a too hasty and partial reading of the first *Logical Investigation* could give the impression that the expressive fact proceeds from a single intentional act: the significatory aim. Certain passages discussing the essential characters of the "meaningful sign" emphasizes this specific act which, by infusing an acoustic or graphical material, confers meaning to it. We can read, for instance (as already quoted in Chap. 2), that "the 'meaning-intention' [...] characteristically marks off an expression from empty 'sound of words'",[1] or that "the essence of an expression lies solely in its meaning".[2] It then suffices to misconceive of the nature of the material substratum animated by the significatory aim, and, correlatively, to misconceive of the character (complex rather than simple) of the act instituting the

---

[1] *RL1/F*, p. 194.
[2] Ibid., p. 199.

© Springer International Publishing AG, part of Springer Nature 2018

D. Piotrowski, *Morphogenesis of the Sign*, Lecture Notes in Morphogenesis, https://doi.org/10.1007/978-3-319-89848-3_4

sign-phenomenon to conclude it is inadequate to analyze the sign in terms of intentionality. Indeed, if we situate the concreteness of the sign on an immanent level, that is, if we relate it to "an experienced complex of sensations",[3] that is, if we ignore the participation within the sign of a simple perceptive intention instituting a concrete phenomenon, then given that "the sensational complex is as little perceived as is the act in which the perceived object is as such constituted"[4] and given that the object which presents itself to consciousness is precisely that which the meaning-intention aims at, then, the sign would reveal itself to be discarnate, that is, to be a pure object of sense, an ideational entity of targeted meaning devoid of any sensorial anchoring.

But this situation is radically contradicted by the phenomenological examination of the semiolinguistic fact, even while conducted in the most rudimentary manner.

Indeed, that which is targeted in and by an expression, that is, an object of meaning, does not exhaust the contents of the field of semiolinguistic consciousness. It most likely the point of attraction for consciousness, what consciousness mainly has in sight through the prism of this "expression", but another component cohabits within the consciousness of signification: It is the ambiguous consciousness of a sensible, graphical, or acoustic complex, not as a simple perceptive object closed upon itself, but as if traversed by a tension which modalizes its appearing.

For, while reading a word, and with my thoughts oriented towards its meaning, this meaning is not alone in my consciousness: Before my eyes are also present, although in another manner and to a lesser degree, that which I recognize precisely as being a signifier, that is, a perceptual complex infused with a significatory aim, but, and we should insist, not merging with the object of the significatory aim, as may be the case, for example, with the adumbrations (apparent contours) of the flux of immanent experiences of spatial perception which merge within the unity of a spatial body.

The trial of phenomenological substitution attests to the involvement of an act of perception within the expressive fact. Indeed, the alternation from a meaning-intention to a perceptual intention has the effect of "uncovering" the underlying physical word rather than of transfiguring the word-expression phenomenon as initially apprehended. There is indeed in expression something of the availability of a perceptual object, one already recognized, and of which the existence finds itself to be confirmed at the moment of intentional substitution.

Dually, the recognition of the semiolinguistic value of a marking already perceived as a sensible phenomenon does not abolish its concrete presence. Without a doubt, the presence of the first object of perception then appears to be diminished; however, despite being relegated in consciousness to a secondary plane of existence, it preserves the attributes which had been recognized at first to present.

This, however, is not devoid of ambiguity, as shown by the following citations which recognize within the expressive phenomenon the persistence of attributes

---

[3]Ibid., p. 214.
[4]Ibid.

characterizing the sign as a concrete marking, all the while emphasizing the essential modalization that it receives.

> It may also be the case that some sensible feature first arouses interest on its own account, and that its verbal or other symbolic character is only then noted. The sensuous habit of an object does not change when it assumes the status of a symbol for us, nor, conversely, does it do so when we ignore the meaning of what normally functions as a symbol [...] One and the same content has rather altered its psychic habits: we are differently minded in respect of it, it no longer seems a mere sensuous mark on paper, the physical phenomenon counts as an *understood* sign.[5]

> [If] we turn our attention to the sign *qua* sign, e.g. to the printed word as such. If we do this, we have an external percept (or external intuitive idea) just like any other, whose object loses its verbal character. If this object again functions as a word, its presentation is wholly altered in character. The word (*qua* external singular) remains intuitively present, maintains its appearance, but we no longer intend it, it no longer properly is the object of our 'mental activity' [...] the intuitive presentation, in which the physical appearance of the word is constituted, undergoes an essential phenomenal modification when its object begins to count as an expression. While what constitutes the object's appearing remains unchanged, the intentional character of the experience alters.[6]

4.2   The Cardinal Difficulty

It should be noted that the trial of substitution considered here alternates the intentional orientations in a binary manner: Either the "word" is "understood", meaning that it responds to a significatory orientation, or the word is "perceived", meaning that it is grasped in its nature as a physical phenomenon. Correlatively, the alternation involves only the "terminal" objects of the two kinds of aimings at work in the phenomenon of expression. There is therefore no place here for an intermediate state or object position, that is, for a kind of hybrid or medium term between the two concurrent intentions.

Now, this binarity will prove to be the source of insurmountable difficulties. This is because the perceptual and meaning-intentions, which are conjugated within the semiolinguistic aim, each institute objects that are mutually exclusive and which pose an obstacle to the constitution of a unitary although complex semiolinguistic intention.

It will therefore be necessary to surpass the binary play of intentions recorded as components of the semiolinguistic act all the while acknowledging it, hence allowing for the existence of an intermediate phenomenological state, one articulating the perceptive and significatory aimings. We will see that it is precisely towards the recognition of a median phenomenological status and towards the elaboration of a specific position of consciousness that Husserl's reflection leads, finally overcoming numerous deadlocks. But, even if it causes us to anticipate, we

---

[5]Ibid., p. 209.
[6]Ibid., pp. 193–194.

will also present its principle, one which Husserl will transpose onto the issue of the attentional field.

In order to uncover and to reach this intermediary state, formed by a sort of significatory and perceptual hybrid, it is necessary to define a variational procedure which does not "discontinuously" skip from a meaning-intention to a perceptual intention, but which comprises a transitional stage between the two. This can be achieved, albeit here in a rather artificial manner that will nevertheless be reconstructed within the apparatus of the attentional field, by distinguishing within the significatory aim that which pertains to the animation of the perceptual substrate from that which pertains to its objective, which is the intentional object of meaning. In order to pass from meaning-intention to perceptual intention, we will first proceed, by means of abstraction, to the neutralization of the object of meaning. In doing so, we will institute a kind of transitory state, one which is incomplete and highly unstable, in which a perceptual basis is intentionally infused but in which the targeted object (of meaning) is not consciously espoused. This state is in a way the one delivered by "generally" significatory construings of a sensible material, that is, construings which, by not targeting specific objects of meaning, only inform their matter towards an indeterminate horizon of meaning: the horizon of possible meaning. The following stage will consist in canceling any significatory construing of the object of perception, so as to solely recover the perceptual intention governing the phenomenological installation of a sign as being concrete.

In short, what phenomenological examination and intentional substitution show is that despite that "we live" in the object of meaning, we cannot entirely escape the presence, as if residual, of an object pertaining to a perceptual act. It is as if in the expression, of which the phenomenological characteristics imprint the relation to the meaning, would reside, in a diminished yet persistent form, the trace of an object as simply perceived. The consciousness of expression, despite its being preferentially involved in a moment of signification, always comprises the underlying albeit modulated consciousness of an object pertaining to a simply perceptual act—in other words: "The expression is indeed perceived, but our interest does not live in this perception."[7]

4.3   The Two Options

It is therefore necessary to conceive of semiolinguistic intention as a complex act, one to which the act of perceiving an underlying physical phenomenon contributes: "[When] we reflect on the relation of expression to meaning [...] we break up our complex, intimately unified experience of the sense filled expression, into the two factors of word and sense." It is precisely this dual yet unitary complexion of the phenomenon of expression which preoccupies Husserl and which he will

---

[7]*RL5/F*, p. 116.

successively address from various angles. Since it is a matter of conceiving the principle and the mode of the connection of a perceptual act and of a significatory act, two main options present themselves, options which Husserl will approach more or less distinctly.

4.3.1 First Option

The *first option is that of layering*: There would first be the moment of a perceptual act producing a sensible phenomenon, an acoustic or visual perceptual object for instance, and then there would be a significatory act which would grasp this primary material and, by infusing it with a significatory aim, would institute for consciousness an object of meaning. Such a conception is apparent for example in the following passages: First, we have "the sensuous acts in which the expression, *qua* mere sound of words, makes its appearance [and] on the other hand [...] acts essential to the expression if it is to be an expression at all, i.e. a verbal sound infused with sense. These acts we shall call the *meaning-conferring acts* or the *meaning-intentions*",[8] and "[T]he function of a word (or rather of an intuitive word-representation) is to awaken a sense-conferring act in ourselves [...] to guide our interest exclusively in this direction."[9] We also find such a conception in the following citation: "If we seek a foothold in pure description, the concrete phenomenon of the sense-informed expression breaks up, on the one hand, into the *physical phenomenon* forming the physical side of the expression, and, on the other hand, into the *acts* which give it *meaning* [...] In virtue of such acts, the expression is more than a merely sounded word."[10]

But this view is not admissible, because we do not see why a perceptual act should trigger or open onto a significatory act: The perceptual act is an accomplished act and the perceptual phenomenon is self-sufficient—it does not indicate, nor does it even suggest the necessity or even the possibility for having an additional act of signification to provide meaning. Reciprocally, according to such a logic of layering, the significatory engagement would be conditioned by a perception which actually is indifferent to it: Everything is as if the possibility for "signifying" was subordinate to the completely random availability of a preceding perceptual moment, such availability being contingent, because the perceptual moment is by no means oriented towards an eventual meaning. Regardless, this manner of conceiving the articulation of perceptual and significatory acts was rejected by Husserl himself, notably when he observed that in the semiolinguistic complex, "no new, independent content [the meaning] is added to the old [the

---

[8] *RL1/F*, p. 192.
[9] Ibid., p. 193.
[10] Ibid., pp. 191–192.

sound complex]: we do not merely have a sum or association of contents of equal status."[11]

4.3.2   Second Option

The second option, temporarily defended by Husserl, consists in a *global act composed of partial acts by virtue of* foundational relations, that is, composed in function of "a law of essence, according to which the existence of [certain characters or acts] presupposes the existence of [some others characters or acts]."[12]

More specifically, the significatory act is a founded act, that is, it presupposes a founding act which is sensible perception. No longer do we have a logic of succession inasmuch as the founding and founded acts participate in a unitary global complex: The founding and founded acts are partial acts and as such they carry the value of their involvement in the total act to which they participate. Thus, the founded act does not occur following the founding act: In the case which interests us, meaning-intention does not add itself to the perceptual act delivering a phonic or graphical complex, but constitutes the necessary complement in the accomplishment of a global act comprising, as a founding act, the perception of the "word form" (the word as a phonic or graphical phenomenon).

This conception of a semiolinguistic intention as a complex of "founded" acts was introduced by Husserl towards the end of the first *Logical Investigation*, and it was reinvoked and discussed extensively in the fifth *Logical Investigation*.

At a first level which serves as a "foundation", by virtue of an act of apprehension aiming towards a concrete entity, "the mere sign appears before us as a physical object, e.g. as a sounded word, given here and now."[13] This level of phenomenological elaboration, in which the concrete mark of the sign therefore appears in its configuration as a physical object, is fully autonomous: It suffices to institute a sphere of phenomenalities. But "on this first conception [...] a second is built, which goes entirely beyond the experienced sense-material, which it no longer uses as analogical building-material, to the quite new object of its present meaning. The latter is meant in the act of meaning."[14] It is therefore based on the sign in its nature as a material entity that, subsequently, the act of significatory aiming takes place, an act by which an object of meaning is instituted within consciousness and which will have the collateral effect of operating a phenomenological mutation of the sign; first seen as a concrete phenomenon, and then as a signifying side of an expression (the signifier). The signifying act is therefore indeed a "significatory" act which confers a meaning to a word form. In short,

---

[11]Ibid., p. 209.
[12]*RL3/F*, p. 12.
[13]*RL1/F*, p. 214.
[14]Ibid.

"meaning is a variously tinctured act-character, presupposing an act of intuitive presentation as its necessary foundation. In the latter act, the expression becomes constituted as a physical object. It becomes an expression, in the full, proper sense, only through an act founded upon this former act."[15]

If the first *Logical Investigation* specifies the type of relation which exists between the act of "signifying" and the intuitive apprehension of the sign as a concrete phenomenon, that is, as a "founding" relation, nothing, at this stage of the phenomenological investigation, has been said about the nature of the phenomenal modulation which the concrete sign undergoes when it is grasped as an expression, nor, correlatively, has anything been said regarding the nature of the relation which institutes itself between the signifying side of the expression (the signifier) and the object of the significatory aim, i.e. the signified.

4.3.3   Second Option: Difficulties

It is in the fifth *Logical Investigation* that Husserl specifically addresses these questions and that he is led to modify his analysis into founding and founded acts.

He begins by recalling that a sign is "as much a physical object as any pen-scratch or ink-blot on paper",[16] and that through acts of perceptual representation, "it is 'given' to us in the same sense as a physical object."[17] The phenomenon of "expression" is then described as "a union of the acts in which an expression, treated as a sensuous verbal sound, is constituted, with the quite different act constitutive of its meaning".[18] In other words, the complex act presiding over significatory phenomena articulates and synthesizes into a global act, a foundational act, by which the phenomenon of the physical sign is constituted, and a founded act, which turns towards an object of meaning.

But breaking down the phenomenon of expression in this manner leads to a problem regarding its constitution, because such a conception of a meaning-intention participating in a complex of partial acts that are in foundational relationships does not account for the phenomenological modulation of the sign as a sensible object. *Indeed, partial acts leave unchanged the phenomenological characters of the objects which they respectively institute.*

Indeed, even if the objectivity intended in the global act has its own phenomenological characters, objects which specifically proceed from the partial acts, as so many constitutive parts of the global objectivity, are recognizable as such within the totality they compose: "[T]he unity of what is objectively presented, and the whole manner of the intentional reference to it, are not set up *alongside* of the

---

[15]Ibid.

[16]*RL5/F*, p. 117.

[17]Ibid.

[18]Ibid., p. 116.

partial acts, but *in* them, in the way in which they are combined [...] The object of this total act could not appear as it does, unless the partial acts presented their objects in their fashion: their general function is [in particular] to present parts [...] of the object."[19]

However, the phenomenological examination of an expression, in the strong sense of the term, would be unable to distinguish, as a constituent of global significatory objectivity, the correlate of the founding act, that is, the expression as a simple graphical or phonic complex.

Moreover, Husserl already observes in the first stages of the first *Logical Investigation* that "the word comes before us as intrinsically indifferent, whereas the sense seems the thing aimed at by the verbal sign and meant by its means."[20] He continues by stating that "the expression seems to direct interest away from itself towards its sense, and to point to the latter"[21]: "[T]he word only ceases to be a word when our interest stops at its sensory contour, when it becomes a mere sound-pattern."[22] Thus, the articulation between the signifier and the signified is viewed here in the mode of a shift of interest.

Regarding this point and immediately, Husserl prevents any confusion between this "indirection of interest" and the referral mechanism which characterizes indicative signs: "[B]ut this pointing is not an indication in the sense previously discussed"[23] as a regulated succession of intentional acts. Indeed, if expression "directs interest away from itself" towards an object of meaning, that does not mean that it configures itself first as the object of a perceptual aim which it would then have the power to reorient towards an object of sense. In fact, "disinterestedness" is the property of intentional experience and it marks a kind of *weakening of the presence* of the word's form to the benefit of the object of meaning aimed at.

In the end, we will consider the following: When faced with a graphical sign, apprehended during a natural act of reading, our consciousness is fully invested in the object of sense which this sign makes present. "[O]ur interest, our intention, our thought [...] point exclusively to the thing meant in the sense-giving act."[24] It follows that the phenomenal constituency of the signifier is not that of the sign in its conformation as a physical object: It proceeds from the very act of meaning intending. Also, "the manifest verbal sound or written sign etc. is not seen as part of the object meant in the whole act, nor even as really determining it, nor as having really to do with it."[25]

Even more explicitly: Whereas in the complex act which merges partial acts by virtue of founding relations, the partial objectivities, as regards their "importance"

---

[19]Ibid., p. 115.

[20]*RL1/F*, p. 191.

[21]Ibid.

[22]Ibid., p. 190.

[23]Ibid., p. 191.

[24]Ibid., p. 193.

[25]*RL5/F*, p. 117.

or their "degree of presence", are equally configured as parts of global objectivity, such is not the case in "the compound act which includes both appearing-expression and the sense giving act." It is then "plainly the latter [which] are peculiarly prominent"[26]—in the sense that it is in them that consciousness fully invests itself: "[W]hen we normally express something, we do not, *qua* expressing it, live in the acts constituting the expression as a physical object—we are not interested in this object—but we live in the acts which give it sense: we are exclusively *turned* to the object that appears in such acts, we *aim* at it, we *mean* it in the special, *pregnant* sense."[27] Thus, "[t]he contribution made by the acts constituting the verbal sound [...] differs characteristically from the contribution of the underlying acts illustrated and discussed above."[28]

The foundational relation between the acts instituting first (physical object) and second (object of sense) objectivities therefore prove inappropriate in the analysis of the significatory phenomenon—and even doubly inappropriate when taking the following difficulty into consideration: Whereas "generally, the greatest energy will be displayed by the act-character which comprehends and subsumes all partial acts in its unity [...] in this act, we live, as it were, principally",[29] it is a whole other matter in what concerns the global act of signification, which, as we have seen, confers a major prevalence to the act of sense presentation.

In order to overcome the preceding inconsistencies stemming from a factorization of the fact of expression according to relations of foundation, Husserl introduced in the fifth *Logical investigation* the problematics of the attentional field, where the inequality of founded and founding acts with respect to the engagement of consciousness can be adequately reunderstood. But it is only in the *Lectures* that this option will be developed in detail. It must be noted that the importance given by Husserl to the attentional modality in phenomenological analysis will grow, and in *Ideen* it will be integrated into the noematic structure (cf. *Ideen* §92). But the rejection of a psychological interpretation of the attentional act already presents itself in the fifth *Logical Investigation*, where Husserl insists on the attentional act's phenomenological content: The attentional modality does not simply consist in emphasizing well constituted objects; it determines the very forms of the phenomenalities which it grasps. "[A]ttention is an emphatic function which belongs among acts in the above defined sense of intentional experiences."[30]

But before introducing the "attentional" solution, it is necessary to complete the review of the difficulties introduced by a description of the fact of expression as a complex of acts of perception and of signification. Moreover that the additional difficulties which we will now examine clearly reveal the cause of the inadequacy of the previous layering or foundational approaches, and, correlatively, that they

---

[26]Ibid., p. 118.

[27]Ibid.

[28]Ibid., pp. 117–118.

[29]Ibid., p. 116.

[30]Ibid., p. 118.

indicate the means by which to overcome them, including that of the possible formula provided by the attentional field.

4.3.4 Shared Difficulties and Consequence

Regardless of the solution contemplated, be it the layering solution or the foundational one—and Husserl retains neither of these—the following cardinal difficulty arises: The word forms, as perceptual complexes, do not present the internal forms for specifying ("polarizing") a general meaning-intention into a particular object of meaning, one which would be specifically attached to them. In order to approach and to illustrate this state of affairs, we may consider the example of visual perception. We know that immanent visual experiences are adumbrations (two-dimensional apparent contours), and that the intentionality of visual perception imbues them in such a way that each aims towards the unity of the flux to which they participate. What should be emphasized here is that the intentionality of visual perception may be qualified as generic in that it does not animate the adumbrations in view and in function of a previously specified object form (cube, sphere…), but animates them according to a general aim (that of three-dimensional reconstruction) which finds in the forms of these adumbrations (essentially the singular points) the information for its specific polarization, which then determines the linkage of the apparent contours belonging to the three-dimensional object thus aimed at (cf. Petitot 2002).

In what concerns meaning-intention, it goes without saying that it must also be considered as aiming at sense in a generic manner, as an act of sense presentation. And, also, it is the form apprehended by virtue of such an act which bears the function of specifying the significatory orientation towards such or such particular content. Thus, the graphical complex *apple* would target the meaning "apple" not because such a word form would be grasped and infused with meaning-intention directed towards the meaning "apple", but because it would polarize and specify a general semantic engagement towards the particular semantics of "apple". But, as far as meaning-intention is concerned, it is not the case. Because unlike perceptual adumbrations, *word forms are amorphous as regards their capacities to orient a horizon of sense towards a specific content in terms of meaning*. Nothing indeed in the internal configuration of *apple* or *pear*, when animated by an act of signification, is able to orient the aim towards their commonly attached contents.

From this, two conclusions must be drawn.

The first is that word forms do not possess as such the configuration or the characteristics likely to particularize a general significatory aim. Therefore, it will be necessary to seek this power elsewhere, but without however leaving the sphere of word forms. And since the specific significatory aims do not proceed from the word forms while requiring to remain attached to them, it is naturally towards the system they constitute, towards the relations to which they participate, that it will be necessary to turn in order to highlight the signifying power of concrete signs. But to achieve this, we will need to refer to Saussure and to structural morphodynamics.

The second point is that the phenomenological modulation which the word form undergoes, when it is promoted to the status of sign, does not pertain to a precisely semantic register of functioning. The reason being that meaning-intention is not achieved by means of isolated signs, but in relations between signs. Also, the phenomenological traits which characterize a sign are to be referenced to a semiolinguistic plane of organization which, without being unbound from the order of semantic operations, is partly independent from it. We will see right away that it is this path which Husserl will pursue, precisely by requalifying the phenomenological characters of the sign, not only following a basis having significatory and perceptual acts for vectors, these being orthogonal and hence mutually inassimilable, but rather on the basis of the stratification of an attentional field of consciousness. Moreover, we will see how this descriptive option which abandons semantic analysis finds itself structurally involved in a global semiolinguistic architecture allowing to reestablish ties with a plane of semantic operations.

4.4   The Attentional Conformation of the Sign

In the 1908 *Lectures on the Theory of Meaning*, and comparatively to the exposition of the first *Logical Investigations*, the phenomenal constitution of the linguistic sign was subjected to thorough reorganization.

Reading the *Lectures* does not at first reveal what new balances are at work in the fact of expression. On the whole, the phenomenological analyses which are developed in the first pages largely intersect with those of the *Logical Investigations*.

For example, just as he did in the first *Logical Investigation*, Husserl begins the analysis of the phenomenon of signification by making a distinction between "the expression itself, that is, without taking its meaning into account, and meaning."[31] But he does so while insisting on the insufficiency of an analysis of the sign in terms of a "physical side of expression: [T]he phenomenon of the sensible sign on the paper" and of a "mental side, that is, certain mental experiences which [...] joined by association to the word's sound, confer meaning to it."[32]

Likewise, he observes that the "word's sound", as a phenomenon of the sensible world, conduces towards nothing but itself—it is deprived of significatory power: "[W]ith the consciousness of the word's sound by means of which the simple sound of the word is objective, we are not yet conscious of its meaning."[33] Furthermore, it is beyond the sound of the word that the significatory orientation should be sought,

---

[31]*Leçons*, p. 32.
[32]Ibid., p. 31.
[33]Ibid., pp. 35–36.

the one that will promote it to the status of authentic expression: "Therefore, new acts intervene *in unity* with the consciousness of the sound of the word, these being the acts of 'aiming at this and that with the word.'"[34]

The fact of expression therefore composes two acts: an aiming by which a word's sound establishes itself in consciousness, and an aiming by which this word's sound directs towards a meaning. And it is clear, in accordance with the assertions reiterated by Husserl, that the unit formed by these two acts cannot have the character of a juxtaposition or of a sequencing: "[The] consciousness of the word's sound and the consciousness of the word's meaning [...] are manifestly not simply given together in the manner of a simple sum."[35]

It is therefore indeed with the idea of the *unit formed by the acts* that Husserl locates the crux of the phenomenological analysis of the fact of expression—hence the importance of the paragraph devoted to the "[p]henomenological characterization of the particular type of connection between the consciousness word's sound and the consciousness of meaning."[36]

In order to approach the nature of this particular link between the two acts involved in the unit formed by the act of signification, Husserl observes that the "phenomenological link [...] bases itself on the essence of the two acts so as to bring them under the unity of the consciousness of an act."[37]

This observation is obviously problematic: How indeed can the self-sufficiency and closedness of the "phenomenon of the word's sound [...] as an intuition, for example as a perception",[38] be reconciled with its hypothetically essential unity with meaning-intention? But formulating this problem in another manner enables to foresee an answer: How may the sign be maintained within consciousness as a simple sensible object when it pertains to a semiotic intention which reconfigures its form of appearing?

Formulated this way, it is no longer question of integrating two acts within a complex intentional unity, but of interrogating the fact of the "signifier", that is, of questioning the ambiguous nature of the "word" which is more than a mere sensible entity since the forms of its appearing (its phenomenal identity) are configured by a certain orientation towards a signified objectivity, but which persists nevertheless, in a certain measure to be specified, in its identity as a "concrete" phenomenon.

Viewed from the standpoint of this other angle of reflection, the issue of unity is realigned: It no longer focuses on the impossible integration of two mutually exclusive acts, but rather on the nature of the unit formed by significatory and perceptual orientations *as they reveal themselves in the phenomenologically ambiguous character of the word*, as a unitary factuality of the sensible and of the intelligible. In doing this, and as anticipated, the focus shifts towards the question of

---

[34]Ibid., p. 36, our emphasis.
[35]Ibid., p. 39.
[36]Ibid.
[37]Ibid.
[38]Ibid.

the nature of a third term serving as a junction between significatory and perceptual acts, that is, the pure "signifier" as a concrete sign infused with an intended generic sense, or, dually, as an expression from which the specific intended object of meaning will have been abstracted.

As mentioned by Husserl, in keeping with what was set forth in the *Logical Investigations*, the phenomenological characters of "words" are not those of "sounds of words", but in words, it is possible to find a recomposition of sensible entities: "[The] consciousness of the word's sound is manifestly not the consciousness of the word. [But] In the apprehension of the word, the first is contained; the sensible sound of the word indeed appears; but only as a founding basis."[39]

Nonetheless, the notion of "foundation" remobilized here cannot have the same content as what it originally had when it was a matter of relations between founding and founded acts. Hence, attention should be given to the new sense conferred to the notion of "founding basis."

In order to do this, the second character of the phenomenological link between the consciousness of the word's sound and the consciousness of meaning will serve as a starting point given that the consciousness of meaning is built upon the word's sound. What must be remembered from this observation is the "inequality of value"[40] between the two moments of consciousness: The consciousness of sounds of words and the consciousness of meanings do not form symmetric poles of a two-sided entity, the one calling for the other and reciprocally as if in a mirror game. The consciousness of an expression (in its full acceptation) is not the equal merging of a consciousness of sound and of meaning. Phenomenological analysis must rather recognize within the unit formed by the expression, as it is constituted in consciousness, a hierarchical distribution between signifier and signified, a distribution which is achieved according to an attentional scale.

This is what can be observed when we examine the natural attitude of consciousness while apprehending signs in their qualities as words: "Consider the phenomenon which is the sound of a word [...] as an intuition, for example, as a perception, [...] with the normal consciousness of the word's meaning (in this case during reading). It is not towards the word's sound that our "attention" is directed. We cast our gaze upon it; and yet, we do not perceive it in the sense [...] that we turn our attention towards an object as it is perceived [...] we see the written sign, we even focus upon it, but it is not our target [...] what we aim towards is something quite different: We aim at signified objectivities, we 'live' in the consciousness of meaning."[41]

And Husserl is quick to point out that it is not a matter of psychological events, but indeed of "specific phenomenological relations"[42]—relations to which he will

---

[39]Ibid., Appendice II, p. 175.

[40]Ibid., p. 39.

[41]Ibid., pp. 39–40.

[42]Ibid., p. 40.

then devote an entire section in view of specifying the various attentional regimes which participate in the elaboration of phenomenal identities.

Thus, the "word's sound" is more than a sound but it is not quite a sign. Rather, it is as if what is sensible within it, i.e. the acoustic phenomenon, is immediately dismissed as soon as it appears when consciousness apprehends it from the angle of a significatory aim—and it is in this sense that it must be considered as a "medium" or a "foundation."

In order to account for this *intrinsic* ambiguity of the "word's sound" *without having recourse to specific meaning-intention*—which, as we have seen, as far as it orients consciousness beyond the simple sound of words, introduces a semantic qualification which cannot be integrated to the concrete sign—the solution, already evoked, will consist in disregarding the intended object of meaning in order to only retain the tension towards the "something" of a generic directedness. The phenomenological examination of a word-form infused with such a generic directedness, one of which the object would have been disregarded, will therefore be faced with the fact of a pure indirection, the fact of a referral to something else, regardless of the identity of this other thing. This phenomenological character can be accounted for in terms of layering of interest or of attentional modalization—the phenomenological character of the sound of words being that of disinterest—and it is to this conception that Husserl's phenomenological analysis of the sign will lead.

4.5 The Strata of Verbal Consciousness

In short, the organic and holistic structure of the attentional field, as described in the *Lectures*, is articulated according to four modalities of "directedness": the "backdrop" mode, the secondary "noticing", the primary "noticing", and the "thematic aim". Furthermore, these modalities "intersect"[43] with the acts which institute certain phenomenological genres (as perceived, imagined, signifying…).

The backdrop's manner of being present is somewhat analogous to that of the "ground" in gestalt theory. As the ground exists and serves only in contrast to a figure, the backdrop is intrinsically coordinated with the secondary mode of noticing, which is itself articulated with a primary mode of noticing. To take this gradation in reverse, primary noticing is the mode of "paying attention" which directs consciousness towards and object in order to confer it some privilege. But the attention thus given can be diverted from the object which, though it may remain "noticed", will only be noticed accessorily. The object remains present as a perception, it is always there in the forms which configure it as a phenomenon (of primary directedness), but it no longer fully occupies attention; it only holds in a secondary capacity, as being noticed but in a less decisive, more floating manner. Such is the mode of secondary noticing which, in contrast to primary noticing,

---

[43]Ibid., p. 41.

institutes a specific manner of being in consciousness. But the objects of primary and secondary noticing which, albeit unequally so, have the characteristics of a "distinct presence", do not appear in a void: They present themselves against a background of objects of which the individuations are uncertain (configured according to the "backdrop" mode) and subject to be promoted to a higher position in the hierarchy of the attentional field, so as to be configured at such level of appearing. Thus, considering an object group, "while the first object [...] detaches itself in a primary manner, the favor of being noticed is only granted to the others in a secondary manner [...]; but then there still remains an objective backdrop from which, so to speak, what is conscious in a primary and secondary manner is extracted and from which it detaches itself."[44]

Again, we must insist on the phenomenological nature of these various modalities. Thus, the manner of being in consciousness, that is, the forms of appearing of an objectivity grasped in a backdrop are distinct from the forms of appearing when such objectivity makes itself present through primary or secondary directedness. Likewise, "phenomenologically, the consciousness of perception alters if we pay attention in a primary manner to that which is perceived, instead of simply noticing it in a secondary manner."[45] These different characters of appearing that are consubstantial to the ground or figure positions have been minutely described in the works of gestalt theory. For the visual field, for example, figures manifest superior properties of resistance, stability and coherence than the elements forming the ground. Likewise, they reveal themselves to be transposable and better circumscribed and, correlatively, they attract the gaze and focus attention. The ground elements, for their part, appear as latent and lacking salience: They give themselves in a mode of "being prepared" to be seen, and we can indeed speak of a "certain form of *paradoxical presence*, one which is vague and relatively indeterminate."[46]

But the modes of "noticing" do not constitute the only dimension structuring the attentional field: There is another way to be attentive than in the manner conferring a more or less great privilege to the object. In other words, the investment of consciousness along a set of objects is not only described by means of a tripartite distribution of attention. Specifically, a separation must be established "between the fact of being oriented towards and object and the fact of being occupied by it."[47] As regards the functioning of consciousness in a manner that is distinct from what occurs in the case of primary, secondary, or backdrop aimings—these being different manners of being *directed* towards an object, in other words, manners of situating the object with respect to the focus of consciousness—the manner of *living* in the object must also be considered. When consciousness is fully involved with the object as it focuses on it so as to penetrate its forms and to invest its matter, in

---

[44]Ibid.

[45]Ibid., p. 42.

[46]Cadiot and Visetti (2001, p. 58).

[47]*Leçons*, p. 43.

sum, to reside in it, what we have is "thematic" directedness. Thus, a "perception" can be the object of primary directedness without however being made into a theme. For example, such is the case when, while busy with something, "we grasp a few sentences from a conversation being held in our midst. Its sense is grasped in a primary manner; but we continue to live within the sphere of our own thoughts."[48] A distinction will therefore be made between the mode of noticing which only favors the object and the mode of noticing which makes the object into a theme, insofar as consciousness invests itself within it.

It is the latter distinction, between primary and thematic directedness, that Husserl will use to accomplish the description of facts of signification. Thus, as regards the normal activity of reading or listening, phenomenological analysis will record the stages of primary and thematic directednesses. The grasping of the sounds or words is only achieved in a mode of primary directedness, as it is towards the sensible sign that attention is directed: "Word perception has the distinction which forms the character of a perception which notices in a primary fashion, but does not have that of a thematic perception."[49] This is because, evidently, it is not in the sensible sign that consciousness invests itself: "The sign-impression is not the object of 'interest'."[50] What consciousness targets as a "theme" is meaning: "It is on the signified that [...] we focus",[51] or "we must live in the consciousness of meaning."[52]

Furthermore, the interlacing of the consciousness of the sound of words with the consciousness of meaning can now be generally described as the necessary passage from primary to thematic directedness: "The function of the consciousness of the word's sound is manifestly not to retain in the primary mode of noticing which is accomplished within it, but to drive it towards a consciousness of meaning."[53] But this formulation is unsatisfactory in two respects. First, it is misleading. We could understand that the connection between signifier and signified operates in the mode of indirection, the word's sound having, in addition to its identity traits, a propensity for orienting consciousness towards the signified theme. But we would then find ourselves with an indicative sign structure. Moreover, it is not without some redundancy. Because from the moment the sign is considered to be a connection, in this case without being further specified, between a primary directedness and a thematic directedness, it is clear, from the essential characters of thematic directedness, that consciousness will not limit itself to the primarily targeted object, but will invest itself in the thematically constituted one. In other words, the simple adjoining of a thematic object to a primary object confers to the latter the appearance of a power to direct attention beyond itself. But the organizational

---

[48]Ibid., pp. 43–44.

[49]Ibid., p. 44.

[50]Ibid., pp. 44–45.

[51]Ibid., p. 45.

[52]Ibid.

[53]Ibid.

regime which commands the sign as an asymmetrical adjoining of the signified to the signifier does not confer primacy to the referral function nor to the status of theme. Fundamentally, this regime has a holistic nature. It is by virtue of their respective positions in the global structure of the attentional field of consciousness, and, therefore, by virtue of the relations of indetermination that take place within, that the word's sound and the signified object constitute themselves into their specific phenomenal identities as signifier and signified: "Since the whole [of the sign] is a unit formed by acts and that, as a whole, it constitutes [...] a correlative objectivity, we then understand that the objectivities on both sides acquire characteristics pertaining to their position with respect to one another."[54] Hence, just as the thematic status of the signified attributes to the word's sound a sort of transitory existence in consciousness, one which constitutes its specific character as a signifier-phenomenon, likewise it is the necessary limitation of the consciousness of the word's sound to the level of primary directedness which, so to speak, compels consciousness to produce a thematical investment beyond the primary level: "In having a tendency to refer belonging to its phenomenological essence, the task of reference to the signified which finds within it its term also participates in conferring its thematic dignity."[55]

In short, such is how the phenomenological conformation of the word-sign takes shape: The organicity of the constituents of the sign proceeds from their modalization within the unity provided by the attentional field of consciousness. The acts of semiolinguistic intention institute the consciousness of sounds of words and the consciousness of meanings into interdependent positions of objects of primary and thematic intention—positions which exhaustively expose their respective phenomenological characters and which enable not only to account for the doubly fusional and dissymmetrical unity of the signifier and signified, but also to explain the phenomenological ambiguity of the signifier, which in its appearing as a phenomenon of a significatory nature, maintains through an allusion the presence of the word's sound as a sensible phenomenon. Because the word's sound constitutes itself as an object of primary noticing, therefore as a sensible phenomenon and, moreover, being intrinsically bound to this level of existence in consciousness, it gives itself to be seen, in its full phenomenal identity, as compelling consciousness to divert from it in order to rather invest itself in its structural counterpart in the attentional field, that is, the signifier as an object of a thematic intending.

Equipped in this manner with a determination, albeit partial, of the forms of linguistic phenomenality, it will now be a matter of establishing the empirical relevance on other bases than phenomenological examination. To this end, we will have recourse to EEG observation data regarding neural processes taking place during linguistic tasks. Specifically, it will be a matter of homologizing the phenomenological analysis of the sign with respect to the generation circumstances of the N400.

---

[54]Ibid., p. 39.
[55]Ibid., p. 45.

But first, we will need to present the Saussurean theory of the sign—a theory which we will establish in morphodynamical writing and which, moreover, we will connect to the Husserlian system (exposed above) of verbal strata of consciousness in order to confer it a phenomenological meaning.

References

Cadiot, P., & Visetti, Y.-M. (2001). *Pour une théorie des formes sémantiques: motifs, profils, thèmes*. Paris: PUF, coll. Formes Sémiotiques.

Petitot, J. (2002). Eidétique mophologique de la perception. In: Petitot, J. et al. (éds). *Naturaliser la phénoménologie: essais sur la phénoménologie contemporaine et les sciences cognitives*. Paris: CNRS Éditions, coll. CNRS Communication.

Chapter 5
The Saussurean Analysis

5.1 Introduction

5.1.1 Foreword

The "semiological" definition of language, which can be found in the third chapter of the introduction to the *Course*—that is that "[language] is a system of signs in which the only essential thing is the union of meanings and sound-images"[1]—is of particular interest to us in that, as we will immediately see, it promptly makes the dimensions of objectivity and of phenomenality coincide, thus anticipating the result towards which we will progressively advance (cf. 1.2.2)

Imaginably, such an approach could be criticized for reducing and crystallizing Saussurean thought into a single angle, one which may be deemed to be the least enlightening one among those from which to apprehend and conceptualize the object which is "language"—angles which Saussure, over the course of his relentless research work, systematically uncovered, explored, and recognized in their mutual equivalencies.[2]

The semiological definition, indeed, breaks the equilibrium in the relation between system and signs, unduly favoring the latter. Already, appearing in the

---

[1] *CLG/B*, p. 15.

[2] "It seems impossible in practice to give priority to any particular truth in linguistics so as to make it the key starting point. However, there are five or six basic truths which are so interrelated that it is equally possible to use any one as the starting point, and to arrive logically at all the others and at every minute ramification of the consequences, starting from any one of them" (Saussure 2006, p. 3). "[G]eneral linguistics appears to me as a system of geometry. We arrive at theorems which need to be demonstrated. However, we observe that theorem 12 is, in another form, the same as theorem 33" (Godel 1969, p. 30).

© Springer International Publishing AG, part of Springer Nature 2018
D. Piotrowski, *Morphogenesis of the Sign*, Lecture Notes in Morphogenesis,
https://doi.org/10.1007/978-3-319-89848-3_5

pages where the notion of *system* had not yet acquired its full epistemic scope—that is, as an integrated complex of connections instituting at their articulations identities of a purely relation nature—the semiological definition, in the first passages of the *Course*, leads to think about the linguistic system in the mode of a combinatorics of signs otherwise constituted. Moreover, as if to further the misunderstanding, the semiological definition asserts that these signs *essentially* proceed from the union of a signifier and a signified, suggesting that they are established beyond any system.

It is clear that such a prevalence conferred to the sign is in flagrant contradiction with Saussurean thinking. To begin, if a reminder were necessary, primacy is attributed to the system rather than to its terms, and, as Saussure emphasizes, giving precedence to the union of sound and meaning amounts to relegating the system to a status of second instance: "[T]o consider a term as simply the union of a certain sound with a certain concept is grossly misleading. To define it in this way would isolate the term from its system; it would mean assuming that one can start from the terms and construct the system by adding them together when, on the contrary, it is from the interdependent whole that one must start and through analysis obtain its elements."[3,4] Then, as we have stated earlier, the fundamental notions and principles of linguistic theory, including the theory of the sign, and as they correspond to the various angles of intelligibility of a sole and same structural reality, essentially refer to one another. Neither of them would therefore be able to claim a status of preeminence.

Such, for example, is the case with the notions of *value*, of *unity*, and of *concrete entity* which can be "mutually blended."[5] Likewise, the principle of arbitrariness, which Saussure nevertheless made into the cornerstone[6] of the edifice of language, is indeed ultimately to be assimilated to the regime of differentiation: "*Arbitrary* and *differential* are two correlative qualities."[7,8] Arbitrariness and differentiability reciprocally imply one another, and, with respect to the constitutive role of the relations of difference in the establishment of *units* and of *values*, the principle of

---

[3]*CLG/B*, p. 113.

[4]Or, furthermore "On the contrary, one must start from <the system>, the interconnected whole; this may be decomposed into particular terms" (Saussure 1993, p. 134a).

[5]"[I]n semiological systems like language […] the notion of identity blends with that of value and vice versa. In a word, that is why the notion of value envelopes the notions of unit, concrete entity, and reality. […] there is no fundamental difference between these diverse notions […]. Whether we try to define the unit, reality, concrete entity, or value, we always come back to the [same] central question that dominates all of static linguistics" (*CLG/B*, pp. 110–111).

[6]The "[p]rinciple [of arbitrariness] dominates all the linguistics of language; its consequences are numberless" (*CLG/B*, p. 68), or: "The hierarchical position of this truth is at the very top" (Saussure *in* Amacker 1975, p. 52). This axiomatic status conferred to the principle of arbitrariness defined a school of thought. For example, De Mauro states that: "The arbitrariness of the sign is first in the order of things: It is the foundation upon which rest the edifice of language as a form" (De Mauro *in* Amacker 1975, p. 85).

[7]*CLG/B*, p. 118.

[8]We will note that the sentence preceding this excerpt even attributes primacy to the regime of differentiation: "[I]t is evident, even a priori, that a segment of language can never in the final analysis be based on anything except its noncoincidence with the rest" (*CLG/B*, p. 118).

arbitrariness is therefore to be situated on a same plane of conceptual operativity as these latter notions.

Finally, and more radically, as this study will demonstrate, the installation of a signifier with respect to a signified, be it within an integrated sign-unit, or be it in the mode of a correspondence between word-forms and concepts, is never but a "secondary" state, one derived by means of semiogenesis, of the underlying primordial phenomenon of expressivity. In what concerns the "secondary" character of the signifier/signified association, both Saussure and M.-P. strongly emphasize this: Saussure notes that "If I state simply that a word signifies something when I have in mind the associating of a sound-image with a concept, I am making a statement that may suggest what actually happens, but by no means am I expressing the linguistic fact in its essence and fullness",[9] whereas M.-P. observes that "if [language] sometimes signifies a thought or a thing directly, that is only a secondary power derived from its inner life."[10] It is because for both Saussure and for M.-P., the sign, having been polarized and stabilized into a signifier and signified, is the "belated" product of underlying differential operations; in short, words are but "secondary realities, the results of a more originary differentiation."[11]

Detached, therefore, from the global theoretical complex and from a logic of exposition of which it is but a stage, the "semiological" definition of language fragments, distorts, and restricts Saussurean thought. It remains that it deserves special attention, because the concept of the sign, if it does not stand at the *center* of the Saussurean undertaking, nevertheless occupies a particular position in that it touches the *empirical ground* of linguistics. There is indeed an experience and an intuition of signs, these being subject to immediate acknowledgement in two respects. First, in what concerns their identity and their unity,[12] and this despite that the forms which determine them and which, notably, circumscribe them as parts of integrated totalities remain widely ignored, it is because "it is one thing to feel the quick, delicate interplay of units and quite another to account for them through methodical analysis."[13] Then, signs allow themselves to be directly apprehended in their specific characters as semiotic objects: Signs manifest and deliver themselves in the form of a certain undivided sound/meaning duality (cf. 2.2 and 2.4). Thus, signs make themselves present following the forms which specifically condition a semiotic "being", and, configured by virtue of this particular regime of appearing, they constitute as authentic *phenomena* the most flagrant experiential matter of linguistic reality: "The signs that make up language are not abstractions but real objects [...]; signs and their relations are what linguistics studies; they are the concrete entities of our science."[14] It will be

---

[9]*CLG/B*, p. 117.

[10]Merleau-Ponty (1993, p. 82).

[11]*PW*, p 32.

[12]Consciousness of the repetition of a same sign which is variable in its material characteristics: acoustic or psychological, cf. Amacker (1975, p. 30).

[13]*CLG/B*, p. 106.

[14]*CLG/B*, p. 102.

judicious to more specifically return to the sense that Saussure gives to the term *concrete* (cf. 5.2.3.2); for the moment, and adopting the views of classical philosophy,[15] we will maintain that signs are *phenomena* in the sense of being indeterminate objects of an empirical intuition (in this case, semiotic).

Let's then return to the semiological definition of language and let's see which lines of inquiry it opens up to investigation.

5.1.2   Precisions

According to the semiological definition, the sign lies at the intersection of two orders. These are, respectively, that of the play of oppositions in language, and that of the signifier/signified connection. As such, the sign therefore doubly pertains to inter- and intra-sign regimes and, correlatively, it asserts their overlapping. But the principle of a reciprocal assimilation, on the one hand, between the forms of language as a system of connections which institute signs, and on the other hand, through the integrative junction between signifier and signified characterizing the sign in its essence is, as must be acknowledged after Saussure, opaque, to say the least. As a reminder, here is how the problem is exposed in the *Course*:

"[H]ere is the paradox: [O]n the one hand the concept seems to be the counterpart of the sound-image, and on the other hand the sign itself is in turn the counterpart of the other signs of language. Language is a system of interdependent terms in which the value of each term results solely from the simultaneous presence of the others, as in the diagram:

How, then, can value be confused with signification, i.e. the counterpart of the sound-image? It seems impossible to liken the relations represented here by horizontal arrows to those represented above [...] by vertical arrows. Putting it another way—and again taking up the example of the sheet of paper that is cut in two [...] —it is clear that the observable relation between the different pieces A, B, C, D, etc. is distinct from the relation between the front and back of the same piece as in A/A', B/B', etc."[16,17]

We know that in order to provide himself with the means for clarifying somewhat this "strange coincidence"[18] between two orders of determination which are

---

[15]Cf. Kant (1944 (1781/1787), p. 53).

[16]*CLG/B*, pp. 114–115.

[17]In short, following the terms of M.-P., how to understand "the lateral liaison of sign to sign as the foundation of an ultimate relation of sign to meaning" (Merleau-Ponty 1993, p. 77).

[18]Godel (1969, p 240).

respectively external and internal to the sign, Saussure has recourse to the concept of *value*.[19] Reformulated in terms of *value*, the initial question is displaced towards a more general framework which does justice to the equivalency of two hetero-geneous modes of determination: "[V]alue is the counterpart of the coexisting terms. How does that come to be confused with the counterpart of the auditory image? (D 271); the solution will be given by a threefold argument [...]. This argument is founded on the analogy between linguistic value and value in the general sense."[20] But if the concept of *value* brings any intelligibility to the "strange coincidence" of which it has been question, we see that it is by means of a process which consists in covering the difficulties this concept conveys by situating it within a context which gives it meaning and which extols it. Now, by transposition, such difficulties are comparable to those which this concept is meant to help surmount. Indeed, *value* is defined following two relational modalities that are foreign to one another. As a reminder: "[V]alues are apparently governed by the same paradoxical principle. They are always composed: (1) of a *dissimilar* thing that can be *ex-changed* for the thing of which the value is to be determined; and (2) of *similar* things that can be *compared* with the thing of which the value is to be deter-mined."[21] The concept of value therefore also comprises the problem of a reciprocal assimilation of two regimes of distinct natures. And we know that this problem, regardless of what its angle of approach will be in the end, resisted Saussure's efforts at clarification: "[I]n a note, in which [Saussure] asserts that value has the specificity of relating two things which display no relationship ('if we consider on the one hand the exchangeable terms, and on the other the co-systematic terms, no relationship is perceptible'), he writes: 'It relates them in a way *which defeats the mind, it being impossible to tell whether it considers these two sides of values to differ, or how.* [...] The relation *simile: dissimile* is something quite different from the relationship *simile: similia*', and yet this relationship nonetheless goes elusively and profoundly to the heart of the notion of 'value'."[22]

We thus see that the problem posed by the semiological definition of language is similar to that carried by the concept of *value*, and it may be foreseen that the solution brought to the one will function for the other. More generally, it can be expected that the eventual answer to the "semiological" problem will lead to a series of readjustments regarding the set of concepts belonging to the Saussurean per-spective and regarding their theoretical composition.

But there is more: As a linguistic *phenomenon*, the sign may be examined from two perspectives of examination. The first has for object the specific characters of semiotic phenomenality, that is: What are the forms which configure and condition

---

[19]Thus, in the *Course*, the definition of value is introduced with the words "[t]o resolve the issue" (*CLG/B*, p. 115)—this question being precisely that of the equivalence of relations external and internal to the sign.

[20]Godel (1969, p. 238).

[21]*CLG/B*, p. 115.

[22]Saussure (2006, p. 240).

semiotic appearing? The second perspective leads to questions regarding objective knowledge: The sign-phenomena being given, to what conceptual and categorical regimes do they pertain as they constitute themselves as objectivities? In other words, what regimes of structure deliver an objective qualification of the semiolinguistic phenomena and also determine their meaning as objects?

Reconsidered from these angles, the coincidence between inter and intra-sign relations postulated by the semiolinguistic definition can be translated as an equivalence between, on the one hand, the phenomenal conformation of the sign, of which the signifier/signified unity constitutes an essential feature,[23] and, on the other hand, the system as a relational complex determining sign-phenomena and correlatively accounting for their identities as linguistic objects.

This hypothetical and highly problematic equivalence is remarkable in that it carries an immense heuristic value. Indeed, the mode of a reciprocal assimilation between the signifier and signified, as an essential character of semiolinguistic phenomena, still remains to be clarified. And the rigorous, explicit, and meticulous recognition of the structural schemas which administer such an order of phenomena constitutes a primordial issue for linguistic knowledge. Now, if we give credit to the semiological definition, we find a key for advancing towards an explanation of the forms of linguistic phenomenality. Indeed, since these would be fundamentally the same as those which govern the linguistic system, the determination of the ones could be acquired following the route towards a sufficient recognition of the others.

In other words, as far as language is concerned, the phenomenological issue would be a constitutive part of the (linguistic) system knowledge to the extent that the order of language would preside as much over the objectivity of its phenomena as over the phenomenalization of its terms (relational identities).

It is precisely this conjecture which we propose to examine here.

We will first address the order of the system (Sect. 5.2). And without any other purpose than to produce a detailed characterization of it, we will find ourselves endowed with a model of *semiosis* (unity of expression and content). In order to do this, it will be necessary to take position regarding several floating aspects of Saussure's thought, to solidify certain concepts, and, moreover, to have recourse to a morphodynamic apparatus. However, it should be emphasized that all of this will be accomplished without leaving the perimeter of the Saussurean exegeses of reference.

But having established a bridge between structure and phenomenality, and given a structural characterization of the appearing of the sign, we will immediately be required to confront these results to those stemming from a specifically phenomenological analysis of the sign, such as that developed by Husserl (cf. 4.5). It may then be observed, thus reinforcing the plausibility of the conclusions obtained,

---

[23] As a reminder: "[T]here are no linguistic facts apart from the phonic substance cut into significant elements" *CLG/B*, p. 110) or "concrete [...] signifies that the idea has its unity in the acoustic medium" (Saussure *in* Godel 1969, p. 211).

that the forms which preside over the sign as a term of the linguistic system are homologized by specifically phenomenological analysis.

It is necessary to emphasize that the result towards which we propose to steer is by no means heterodox.

Already, it covers the most fundamental conceptions of the structural episteme. This is because what is also at stake in the semiological definition is the question of the *nature* of structures, *formal* and *abstract*, versus *organic* and *material*. And indeed, if we consider that the qualifications which proceed from a position in the system are of a purely formal order (ideal and abstract) as is the case in a logico-symbolic axiomatics, then the equivalence given by the semiological definition seems incomprehensible—simply because it amounts to inconsiderably reducing the formal to the substantial. But we know that the conception defended by structuralism, and thereby opposed[24] to the formalist approach, recognizes an empirical quality in structures and in their objective value—The objective content of structures, in other words, the intelligibility of the material configurations they govern, delivers itself in the intuitive apprehension and in the concrete experiences of such configurations: "For [formalism] form alone is intelligible, and content is only a residual deprived of any significant value. For structuralism, [...] content draws its reality from its structure and what is called form is the 'structural formation' of what the content is about".[25]

Moreover, the unity of the questioning put forth by phenomenology and structuralism, in what is most manifest in it, is, so to speak, an analytical truth: While phenomenology concerns itself with the structures *of* appearing, structuralism sees structures *as* appearings—in other words, it sees structures as they participate in the empirical, that is, as they offer themselves to be seen, and, more broadly, to be intuited and reflected in the experience of the world. Regarding this point, of the empirical reality of structures, and where structuralism meets phenomenology as a science of the regimes of appearing, it suffices to quote Lévi-Strauss once more: "[Whereas] *Form* is defined by opposition to material other than itself, [...] *structure* has no distinct content: it is content itself, apprehended in a logical organization conceived as property of the real."[26] Structuralism is therefore radically opposed to formalism. Whereas formalism conceives of structures in terms of an abstract system, both in what concerns its pure ideal and atomical elements, and in what concerns its relational regimes as defined through axiomatic schemes, for structuralism, structure is to be conceived in terms of hylomorphism: organization of matter (Saussure: shapeless mass of ideas or sounds) by the actualization or by the emergence of a form (Saussure: network of reciprocal delimitations) establishing such matter into substances (Saussure: signifiers and signifieds). That is to

---

[24]"Formalist dichotomy, which opposes form and matter and which defines them by antithetic characters, is not imposed on him by the nature of things, but by the accidental choice which he made in a domain where form alone survives" (Lévi-Strauss 1983, p. 131).

[25]Lévi-Strauss (1983, p. 131).

[26]Lévi-Strauss (1983, p. 115).

say that according to the structuralist perspective, the form has absolutely no existence separately from the matter through which it deploys and expresses itself.

5.2 The System

Approaching language from a structural standpoint, and correlatively considering signs as relational identities, the question immediately arises as to the nature of the connections which form the web of the linguistic system.

We know that among the main relational concepts which Saussure mobilizes in order to approach and to account for the linguistic system, there are the concepts of *difference, opposition, negativeness, limitation, proximity, exchange*. Let's address them one at a time.

5.2.1 Difference and Opposition

Concerning the concepts of *difference* and *opposition*, notably in *Course II* and while discussing and comparing the main features of graphical and linguistic signs, Saussure successively mentions the properties of (i) *arbitrariness*, (ii) *difference* or *negativeness*, and (iii) *opposition*. In this passage, of which the main points can be found in the second and third paragraphs of the 4th chapter of the *Course*, Saussure distinguishes on the one hand the regime of *differentiation* which, *modulo* an approximation needing to be explained (cf. 5.2.2), he assimilates at that point to *negative differentiation*, and on the other hand, the regime of *opposition*.

In order to establish in strict conformity with Saussure's thought the specificity of each of these regimes and, correlatively, how they differ from one another, it is necessary to take the Saussurean definition of *opposition*[27] into account, that is, "difference joined with a relation."[28] This should be done keeping in mind that two sorts of relations exist: The paradigmatic and the syntagmatic (henceforth S&P): "[T]he play of differences is apparent in two orders of relations: Those of syntagmatic and those of associative relations."[29]

---

[27]"The Saussurean notion of opposition thus implies both difference and relations (III 142)" (Godel 1969, p. 197).

[28]Godel (1969, p. 200).

[29]Godel (1969, p. 200) based on Saussure *in* Godel (1969, (*Cours II*, note 74) p. 72 & (*Cours III*, note 142) p. 89).

5.2.1.1 The Relation of Difference

In what concerns *difference*, we know that Saussure means a relation of "noncoincidence" in terms of identity: The differential character of the sign expresses the fact of a "non-identity" of this sign with another. For example, "what we require [from a *t*] is that it not be completely identical [to an *l*]"[30] or "The only requirement is that the sign for *t* not be confused in his script with the signs used for *I*, *d*, etc."[31] Finally, to say that *a* is different from *b* "simply amounts to saying that *a* is not *b*"[32] and, by extension, linguistic unity is never founded "on anything except its non-coincidence with the rest."[33] Dually, "[the] most precise characteristic [of a sign] is in being what the others are not."[34]

Now, a few comments would be of order.

First, we must recognize with M.-P. that the idea of a differential identity, that is an identity which proceeds from differences between terms themselves devoid of identity ("in language there are only differences without *positive terms*")[35] is a "difficult idea, because common sense tells us that if term A and term B do not have any meaning at all, it is hard to see how there could be a difference of meaning between them."[36]

And we have lengthily discussed the circularity of the relation of difference, and concluded that it was sterile in terms of its ability to provide a definition. For instance, Itkonen asserts: "Suppose that I attempt to define A, B, and C in 'purely negative terms,' that is, by what they are not. The result, apart from not being very workable, will always contain two positive terms and, circularly, B and C will be implied in the definitions of B and of C."[37]

This is probably true, but *a contrario*, two things must be acknowledged which, taken together, attest to the soundness of the primacy of difference.

On the one hand, there is the fact that it is practically impossible to characterize a semiolinguistic identity by "positive" means, for example, in the form of a list of predicates—a positive characterization on the basis of which a judgment in terms of identity or of difference could then be produced.

On the other hand, and despite the preceding, an empirical judgment concerning the identity or the difference of semiolinguistic occurrences is undeniably possible[38]—and such judgment pertains to intuition inasmuch as it is the expression of

---

[30]Saussure *in* Godel (1969, p. 193).

[31]*CLG/B*, p. 120.

[32]Saussure *in* Godel (1969, p. 197).

[33]*CLG/B*, p. 118.

[34]*CLG/B*, p. 117.

[35]*CLG/B*, p. 120.

[36]Merleau-Ponty (1993, p. 76).

[37]Itkonen (1991, pp. 298–299).

[38]"Be it as it may regarding the justification or origin of identity judgments which enable speaking subjects to recognize, for example, in always physically differing concrete utterances, that they are

an immediate consciousness of difference or, dually, of the identity of the terms examined. As Amacker insists: "[T]he linguistic units are the object of research [i.e. they pertain to conceptualization], whereas identities are acknowledgeable [i.e. pertain to intuition]."[39]

A differential conformation of the semiolinguistic field must therefore indeed be acknowledged, as well as the fact that this conformation does not operate on what we could call "manifest" units, those constituted and completed units delivered by the semiolinguistic "reality", that is, signs as "delimited" entities (cf. 5.2.3.2), but that it works, so to speak, "underneath" of its most apparent elements.

Indeed, as we have said, the difference which arises between signs is not grasped in its own order, as in a consciousness of pure form, but reveals itself, through terms to the production of which it contributes, as a judgment regarding its identity or alterity. The "pure difference", the specific regime of differentiation, works at a level lower than that of the sign-phenomenon, a level to which semiolinguistic intuition does not directly have access, one which can only be attained by abstraction: "A negative [i.e. differential] character can never be observed in a pure state: It could only be observed as such if, by abstraction, we were to consider only one side of the sign."[40] This level would be a place, as we will also see with M.-P., which linguistic consciousness has the vocation of leaving from the very moment it is formed (cf. 7.6). This is why, nonetheless, "Saussure may show that each act of expression becomes significant [...] as it is differentiated from other linguistic gestures. The marvel is that before Saussure we did not know anything about this, and that we forget it again each time we speak."[41]

5.2.1.2 From Difference to Opposition

According to Saussure's definition of the regime of opposition, the "relational" dimension (the S&P relations) operates as a principle which commutes differences into oppositions.

In this respect, the sources leave no room for a shadow of a doubt. For example, and among many other references, in the presentation of the characteristics of the graphical sign, Saussure notes "the negative or differential character [...] the purely negative and differential value of the sign", and in the following paragraph which introduces the oppositional character, we can read that "[in] writing, values only work in opposition within a defined *system* [...] the value of the sign is oppositive

---

perceiving occurrences of a same word, such identity judgments are observable" (Amacker 1975, p. 30).

[39]Amacker (1975, p. 64).

[40]Godel (1969, p. 197).

[41]Merleau-Ponty (1993, p. 76).

and only holds within a *system*."[42] Clearly, therefore, by inscribing itself within a system, the regime of differentiation is transformed into opposition. And since it is a matter of linguistic signs, we find the same thing further down: "[...]2. The negative value of the word is obvious. Everything consists in differences", and then: "[...]3. Value becomes positive by means of opposition, by proximity, by contrast."

The definition of opposition therefore establishes *opposition* as a *systemic (and linguistic) promotion of difference*, as a structural assimilation *into language* of an underlying order of negativity.

5.2.1.3   Consequences for the Architecture of the Linguistic System

The "incorporation" of differences within the linguistic system, as evoked by Saussure's text, as given in the equation "opposition = difference + S&P relations", requires two things, which follow one another.

It is first necessary to determine the nature and the formal characteristics of the differential connections and, ideally, to do so using mathematical writing (cf. Chap. 6).

It is also necessary to ensure that this determination is "incomplete" with respect to the semiolinguistic order—in that it is the "S&P relations" that are to accomplish the conversion of differences (these being of a proto-semiolinguistic nature) into fully semiolinguistic oppositions.

This incompleteness already gives some indication regarding the nature of the contacts which the language system establishes with the substances (summarily: mental and sensible) which it invests to its own ends. Indeed, seizing substrates which are external to the sphere of semiolinguistic objectivity so as to configure within them linguistic identities which are "inchoate" but nevertheless adjacent to the linguistic system in that their structural reason ultimately resides within this very system, the differential relations thus operate, so to speak, at the *periphery* of language.

The distinction between opposition and difference therefore determines a topology for the language system in its connections with the substances which it informs. Thus, we will firstly distinguish an "inside" in which the opposing forms constitute the regime of unity and of homogeneity. Secondly, we will distinguish an "exterior", a sphere of substances towards which language ultimately conduces and upon which language rests, and finally, we will distinguish "boundaries", that is, interfaces between the system of oppositions and its externalities, boundaries which we have seen to consist in a differential structuration of these externalities (cf. Fig. 5.1).

To summarize, we have thus distinguished two sub-problems and one constraint. The sub-problems can be formulated as follows: (i) Which formal characterization should be given to the relation of difference?—and in a manner where (ii) joined with the "S&P relations", it is achieved as an opposition. As regards the constraint,

---

[42]Saussure *in* Godel (1969, p. 193, our emphasis).

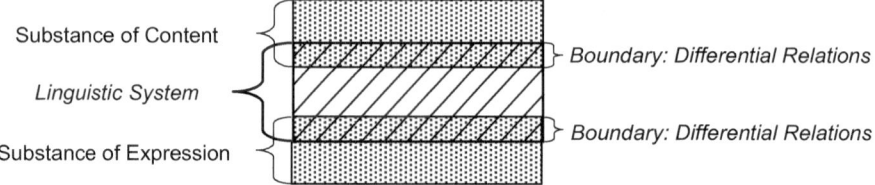

Fig. 5.1 Topology of the linguistic system

it weighs upon the topology of the linguistic system which will therefore be required to comprise "boundaries" where relations of difference, taking hold of extralinguistic materials, institute differential identities to be promoted into sign-units.

In order to advance in the resolution of these two sub-problems, and to satisfy the aforecited constraint, it will suffice to return to Saussure's texts and to find the necessary elements.

5.2.1.4   Detail and Confirmation of the Preceding

The first thing we will need to inform and homologate in a strict Saussurean perspective is that differences and oppositions do not have the same field of functioning. We have had a sense that oppositions function *in language* as relational modalities which institute signs, whereas differences, which invest empirical substances, operate *at the margin* of the linguistic system.

This distinction in the sphere of effectivity of each of these two relational modalities is fully corroborated by Saussure's texts, which clearly indicate that differences concern either signifiers alone, or signifieds alone, whereas oppositions concern signs in their essentially indivisible unity. Thus, as stated in the *Course*: "When we compare signs [...] with each other, we can no longer speak of difference; the expression would not be fitting, for it applies only to the comparing of two sound-images, e.g. *father* and *mother*, or two ideas, e.g. the idea 'father' and the idea 'mother'; two signs, each having a signified and signifier, are not different [...]. Between them there is only *opposition*."[43] This is a citation which is supported namely by the following: "in a language-state, there are only differences [...] either in the signifiers, or in the signifieds. Once we get to the terms themselves, as a result of the relation between signifier and signified, it may be possible to speak of oppositions",[44] or: "[T]wo signifiers or signifieds are different, two signs are opposed." Likewise, in the third *Course*: "[I]n language there are only differences [...] There are only differences when it is a matter of [...] signifiers or signifieds.

---

[43]*CLG/B*, p. 121.
[44]Saussure *in* Godel (1969, p. 92).

<Once we get to the terms themselves, as a result of the relation between signifier and signified>, it may be possible to speak of oppositions."[45]

Correlatively, it is the difference of status between the relational identities respectively established following the relations of opposition and of difference which are regularly asserted in Saussure's texts. Indeed, to situate oneself on the sole plane of the signifiers or of the signifieds, hence, to situate oneself at the "boundary" of the system in which only differential relations operate, is as much to partially escape the regime of oppositions which govern linguistic objectivity as to break the essential unity of the sign and to leave the field of "concrete" linguistic existence: "The linguistic entity exists only through the associating of the signifier with the signified [...]. Whenever only one element is retained, the entity vanishes; instead of a concrete object we are faced with a mere abstraction",[46] or: "If we unwittingly take only one of the elements [of the sign], one of the parts, we have straight away created a spurious linguistic unit. We have made an abstraction ['abstract = not linguistic'[47]] and it is no longer a concrete object that we have before us".[48]

But if, leaving the terrain of signs and the order of the oppositions which shape them so as to artificially limit ourselves, at the boundary of the system, to the sole relations of difference, we indeed leave the sphere of linguistic objectivity and empiricity, we do not however fall beyond language. This is because the signifiers and signifieds do not fully pertain to the planes of materiality, be they sensuous or ideal, which language invests and informs: They specifically result from the differential structuration of such planes. Also, we will recognize a hybrid status for signifiers and signifieds taken in isolation: Not fully pertaining to forms of linguistic objectivity nor to the planes of materiality which language takes hold of, they occupy an intermediate position which confers them a doubly transitional sense. On the one hand, they appear to prepare for language inasmuch as they constitute a structural moment turned towards linguistic accomplishment, on the other hand, they give access to what lies beyond it, by being partly unlinked from the linguistic order and by touching extralinguistic empiricities.

The foreseen gap between the modes of difference and of opposition, as they do not regulate nor institute identities of a same nature, is thus ratified in Saussurean thought. But such a conclusion is not terminal. We must still, in keeping with the program defined, clearly identify the formal nature of the relations of difference, as regimes administering the non-void intersection between the systemic interior of language and its substantial exterior. To do this, it will be necessary to successively address the question as it may apply to the plane of the signifiers and to the plane of the signifieds, and to produce each time a formal determination of the mode of difference.

---

[45]*Cours III*, p. 288.

[46]*CLG/B*, p. 102.

[47]*Cours III*, p. 229.

[48]Saussure (1993, p. 79a).

5.2.2   Planes and Types of Difference

5.2.2.1   The Plane of Content: The Signifieds

To address the question of the determination of the relations of difference, we may first observe that the mode of "noncoincidence" (of identities) is likely to receive various formal translations. The first, having a taxonomical aim, consists in reducing each identity to a specific feature, generally related to the predicative format, which retains the "essence" of the examined occurrence by correlatively disregarding those among its characters which are deemed to be irrelevant. The second, which introduces a relational dimension, establishes differences in the mode of binary oppositions: The positive and negative values of a binary opposition carry over the differential structuration from a dimension of which they constitute the poles.

The third, which Saussure will retain in order to account for the differences between signifieds, is not of a logical making, but of a *topological* one: Difference is conceived in the form of a system of discontinuities (network of boundaries) categorizing a presumably homogeneous space into adjacent subdomains. These structures of discontinuity operating, in a same conjunctive and disjunctive moment, the division of an intrinsically undifferentiated substrate into adjacent fragments, constitute the fundamental forms of a "pure"[49] structural intuition—and it is incontestably these forms which are at work in Saussurean analysis: "The linguistic fact can therefore be pictured in its totality—i.e. language—as a series of contiguous subdivisions marked off on both the indefinite plane of jumbled ideas (A) and the equally vague plane of sounds (B). The following diagram (Fig. 5.2) gives a rough idea of this."[50]

But this topological intuition is coupled with a dynamic dimension, and here the descriptions which Saussure gives of the relations between signifieds leave no room for doubt: The network of boundaries fragmenting a substratal space into subdivisions is fundamentally the actualization of a configuration of equilibrium to which underlying dynamics, *spatially* expressing themselves as expansionist propensities, arrive by mutually limiting one another.

This topological *and* dynamic conception is flagrant in the passages concerning synonymic relations, notably as restituted in the *Course*: "[A]ll words used to express related ideas limit each other reciprocally; synonyms like French *redouter* 'dread,' *craindre* 'fear,' and *avoir peur* 'be afraid' have value only through their opposition: if *redouter* did not exist, all its content would go to its competitors"[51] and "[I]f, by any chance, we had chosen only two signs to begin with, all meanings would have been distributed among the two of them."[52] Or, regarding the context as

---

[49]cf. Petitot (1985b, p. 62 *sq.*).

[50]*CLG/B*, p. 112.

[51]*CLG/B*, p. 116.

[52]Saussure *in* Godel (1969, p. 199).

Fig. 5.2 Contiguous
subdivisions in the planes of
ideas (*A*) and sounds (*B*)

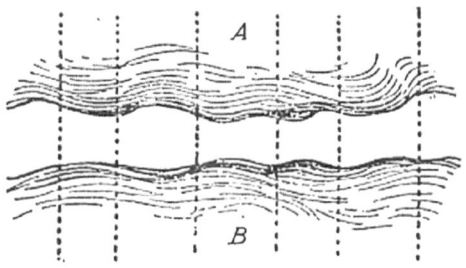

contributing to circumscribe the contents of meaning: "[T]he context, by opposition, determines what is 'enclosed' within the sense."[53] This point not being controversial, it will not be useful to defend it further, if not to conclude with Godel that "signifieds, for their part, [are] differential and their limitation is entirely negative."[54]

The structures of discontinuity, as they proceed from a stabilization of dynamic regimes and as they categorize a substratal semantic space into signifieds, thus deliver the adequate formal expression of the relations of difference at work on the content plane. Correlatively, it is the "border" character of the substances of content thus categorized which is confirmed and elucidated. Indeed, by principle, the differential forms, which we know to constitute the structural periphery of the linguistic system, invest extralinguistic substances, the differential identities thus produced, i.e. the signifieds, pertain as much to the system of the language from which they proceed in part as to the extralinguistic substrates within which they are actualized.

This step being completed, the same question must be addressed with respect to the plane of signifiers. Following this, having obtained a precise characterization of the differential forms on both planes, it will be a matter of operating, via the mode of S&P relations, their promotion into opposing forms.

5.2.2.2   The Plane of Expression: The Signifiers

Introduction

If, in Saussure's view, phonemes conceived as "opposing, relative, and negative entities"[55] proceed from differential relations similar to those which regulate the plane of content, that is, topological and dynamic regimes which categorize an acoustic or articulatory substratal space, it is, however, a whole different matter in the case of signifiers.

---

[53]Saussure *in* Godel (1969, p. 196).

[54]Godel (1969, p. 198).

[55]*CLG/B*, p. 119.

Fig. 5.3 Double-arrowed
schema of the sign

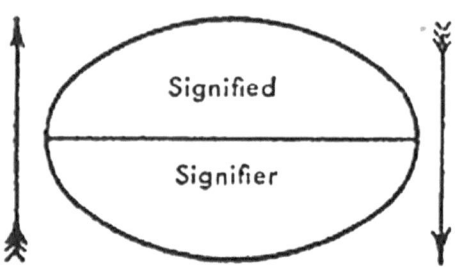

This question of the differential forms operating on the plane of expression is
particularly important inasmuch as it forces to reconsider certain structural configurations which the *Course* privileged and unduly perpetuated but which exegetic
works have long since denounced: These mainly concern the problematic symmetry
between signifier and signified, such as is given to see (Fig. 5.3) by the famous
double-arrowed schema of the sign.[56]

Already, the metaphorical vocabulary (incorporation, assimilation, impregna-
tion…) in addition to epithets of reciprocity (mutual, coextensive…) which are
often used to illustrate the relation formed by the signifier and signified within the
totality of the sign have favored this conception of a symmetry between the two
sides—a conception supported by the double arrow used in the illustration presented in the *Course*.

More fundamentally, the symmetry principle is formally employed and
explained in a number of passages devoted to the structural economy of the linguistic system, in the *Course* as well as in its sources. Thus, the definition of
language as a "series of differences in sound combined with differences in ideas"[57]
leads as well to imply that the signifieds and signifiers occupy structural and
functional positions which are positioned face to face. Even more explicitly: "There
are only differences which belong to the two orders [of signifiers and signifieds] and
which mutually condition each other."[58] And it is indeed also this conception
illustrated by the famous schema (cf. Fig. 5.2) in which cross-cuts operated upon
"jumbled ideas" and "equally vague" sounds locally and symmetrically proceed
from a same differential principle (vertical line) which distributes the signifiers and
signifieds along reciprocal relationships. The explanation which accompanies this
schema is also devoid of ambiguity: "The characteristic role of language with
respect to thought is […] to serve as a link between thought and sound, under
conditions that of necessity bring about the reciprocal delimitations of units."[59]

But the principle of a functional symmetry in the relations of difference, and,
dually, of a symmetrical arrangement of the two faces of the sign, finds itself to be

---

[56]*CLG/B*, p. 114.
[57]Saussure *in* Godel (1969, p. 259).
[58]Ibid.
[59]*CLG/B*, p. 112.

disrupted, firstly due to a distinction regarding the nature of the differential relations operating respectively on the plane of ideas and of sounds, and following which, correlatively to the introduction of the concept of value in the Third course, by a modification of the schema of the sign in which the double arrow is replaced by a single arrow (from the signifier to the signified).[60] Let's begin by examining the first point.

Distinctive Differences

As emphasized by Godel, "when he talks about differences, Saussure usually has the signifiers in mind."[61] But, referring to our framework of argument, the whole question is to know which formal specification is to be attributed to the relations of difference when they pertain to signifiers: Is it a matter of differential connections in the topological and dynamic sense, hence of relations that are structurally homogeneous with respect to those operating on the plane of signifieds, or is it indeed a matter of relations of a different nature? It should be noted that in the first case, we favor a symmetrical conception of connections between the planes of content and of expression, whereas in the second, we introduce a disequilibrium which breaks this symmetry.

To respond to this, we will note, after Godel, that "without recognizing it expressively, Saussure must have thought that the differences in signifiers allow themselves to be observed and analyzed more readily than those pertaining to signifieds. *Two sequences of sounds can only differ in the number, quality, and order of the irreducible units*, and we have seen [...] that the determination of these units, for Saussure, does not pose a real problem."[62] This amounts to saying, since it is a matter of signifiers, that the differences are not *productive* but *resultative*: The difference between two signifiers does not function as a principle of constitution for these signifiers, but proceeds from the identities which these signifiers themselves possess as particular compositions of phonemes. In such case, we will speak of "distinctive" difference.

Indeed, differentiation in terms of *quantity, quality, and order* evidently supposes that we are situated on a plane of object apprehension and qualification which regulates the counting (*quantity*), identification (*quality*), and distribution (*order*) of the phonematic units, hence on a plane "dominating" that of negative differentiations from which the phonematic identities fundamentally proceed, and where these phonemes are approached as units a, b, c... which are autonomous, therefore denumerable, and which are carriers of specific attributes (distinctive features) having the format of predicates $P_i(a)$, $P_j(b)$... that enable their qualitative comparison. In this manner, the signifiers are endowed with an identity

---

[60]*Cours III*, p. 286.
[61]Godel (1969, p. 198).
[62]Godel (1969, p. 199), our emphasis.

(as arrangements of phonematic components) which is not conditioned by their mutual differences but, on the contrary, which founds them (in terms, therefore, of number, quality, and order).

We therefore see that we have abandoned the principle of a correlation of topological differentiations between the plane of ideas and the plane of sounds, and certainly without loss. This is because, as illustrated by the famous schema upon which, incidentally, Hjelmslev cast a severe gaze,[63] the principle of a "direct" correlation between differential identities of sound (phonemes) and negative values of meaning (signifieds) is spurious.

But the possibility for a connection between differences in terms of signifieds and differences in terms of signifiers that would relate signs which are contiguous in terms of their content but not in terms of their expression, such as *night* and *day,* or *dog* and *wolf,* is nevertheless not imperiled. It would only imply transcribing into algebraic terms the distinctive differences instituted at a level lower than the plane of signifiers. Specifically, the situation is the following: The phonematic chains configured outside of the system of language (in the strict sense) and endowed as such with specific substantial (phonematic) identities are to be converted into purely relational identities following "#" relationships of difference which formally translate, i.e. on the plane of the linguistic system, the distinctiveness judgments instructed at a lower level. Moreover, these formal units participate in the system of language as they engage a functional connection (which remains to be determined) with the order of signifieds. The signifiers reveal themselves to be endowed with a complex status. Functionally participating in the linguistic system while being constituted outside of it, they present an intrinsically dual nature which does not go without posing a problem, and which immediately requires clarification.

The Duality of the Signifier

We discussed earlier the hybrid nature of signifieds. As we know, this character comes from their positioning at the boundary of the linguistic system where the differential forms which configure them invest extralinguistic matter. The nature of signifiers is likewise composite, but in a very different manner. On the one hand, just like signifieds, signifiers, in what concerns the forms which regulate them, are situated at the periphery of the linguistic system and therefore do not constitute full-fledged linguistic entities. But, contrarily to signifieds, the differential relations which signifiers establish between themselves do not condition their identities—for the simple reason that the relations of difference between signifiers are a transposition, without modification, of phonematically based distinctiveness judgments into the realm of the linguistic system.

---

[63]"This pedagogical experiment is meaningless, as well designed as it may be, and Saussure must have thought so himself" (Hjelmslev 1971, p. 68).

Thus, whereas signifieds, in their differential identities, are fully suspended to the regimes of discontinuities which institute them, the signifiers, conversely, found their differences upon the "quantity, quality, and order" of the phonematic units which they combine. And the transposition which introduces these phonematic complexes at the margins of the linguistic structure produces nothing more than a formal replication to plays of difference elaborated at a lower level.

More specifically, we have here a three-layered configuration: *First,* there are identities constituted outside of the linguistic system, namely the phonematic chains, and these are endowed with their own identities (for example, the chains /m-a/ and /t-a/ identified as specific combinations of the phonemes /m/, /t/ and /a/); *and,* on a second level, these extralinguistic identities will be instituted in identities we could call "peri-linguistic" inasmuch as they are converted into simple formal units establishing relations of difference that are of a logico-relational nature and which reproduce distinctiveness judgments informed at a lower level (thus, the chains /m-a/ and /t-a/ will be converted into the formal units A and B having a relation of difference "#" stemming from the comparison of these chains as specific compositions of phonemes, that is: A#B). *Finally,* on the third level, these units are engaged in the linguistic system as they participate in a functional connection (which remains to be determined) with the order of signifieds.

The nature of signifiers, in what it specifically comprises, lies in this distribution over three levels. We clearly see indeed that the middle level, which provides the junction between the interior and exterior of language, engages its exterior just as much as its interior: On the one hand, the pure formal units which enter relations of distinctiveness have phonematic materiality for basis and for principle and, on the other hand, these formal units are instituted as such due to their place and to the function they hold within a semiolinguistic system, that is, of serving as a junction with the plane of signifieds.

In this sense, signifiers have more of a *dual* than a hybrid nature: As quantities belonging to the medium level, they pertain to two independent orders of constitution and, as such, can be subject to a double approach. The moment of the signifier therefore discovers itself to be intrinsically and essentially *ambiguous.* The signifier exists as an object to be grasped in two respects: as directed towards meaning and, simultaneously, as a phonematic composition.

We will then note that it is precisely towards the observation of this essential ambiguity of the signifier that the phenomenological analysis of the sign leads, making it into its central point of articulation (cf. Chap. 4). And it is also this state of affairs that a structural description of languages will specifically need to account for. In order to progress in this respect, it will be required to examine more closely, in its principle and in its functional signification, the breaking of the symmetry between signifier and signified.

5.2.3   Signifier/Signified Dissymmetry

5.2.3.1   Functional Architecture: First Approach

As noted above, in the Third course, Saussure breaks the symmetrical arrangement of the signifier and signified within the totality of the sign (double arrow) in favor of an orientation from the signifier towards the signified (single arrow).

This new conformation of the internal structure of the sign is to be juxtaposed to the concept of value, which unevenly arranges "value" and its counterpart in accordance with the relation of "exchange". To give the breaking of symmetry discussed here its full structural signification, we will need to mobilize the question of value, but, beforehand and on the basis of previous results, we can already clarify some aspects.

It can already be noted that the disequilibrium introduced by Saussure within the sign is coherent with the linguistic forms as they are arranged on the basis of relations of opposition and of difference.

Indeed, as we have just shown, if we use "#" to indicate the relation of distinctive difference, "/" to indicate the relation of topological differentiation, and "$\leftrightarrow$" to indicate the functional connection between the differences of signifiers (Sr) and of signifieds (Sd), we pass from the formula "$Sr_1/Sr_2 \leftrightarrow Sd_1/Sd_2$" to the formula "$Sr_1\#Sr_2 \leftrightarrow Sd_1/Sd_2$", and, in doing so, by upholding structural regimes which differ with respect to the planes of expression and of content, the symmetry has clearly been broken.

At this stage of our discussion, we have at our disposal a sufficient amount of information to approach the functional architecture of the sign, which we have nevertheless chosen to address at the term of a process which is more complete and hence more instructive regarding the various notions involved. However, allowing ourselves to use a shortcut, we will immediately establish the construction of the functional and structural wireframe of the Saussurean sign.

At this stage, we thus have at our disposal three pieces of structural information which may be foreseeably integrated within a unitary system—that is, on the one hand, a topology (elementary) of the linguistic system (cf. 5.2.1.3), on the other hand, the "$Sr_1\#Sr_2 \leftrightarrow Sd_1/Sd_2$" formula and, finally, the Sr $\rightarrow$ Sd directionality.

These three structural schemas allow themselves to be easily superimposed, at least diagrammatically, by way of a few arbitrations and adjustments.

To do this, we may consider two signs, noted α/A and β/B (following the Sr/Sd pattern). We already know that the $\alpha\#\beta$ and A/B relations are to be situated at the "boundaries" of the system. In what concerns the $\leftrightarrow$ connection between the differences pertaining to signifiers and signifieds, it would be logically appropriate to make it unidirectional—that is, and in accordance with the Sr $\rightarrow$ Sd orientation and with that of the relation of "exchange" between a sign and its "concrete" counterpart (regarding the exchange relation, cf. 5.2.4 the definition of *value*): "$Sr_1\#Sr_2 \Rightarrow Sd_1/Sd_2$".

But then the expression/content directionality is achieved in two ways: between α (resp. β) and A (resp. B) and between "$Sr_1\#Sr_2$" and "Sd_1/Sd_2", to which must be added the directionality of the relation of *exchange* between α (resp. β) and their counterparts [occurrences of the substance of content—noted "a" (resp. "b")], for instance concepts, as entities that are external to the linguistic system. The situation is therefore somewhat confused, and it is fitting, in order to avoid any redundancy, to retain only the main points of these different accomplishments of a same functional orientation.

To this end, we will observe that the Sr → Sd link can be deduced from the "⇒"relations and from the relations of "exchange". Indeed, the ⇒ relation establishes for its part a directional link between *pairs* of signifiers and signifieds (as a reminder, $Sr_1\#Sr_2 \Rightarrow Sd_1/Sd_2$). In order to move from a link between pairs of signifiers and signifieds to a link between a signifier and its signified taken in isolation, it is necessary to enhance the ⇒ relation with a "substantial" piece of information. The introduction of a substantial indication indeed enables to "liberate" the signified or the signifier from the differential or distinctive relation which establishes it in its unity and its identity, in order to establish an autonomous signifier/signified connection. Now, the relation of exchange, which associates a signifier to occurrences of content substances, enables precisely that.

We will therefore retain "⇒" as structural directionality, and the exchange relations α/a (and β/b) as a principle of "encapsulation" of the α/A (and β/B) relations. The resulting diagram (Fig. 5.4) is the following:

This schema of semiolinguistic structure, which requires numerous clarifications and precisions, will be addressed once more following a more detailed demonstration than the one accomplished so far, and we will do so by resuming the momentarily suspended discussion regarding the dissymmetry of the sign. This dissymmetry asserts itself in yet another manner.

Indeed, being interested in the forms which govern signifiers and signifieds, and supported by Saussure's texts, we are logically led to orient the signifier/signified relation as being directed from the first term towards the second.

We have seen indeed that the signifier and the signified, if they both present a hybrid nature, distinguish themselves in that the signifieds *by no means* proceed, be it in the form of an identitary persistence or of a functional implication, from the qualities of substance which the differential forms invest. Now, such is not the case with signifiers. We have seen that the relations of difference between signifiers rest upon differences elaborated outside of the linguistic system, and to which they are a

Fig. 5.4 Functional architecture of the linguistic system

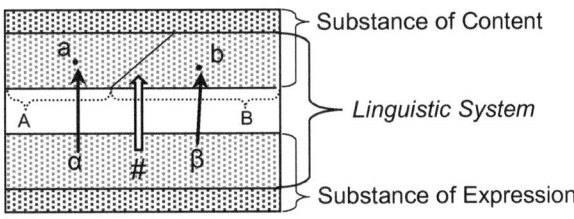

formal transcription. Then, and albeit that these plays of formal differences are correlated with differences in content, they preserve, as though in underlay, the distinctive "external" identities—or, to use a Husserlian expression, the "phonic complexes"—from which they first proceed.

Thus, the phenomenological equivocality of the signifier—which assimilates two distinct moments of intuition into a tiered consciential complex: moments of perception[64] and of signification—considered as regards the plane of structures leads to conceive of a sort of functional participation of the "phonic complexes", as they are simply perceived, within the order of the linguistic system. Indeed, since the signifiers preserve a link to the qualities of the phonematic substance, it is necessary for this link to find within the linguistic system a structural signification, in short, for it to find in one way or another its operational reason in the economy of the linguistic system. In order to address this, it will be necessary to reflect upon the notion of "concreteness".

5.2.3.2 Regarding "Concreteness" in Language

We know that Saussure first relates linguistic concreteness to a consciousness of significativity: "[C]oncrete, as well as real, applies to everything of which speaking subjects are *conscious* in any degree"[65] and, more specifically: "What is concrete or real is what is in the consciousness of speaking subjects, that is, what is *meaningful* in any degree."[66] But this criterion of meaningfulness proves insufficient with respect to the doubts and to the controversies raised by linguistic analysis, for example with respect to the status of several infixes that are poorly determined on a semantic level (and of which we know that their share of meaning is not separable from the syntagmatic and paradigmatic configurations of which they are members, therefore from the relations they form within the system). In short, it is a matter of knowing if "the intuitive analysis of speaking subjects goes as far as that of grammarians."[67] Now, following this first acceptation of *concrete*, and although the principle of a gradation of meaningfulness ("there are degrees of consciousness and of meaningfulness"[68]) enables to cast certain interrogations aside, it seems however difficult to recognize for the entities obtained by abstraction (flexional classes, for example) a concrete character on the basis that they exist just as much in the consciousness of speaking subjects: "[I]n this sense, all units as such [would be] concrete, including paradigmatic series."[69]

---

[64]Specifically: categorical perception of phonemes.

[65]Godel (1969, p. 210).

[66]Godel (1969, p. 157).

[67]Godel (1969, p. 210).

[68]Saussure *in* Godel (1969, p. 233).

[69]Godel (1969, p. 210).

Saussure is therefore led to revise his conception of the concrete, and firstly by bringing it closer to the essential character of signs: "[I]n language, what is concrete is all which is present in the mind of the speaking subject [...]. But it is not in this (justifiable) sense that we have taken the terms: concrete and abstract. Concrete, here, signifies that the idea finds its unity in the acoustic medium."[70] Clearly, therefore, at this stage, the concrete refers to the regime of the signifier/signified interdependence.

Now endowed with a criterion for concreteness, we also have a "method of delimitation": Concrete units respond indeed to the segmentations operated within the chain of discourse inasmuch as "the division along the chain of sound-images [...] will correspond to the division along the chain of concepts."[71] We see that the mode of delimitation is related to concreteness in language, and it is quite logical that Saussure would inscribe it as a property of concreteness—thus: "The concrete unit is a delimitable unit."[72]

But the practice of delimitation will highlight a dissymmetry between the signifier and the signified. Because if both faces of the sign occupied equivalent functional positions, the method of segmentation should have two referentials: that of meaning and that of sound. Thus, in the first case, choosing the measurement (or criterion) of meaning, the procedure will consist in segmenting the sequence of sounds and in evaluating the signifying content, i.e. the connection with a component of meaning, of the segments thus isolated. It is this method, by virtue of which primacy is conferred to meaning, which is implicitly exposed in the following definition of the concrete entity: "a slice of sound [...] linked to a certain concept which serves to delimit the slice"[73] and which we find exemplified in different passages of Saussure's works. In principle, and in first instance, this way of doing things does not seem to encounter major obstructions. This is not the case however when we proceed with regard to sound. It is difficult to imagine, indeed, how to circumscribe, free from any calibration by means of a signifier, a segment in the mass of thoughts so as to later observe that it is assimilated as a signified to a certain fragment of the phonic chain.

From the standpoint of the practice of delimitation, hence from the standpoint of the constitution of the concrete unit, we thus observe a dissymmetry between the two sides of the sign.

We may observe here that the possibility for the practice of delimitation is dependent upon the previous (precedingly acquired) recognition of the ambivalent character of the signifier. This is because, by way of the criterion of meaning, one first thing among two: Either the analytical procedure addresses a sequence of sounds *as a signifying organism*, in which case it will tautologically perform a segmentation into signifiers—which is sterile—or it addresses the sequence of

---

[70]Saussure *in* (Godel 1969, p. 211).

[71]*CLG/B*, p. 104.

[72]Godel (1969, p. 217).

[73]Saussure *in* Godel (1969, p. 140).

sounds as a simple acoustic-phonematic flux, in which case, since the arbitrary segmentations engage no link to meaning, their appreciation as a medium for meaning has no logical or structural *raison d'être*. Such would be the case if the signifier and segment of sound were entities pertaining to distinct and irreducible spheres of objectivity. But we know that the signifier presents a hybrid character, an indecisive composition of a concrete material and a signifying form, which evidently enables to overcome this difficulty—one which is indeed quite real when considered from other perspectives.

Now, not having expressively or previously conceptualized the ambivalent character of the signifier, Saussure cannot fail to be confronted with this difficulty.

The Saussurean solution will consist in inscribing the material dimension (graphematic, phonematic…) within the constitution of the signifier. And it is this inscription which delivers the precise sense of the principle of delimitation. Indeed, as we will see, the method of delimitation, although it may aim to circumscribe the linguistic entities within the flux of discourse, will just as well engage an order of things which is not that of language but that of its medium (substance) of expression: "*[D]elimitation* is an operation that is not purely material but necessary and *possible* because there is a *material element*."[74] This may clearly be read in several passages. Thus, "to exist, for a linguistic entity, is to be delimited *from back to front* or, conversely, by a value (the distinct meaning) the speaking subject attributes to it"[75] and, in the *Course*, the concrete unit is defined as "a *slice of sound* which to the exclusion of everything that precedes and follows it in the spoken chain is the signifier of a certain concept."[76]

It is clear that, in this manner, Saussure introduces within the linguistic system a character belonging to phonematic matter, that is, *linearity* as a property of time, distributing a *before* and *after*, and from which signifiers proceed: "The signifier, being auditory, is unfolded solely in time from which it gets [its] characteristics."[77]

We thus find once more the equivocal nature of signifiers, as they doubly pertain to language's exterior and interior. And regarding this ambiguous character, Saussure is perfectly clear. For example, the following citation: "The concrete unit […] will be defined as a delimitable element: It is the sign inasmuch as its signifiers *coincides* with a certain slice of sound"[78] (and "the word is the most delimited unit"[79]) clearly describes a sort of overlapping (coincidence) between the phonematic matter and the signifier as a linguistic entity. Now, it is also clear that the principle of a "coincidence" necessarily supposes homogeneity among the qualities recognized to be in coincidence, hence it supposes identity between the forms of the simply perceptual medium and of the signifier as such. In this respect, Godel indeed

---

[74]*Cours III*, p. 224, our emphasis.

[75]Saussure *in* Godel (1969, p. 211), our emphasis.

[76]*CLG/B*, p. 104.

[77]Ibid., p. 70.

[78]Saussure *in* Godel (1969, p. 211).

[79]Ibid.

speaks of the sharing of forms between the phonic medium and the linguistic signifier: "the only common trait which may be found [between the concrete unit] and the phonic unit [is precisely to be delimitable]."[80]

The signifier thus presents a complex nature in that its apprehension conjugates two inseparable points of view: On the one hand, the signifier in its purely perceptual aspect, as a simple phonematic complex, and on the other hand the signifier as structurally constituted in and by the linguistic system, that is, as orienting towards a content.

It should be noted that this double nature of the "signifier", which was anticipated above, and of which we find the principle in Saussure's works, though this aspect of things has not been systematically thematized, overlaps with Husserlian phenomenological analysis: The "word's sound" preserves something of the "sound" but, without fully being a "word" (i.e. incorporating a signified), it is already directed towards meaning (cf. Chap. 4).

We may also note that this conception of an intrinsic equivocality of the signifier enables to surpass the difficulties posed by the principle of linearity and to lift a few misunderstandings regarding the question of arbitrariness.

Very schematically, first concerning this second point: Since the signifier, in its phenomenal equivocality, assimilates its coextensivity with the signified to a sort of endurance of the perceptual medium which it mobilizes in its constitution, the relation between signifier and signified reveals itself to be structurally contingent *and* necessary. It is *contingent* (or arbitrary) if we emphasize the phonic medium as such, and *necessary*[81] if we only retain its relation to the signifier. Likewise, concerning linearity this time, favoring one or another of the viewpoints, we will be led to recognize the linear occurrence of signifiers over the time of phonematic perception or the essential unity of the phrastic organicity which comes into existence in a neutralized time: "[I]nterlocutors [...] only have a very obscure awareness of the time taken by the utterance of words and syllables. In practice, such time does not count, although the linear character of discourse can only be explained by its means."[82]

But if the ambiguity of the signifier seems indeed to constitute a *structural fact*, its meaning and principle remain to be understood. In what concerns the first aspect, the key is again to be found with Saussure, in the previously cited passage: "Linguistic entities must be delimited, an operation which is [...] by no means purely material, but necessary and logical because there is a material element."[83] What is being said here, namely, is that the determination of linguistic entities implies the material element, and therefore, dually, that the material components participate in the elaboration of linguistic entities. This amounts to acknowledging for phonematic segments and in their simply perceptual character, a functional

---

[80]Godel (1969, p. 211).

[81]Conception defended by (Benveniste 1966, p. 50 *sq.*).

[82]Godel (1969, p. 207).

[83]Saussure *in* Godel (1969, p. 211).

effectivity in the establishment and functioning of the linguistic system and, correlatively, to supporting the principle of an asymmetry of the two sides of the sign:
The one, due to one of its constitutive features, finding itself to be invested with a
dimension which is operational in the constitution of the other—an operative
dimension which may have received an instrumental interpretation (hence,
Amacker, following Prieto, who describes a "subordinate, instrumental situation of
expression"[84]), but which, in the context of a morphodynamic approach, will lead
us to qualify it in terms of the "control" of differential forms categorizing the
substance of the content.

Evidently, and this is the second point, it is still necessary to precisely explain
the *modus operandi* of the material substrates in the constitution and functioning of
language. To do this, we will operate a problematic shift: Putting aside for a few
moments the considerations we have pursued and the conclusions we may have
reached, we will focus upon the concept of *value*. This shift is quite in accordance
with the thinking of Saussure who, in the Third course, reevaluated the forms of the
object and of the linguistic system using this new conceptual angle.

5.2.4   The Standpoint of Value

The questions raised by the concept of value are highly complex and somewhat
entangled. It will not be a matter in this short contribution of presenting them
exhaustively and even less of providing a comprehensive response. At best, we will
present some elements in order to situate our own approach within a global context
formed by the problematics at hand. Turning our attention towards only one of the
numerous problems posed by the concept of value, it will be a matter of proposing a
manner of addressing it, at the term of which we will then be able to "recuperate"
the set of previously drawn conclusions within a unitary schema.

As we know, value is defined at the intersection of two relational dimensions,
and in this respect, the *Course* and the sources converge. We may recall, to avoid
any confusion, that "values are apparently governed by the same paradoxical
principle. They are always composed: (1) of a *dissimilar* thing that can be *ex-
changed* for the thing of which the value is to be determined; and (2) of *similar*
things that can be *compared* with the thing of which the value is to be determined."[85] The "paradox" proceeds, as we know, from the heterogeneity of the two
relational modalities which value traverses and unifies, that is, a relation of
exchange (or "counterpart") with an entity which is *external* to the set of units with
which the value participates in systemic (*internal*) relations of comparison: "*Value*
is eminently synonymous [...] with an element situated within a system of like
elements, just as it is a [...] perfect synonym for what is exchangeable [...]. If we

---

[84]Amacker (1975, p. 160).
[85]*CLG/B*, p. 115.

consider on the one hand the exchangeable thing, and on the other the co-systematic terms, no relationship is perceptible."[86] The concept of value therefore "paradoxically" integrates relations of internality (constituting the system) and of externality (to the said system). This point is capital and, as widely acknowledged, it constitutes without a doubt the stumbling block of Saussurean thought. But before addressing the problematic unity of the two relational dimensions of value, it is already necessary to specify what are the terms of the relations that are mobilized within—in other words, what is the nature of the term "value" (entity of the system) and of its external counterpart.

Let's recall the main possibilities, knowing that these different options can be "assembled"[87]: (i) value is the signifier, its counterpart is the signified, (ii) value is the signifier, its counterpart is a concrete phonation (a unit of the expression's substance), (iii) value is the sign, its counterpart is a "meaning" (an "idea" or psychological moment, that is, a unit of the content's substance), or (iv) value is the signified, its counterpart is a "meaning".

For example, taking into consideration that "in the association constituting the sign, there is nothing [...] but two values which exist the one for the other",[88] Amacker is led to recognize two planes of value and a layering of relations of exchange. Regarding the plane of expression, "the coexistence with 'similar things' is the set of mutual relations of difference between signifiers" and "exchange with what is dissimilar is the relation to the signified."[89] But the signified is itself a value defined by relations of difference doubled by relations with "'exchangeable things,' in the case of concrete meanings which are exchangeable for value in speech."[90]

This doubling is to be found in Godel's analysis, but in contrast to Amacker who articulates the various strata of value, Godel raises the question of their superimposition and thus transposes the issue which spans the semiological definition of language towards a new field of questioning.

First discussing the difficult distinction between "value" and "meaning"—a difficulty regularly attested but nevertheless muddled by the frequent synonymic use of the two terms—and referring to the exposition of the economic analogy in the second course, Godel is led to raise *the issue [of the] double aspect of the signified:* the value of the auditory image, on the one hand, and at the same time the value generated from the relations between coexisting terms."[91] Indeed, in the economical analogy, "the crown [coin] with its effigy [is compared] to the signifier; the value (5 francs) is compared to the signified."[92] Correlatively, after having

---

[86]Saussure (2006, pp. 239–240).

[87]Cf. notably Godel (1969, p. 231).

[88]Saussure *in* Amacker (1975, p. 159).

[89]Amacker (1975, p. 160).

[90]Ibid., p. 159.

[91]Godel (1969), p. 237, our emphasis.

[92]Ibid., p. 239.

introduced the idea of signification as the referral[93] by the signifier to the signified
—for example: "the arrow [from the auditory image towards the concept] indicates
meaning as counterpart of the auditory image"[94]—Godel notes that Saussure
indicates the insufficiency of this acceptation which he rectifies by stating that "the
counterpart of the auditory image [...] is just as much the counterpart of terms
coexisting in the language"[95]: "The value of a word will be the result only of the
coexistence of the different terms. The value is the counterpart of the coexisting
terms [...] How does that come to be confused with the counterpart of the auditory
image[?]".[96] When supported in such a manner, the signified appears in its double
nature, both as value of a signifier and as a systemic value. Although the signified is
taken to be the counterpart of the signifier, it "cannot, in any case, be anything else
than value",[97] and immediately the problem arises of "the strange coincidence
between the two determinations at play [with respect to value]."[98]

In order to follow Godel's reasoning and to understand his choices, it is necessary to keep in mind the privilege he confers to the systemic acceptation of value:
"[A]lmost always, it is the idea of a system, of the coexistence, of the solidarity of
terms which provokes or motivates the use [of value]."[99] Now, we can see that the
transposition of the question of value, which is that of the integration of heterogeneous relational modalities, as a simple issue of "coincidence" of such relational
modalities, amounts to a systemic qualification of the relation of "signification" (or
of "exchange"). Indeed, it is necessary and sufficient to internalize in the linguistic
system the relation of exchange which participates in establishing values for the
"paradox" of value to be transformed into a problem of relational equivalence. In
this perspective, value therefore finds itself to be entirely situated within the linguistic system, precisely as it is "depend[ant] upon three orders of relations: the
sign's internal relation (the signified is the value of the signifier); the [limitative]
relation of terms *in absentia*; the [limitative] relation of terms *in praesentia*."[100]
Given that primacy is given to the system and to oppositions between terms, the
signifier/signified relation "is not initial, it contains only the summary[101] of
value."[102]

---

[93]In other words: "[M]eaning, that is the evocation of sense by means of the word" or "the word
carries a meaning which adds itself to syllables" (Godel 1969, p. 237).

[94]Saussure (1993, p. 135a).

[95]Ibid.

[96]Ibid.

[97]Godel (1969, p. 241).

[98]Ibid., p. 240.

[99]Ibid., p. 236.

[100]Ibid., p. 244.

[101]"[T]he word does not exist without a signifier and a signified; but the signified is but the
summary of the linguistic value which assumes a play of terms between themselves" (Saussure *in*
Godel 1969, p. 237).

[102]Godel (1969, p. 246).

We therefore understand Godel's uneasiness with respect to the numerous passages where the relation of signification is, unequivocally, presented as a relation between terms among which one is foreign to the system. Thus, concerning the definition which puts into play the two relational modalities, he notes that "the text is not devoid of obscurity [...] we hesitate regarding the interpretation of the words: 'exchangeable thing' [...]"[103] In addressing the "identity of meaning [which Saussure discusses] between Sanskrit plural and German plural", Godel notes that "the identity of meaning [...] can only be that of abstract concepts pertaining to pure psychology [...] We are therefore faced with a dilemma: Either meaning consists in the signified, in which case it is confounded with value [preceding option], or it is the concept taken in abstraction, in which case meaning would be foreign to language."[104] Now, this second option, although it may appear to raise insurmountable obstacles, must nevertheless be maintained—the sources converge too strongly towards it for it to be dismissible. For example, in the *Course*, the relation of exchange (or of "signification") which concerns words (as linguistic entities) indeed extends towards the extralinguistic sphere of *ideas*: "a word can be exchanged for something *dissimilar*, an *idea*; besides, it can be compared with something of the same nature, another word."[105] Moreover, "[the word] can be 'exchanged' for a given concept, i.e. that it has this or that signification: [but] one must also compare it with similar values, with other words that stand in opposition to it."[106] And correlatively: "Modern French *mouton* can have the same signification as English *sheep* but not the same value."[107]

By enclosing the issue of value within the closed perimeter of the linguistic system, we lose two things. On the one hand, we deprive language of its relation to the world of materialities which it invests by means of significations, and, on the other hand, we dispossess the concept of value of its heuristic potentialities which precisely stem from its constitutive, though problematic, heterogeneity. Transposing the problem of the semiological definition onto the plane of values, we will be required to place at the center of our enquiry the modalities of a mutual integration of the order of language with its extralinguistic substances.

In order to do this, and establishing a first connection with Husserlian phenomenology, we will choose to approach the issue in terms of the structural promotion of the "indicative" sign to the rank of "meaningful" sign.

---

[103]Ibid., p. 240.

[104]Ibid., p. 241.

[105]*CLG/B*, p. 115, our emphasis.

[106]Ibid.

[107]Ibid.

5.3   Structural Promotion: From the "Indicative" to the "Meaningful" Sign

5.3.1   The "Indicative" and the "Meaningful" Sign

We should first recall (cf. 2.2) that a phenomenological analysis of the acts con-figuring in consciousness the appearing of signs led Husserl to distinguish the *meaningful* sign (expression) from the *indicative* sign (symbol).

The latter, the *indicative sign*, coordinates two moments of consciousness: There is first a certain experience of consciousness, which is the perception of the sym-bolic marking, and, by its constitutive function, the symbol reorients consciousness towards another content which is the thing, the idea, or the state of things to be communicated (of which the listener is to be informed).

Conversely, the *meaningful sign* inscribes itself within a single and same moment of consciousness: The apprehension of the sensible manifold and its elaboration into a sign-phenomenon (the noetic moment) operates within a single intentional act which is the aim of an object of "content." This is its very principle of constitution: "[T]he essence of an expression lies solely in its meaning."[108]

Assuredly, the "true" signifier, which Husserl thus calls "expression", comprises in its phenomenal nature the orientation of consciousness towards a meaning. It is this intentional directionality which shapes its appearing as a word-sign: "The 'meaning-intention' [...] constitutes the phenomenological character of expres-sion"[109] and it is therefore "By virtue [of the intentional acts] that expression is more than a simple acoustic phenomenon."[110]

The opposition between the *meaningful* sign and the *indicative* sign is not for-eign to Saussurean thought. We may begin by recalling that for Saussure, language is not organized like an index. As a conventional reference of sound-units to meaning-units, each constituted within their own spheres: "The characteristic role of language with respect to thought is not to create a material phonic means for expressing ideas [i.e. the indicative sign]".[111] And in keeping with the meaningful sign, the existence (the character of essence) of the Saussurean sign stems from the indefectible unity between a form and of a meaning.

We have sufficiently emphasized (cf. 2.2) the proximity between Saussurean and Husserlian thought for it to be necessary to insist on it any further. Let's therefore return to the issue of value, which we know to be that of the delicate integration of two relational modalities which articulate values, the one being internal and the other being external to the system they compose. It should readily be noted that this problem could not be resolved as a simple summation, because that of which it is

---

[108]*RL1*, p. 56.
[109]Ibid., p. 47.
[110]Ibid., p. 43.
[111]*CLG/B*, p. 112.

fundamentally question in the constitution of values, is not an additive superimposition of two relational schemas, but indeed of their mutual assimilation and necessary unity, and, as we have said, it is this difficulty which will have, in a certain, way triumphed over Saussure.

5.3.2 The Issue of Value: Towards a Solution

In order to resolve this problem, it appears beneficial to pose it in terms of the structural promotion of the "indicative sign" to the status of "meaningful sign". Indeed, if the values are configured at the intersection of relations of exchange and of opposition, we can reasonably suppose that each of these relations carries a partial semiotic rationality, and therefore, in particular, that the relation of exchange institutes at its sole level a regime of significativity of which the achieved form, proceeding from its structural enrichment following oppositive modalities, conserves its "trace".

It will therefore be a matter of starting with the structure of the indicative sign, that is, with the conventional connection between units of substance of expression and units of substance of content, in order to make it into a "meaningful" configuration where the exchange relationship, though maintained, is functionally requalified. And this transmutation will need to pertain as much to the terms of the relation (units of expression and content) as to the relation as such.

Now, in order to progress regarding this problem, we have elements which pertain to the general architecture of the linguistic system. We indeed know that the system comprises an interior and two exteriors (the substances), and that each exterior is in contact through a boundary of which we have approached the internal forms, that is, on one side, a system of discontinuities separating the substance of content into negative identities, and, on the other side, "equivocal" units of expression, in the sense that their qualifications as substances "persist" in their structural identities in language.

The promotion of the indicative sign to a status of meaningful sign, given these elements, will therefore fundamentally consist in passing from a simple relation of exchange (written $\rightarrow$) between symbols and ideas to a relation of oriented determination (written $\Rightarrow$) between differences on the plane of content (written "/" and established as boundaries categorizing a homogeneous substrate into adjacent subdomains, respectively indexed by means of A and B) and on the plane of expression (written "#"). This therefore requires two relations of exchange, which we could write $\alpha \rightarrow a$ and $\beta \rightarrow b$ so as to produce the schema: $\alpha\#\beta \Rightarrow A/B$ (cf. Fig. 5.5).

This configuration obviously lacks clarity, at least in three respects. *First*, the $\rightarrow$ relation, although it is "persistent" since it participates in configuring the values, cannot be maintained as such: It must be functionally requalified. *Second*, the relation of determination, noted $\Rightarrow$, between differences belonging to both planes

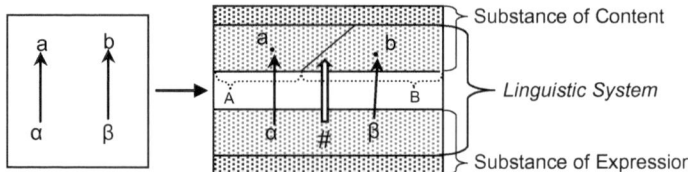

Fig. 5.5 From indicative sign to meaningful sign

must be introduced and explained. *Finally*, this schema lacks the order of S&P relations.

It is through the examination of the latter that we will provide a comprehensive answer to these various questions.

5.3.3   Syntagmatics and Paradigmatics

5.3.3.1   Introduction

Let's briefly recapitulate the situation regarding these two types of relations.

In the Saussurean perspective, and in essence, the syntagmatic relations which link units into discourses have for foundation "the linear character of language", knowing that this "linear character" stems from a "principle of exclusion" by virtue of which the "actual present" of the activity of speech (*versus* the "accomplished present" which maintains the previously uttered elements) defines a locus which can only receive one single linguistic unit at a time, this being "the linear character of language, that is, the impossibility of simultaneously pronouncing two linguistic elements."[112]

The syntagmatic relation therefore concerns elements as they are actualized ("pronounced") and which, mutually excluding one another in the time of their actualization, "are [thus] arranged in sequence on the chain of speaking."[113] "Consecution" (succession) and "combination" are the terms used by Saussure to qualify the syntagmatic relation. In the syntagmatic series, elements succeed one another inasmuch as they are actualized one after the other.

The paradigmatic is opposed to the syntagmatic inasmuch as it does not establish between its terms relations of disjunction (of mutual exclusion), but considers them conjointly on a same plane of existence which can only be virtual. Thus, whereas the syntagmatic relation concerns terms *in praesentia*, the paradigmatic relation "unites terms *in absentia*."[114] Furthermore, the mode of virtual coexistence of terms

---

[112]Saussure, *Course III*, *in* Godel (1969, p. 203).

[113]*CLG/B*, p. 123.

[114]Ibid.

within a paradigm is that of genericity: The elements of a paradigm "are [all] related in some way."[115] The terms of a paradigm therefore enter into relations of "association" on the basis of a shared generic feature, which may be of various natures (founded on the signifier, the signified, or the *undivided* sign), and which "the mind grasps."[116]

From Jakobson's point of view, the paradigmatic and the syntagmatic are the relational modalities underlying the operations at work in the activity of speech, that is, *selection* and *combination*: "Speech implies a selection of certain linguistic entities and their combination into linguistic units of a higher degree of complexity."[117] We therefore have *simultaneity* and *succession* as regulatory principles of the paradigmatic and of the syntagmatic: There is "concurrence of simultaneous entities and the concatenation of successive entities."[118] However, Jakobson modulates somewhat the Saussurean point of view. First, the relations between paradigmatic elements are not approached in terms of association but of *concurrence*, and this concurrence is carried by an operation of substitution. Otherwise, and as is the case for Saussure, the terms of a paradigm enter into relations of genericity/specificity. These few observations are apparent in the citation which follows: The linguistic system articulates "two modes of arrangement. (i) *Combination* [which concerns] the *actual grouping* of linguistic units [...] and (ii) *a selection* between *alternatives* [which] implies the possibility of *substituting* one for the other, *equivalent* to the former in one respect and *different* from it in another."[119] This emphasis on the relation of substitution can be found in the works of Benveniste who defines the paradigmatic relation as "the relation of the element with the other elements which are *mutually substitutable*."[120]

5.3.3.2   Difficulties

These conceptions of the S&P relations pose serious problems.

First, concerning the syntagm as a series of units: The difficulty of this concatenational conception (succession of units) is apparent in the difficulty of conjoining the atomical character of the linguistic items which are successively actualized through the activity of speech with the effective unity of the syntagm of which these items are members.

Indeed, it is necessary to admit that the "present" of the syntagmatic dimension, and more generally the present of speech, extends beyond the present which finds itself to be instituted in the strict moment of an utterance regulated by virtue of the

---

[115]Ibid.

[116]Ibid., p. 125.

[117]Jakobson and Halle (2002, p. 72).

[118]Ibid., p. 73.

[119]Ibid., p. 74.

[120]Benveniste (1971, p. 102).

principle of exclusion. In fact, if the present was but the present of a single locus, a present which is not shareable, then the units initially configured in terms of exclusion would not allow themselves to also serve as parts of an integrated syntagmatic totality. Moreover, upon leaving the present of their actualization, and being devoid of other forms of actuality, they would be immediately disqualified.

The possibility of the syntagmatic fact therefore implies to *overcome the principle of exclusion which configures it* nonetheless, and, correlatively, to conceive of syntagmatic actuality following another order of temporality than that which is established in the moment of utterance. These are all things which Saussure implicitly recognizes, as demonstrated in the following citation: "The syntagmatic relation […] is based on two or more terms that occur [*equally*] in an *effective* series."[121] Within the syntagm, Saussure indeed considers terms which are *equally* present, that is, *conjointly* present *and this with a same degree of presence*—which thus supposes an existence regulated following another mode than that of reciprocal exclusion—and this in an *effective* series, that is, a series of which the totality taken "as a unit" is of the order of factuality.

We thus understand why Saussure approaches linearity in spatial terms: "The signifier, being auditory, is unfolded solely in time from which it gets [its] characteristics: (a) it represents a span, and (b) the span is measurable in a single dimension; it is a line"[122]—since the figure of a line sets up and distributes, below and beyond the simply punctual present, the totality of time.

We also understand why Jakobson, Benveniste and also Hjelmslev include hierarchical relations (with higher-level units) in the definition of the syntagmatic relation. In Beneveniste's words, the syntagmatic relation is: "the relationship of the element with the other elements simultaneously present in the same portion of the utterance".[123] Or Jakobson: "the speaker combines [syntagmatic] words into linguistic units of a higher degree of complexity."[124] We indeed see in these two citations that the presence of the word is not a presence which isolates it, but a presence as a part of a syntagm forming a totality.

It remains that we must acknowledge the linear character of the syntagm: The syntagm is not delivered in a single block; its presence is not without articulations. It arises at the term of an unfolding in which each instant sets one of its items in a position which effectively cannot be shared, in sum, following Saussure, a position which excludes other elements.

There is, in truth, a sort of uncertainty between subjective time as per an individual consciousness inhabiting the empirical world—a subjective time in which the principle of exclusion has its place, but without being reducible to it—and time as per a linguistic consciousness which, though submitted to the principle of exclusion, differently articulates the temporality of units of different levels and

---

[121]*CLG/B*, p. 123.

[122]Ibid.

[123]Benveniste (1971, p. 102).

[124]Jakobson and Halle (2002, p. 72).

different sorts (syntagms and paradigms). In this respect, as we have seen, Godel speaks of a "very obscure awareness of the time taken by the on going utterance of words and syllables."[125]

In order to overcome this paradox of the syntagm, which establishes its units by means of exclusion in order to later integrate them into a unitary complex, it will first be necessary to take some distance from any intuitive or conceptualized representation of physical time, and to instruct the form of semiotic time. Specifically, it will be necessary to establish in their specifically linguistic rationality these "punctual moments" which are governed by the order of exclusion.

5.3.3.3 Solution

In order to do this, the path is clearly set and it is quite direct. Indeed, as we have seen earlier, the paradigmatic, in its intersection with the syntagmatic, administers *concurrence*, therefore *selection* and, furthermore, *exclusion*. The paradigmatic relations, when they encounter the syntagmatic axis, deliver the linguistic form of exclusion, the form which institutes the segments into qualities of exclusive isolates and which we know to be at the source of (the problematic) syntagmatic successivity. Specifically, the paradigmatic axis articulates linguistic linearity as it institutes the exclusive position of a present which is "underway" by means of an operation of selection. This present pertains to "that which is being said" *versus* the present of "that which has been said", which is the syntagmatic present.

But we have nevertheless not resolved the problem of the *equally present units* within an *effective syntagm*, that is, the problem of the simultaneity of elements *equally present* within an *integrated totality which is equally present in a unitary form*. This is because by establishing the principle of exclusion in the paradigmatic format, we indeed confer it a linguistic status, but we open no means to overcome the problems it induces. How to ensure that the units, configured as isolates at the intersection of the syntagmatic and of the paradigmatic, may be associated to previously established or future units, in a way so as to compose a higher-level unitary totality?

The answer to this question, as it is well known, is to be found in the reversal of views operated by structuralism, that is, in the primacy conferred to the relational and which is denied to the substantial. Indeed, in the structural perspective, what is first, are not the identities of substance, supposed to carry their own qualities, and between which we could observe relations of various sorts. The relations do not derive from identities, but it is rather through relations that identities are instituted: It is in the order of relationships that the objectivity of things is configured.

Once the structuralist reversal is taken into account, the form of exclusion weaved by the paradigmatic is no longer to be considered as an operation which would consist in preventing the installation of any element available in its unity and

---

[125]Godel (1969, p. 207).

its identity in a place which is already occupied by another element. It should rather be considered as a form of connection which establishes the existence and the identity of the semiolinguistic objects involved. In other words, the paradigmatic, here through the regime of exclusion, has a constitutive function: It institutes and determines the identities of language.

Thenceforth, it is easy to overcome the difficulties we have been discussing. Indeed, in a very general manner, the integration of various objects endowed with their autonomous characters is only possible and conceivable if they proceed from a common principle. In other words, it is in a motif of shared constitution that lies the possibility for an authentic integration for identities that are themselves endowed with their own identities.

Applied to our case in point, this latter consideration expresses itself as a principle of interdependence between the syntagmatic and the paradigmatic. Indeed, for it to be possible to integrate an entity, for instance A_1, with an entity A_2 when both are governed from the onset by a regime of exclusion which precludes any form of coexistence, it is necessary for the identity of A_2 to be regulated by constraints stemming from A_1. Specifically, it is necessary for the paradigm where A_2 is instituted through exclusion/selection relations with the other members of this paradigm to be determined by the identity of A_1 in its syntagmatic relation to A_2. In its most elementary and general form, the schema of the *S/P* interdependence[126] is the following: Let there be a syntagm coordinating two elements A and B, among which A occupies the "first" place. The actualization of A conditions the paradigm of which B is a member, that is, the terms concurrent to B which institute its *value*.[127] Retroactively, B, actualized in turn, determines the paradigm centered on A, and therefore the *value* of A. Of course, this schema—which we will note to account for the formation of a holistic structure—functions in a more complex mode in hierarchical structures of constituents.

It is therefore by the indetermination of the syntagmatic and the paradigmatic that the possibility of a syntagmatic relation is achieved, at the level of its units as well as at the level of its totality, authentically and fully *in praesentia;* a relation which "is based on two or more terms that occur [equally] in an effective series"[128] and which instructs the compresence of the parts (semiolinguistic elements) and of the whole (syntagmatic set).

This interdependence between the syntagmatic and the paradigmatic, which has been established here from the angle of a structural necessity, participates in the empirical foundations of semiolinguistics.

---

[126]Syntagmatic/Paradigmatic.

[127]In Saussure's sense.

[128]*CLG/B*, p. 123.

5.3.3.4 Illustrations and Precisions

In order to illustrate and discuss this "reciprocal incidence of paradigmatic varia-tions and their contexts [i.e. syntagmatics]",[129] let's return to the famous example from the *Course* which looks at one of the paradigms centered around *enseignement* ('teaching'), that is, *éducation* ('education'), *instruction* ('instruction'), *apprentis-sage* ('*apprenticeship*'), etc.

From the structuralist viewpoint, the sense of *enseignement* ('teaching') is established through relations of reciprocal delimitation between this word and terms which, in certain contexts, compete with it. Thus "the intellectual connotation contained in *enseignement* ('teaching') reveals itself in its possible opposition to *education* ('education') in contexts where both terms compete."[130] Such is the case, for instance, with constructions of the *Adj. + N* type, such as: *arts education, literary education, professional education, religious education, scientific education.* And "from the moment this concurrence ceases, from the moment when *en-seignement* ('teaching') can no longer be replaced by *éducation* ('education')—as in the sentence 'this story holds many teachings'—the vacant area is immediately filled and *enseignement* ('teaching') recuperates the moral acceptation it had otherwise abandoned."[131]

What is apparent with this example is that *two sorts of exclusion* must be considered and their relations examined. The first sort concerns the construction in whole or in part of a paradigm as a set of terms likely to occupy a place in a given syntagmatic context (cf. Benveniste's definition of the paradigmatic relation as a "the relation of the element with the other elements which are mutually substi-tutable."[132]). Exclusion in this case consists in excluding from the paradigmatic field the terms which could have claimed to reside in it under different circum-stances (as in the sentence "*this story holds many teachings*" where, in contrast to *leçon* ('lesson') for example, *éducation* ('education') finds itself excluded from the paradigm of *enseignement* ('teaching')—cf. below). This first regime of exclusion therefore establishes sets of concurrent terms. Hence, it is the second sort of exclusion which must actualize the one or the other: For example, *éducation* ('education') rather than *enseignement* ('teaching') in *religious education*, which then receives a differential value (/morality/) in its contrast to *enseignement* ('teaching') (/knowledge/).

These two sorts of exclusions, respectively "constituting" and "actualizing" exclusions, participate in a shared systemic logic, and, more specifically, opera-tionalize two facets of a same semiolinguistic rationality which should be specified.

First, as we have just seen, these two sorts of exclusions seem to be in a hierarchical relation. Indeed, it is necessary to first construct a paradigmatic class

---

[129]Rastier (1982, p. 11).
[130]Ducrot (1968, p. 60).
[131]Ibid.
[132]Benveniste (1971, p. 102).

(by retaining certain terms and by rejecting others) in order to actualize an item of this class in a context of competition effective as regards the other items. This hierarchical distribution thus confers a certain primacy to "constitutive" exclusion. And this primacy, firstly hierarchical, is asserted with respect to the structural significations specific to each kind of exclusion. When "actualizing" exclusion concerns *semantic effectivity*, that is, the actual production of sense through discourse by selecting terms among several which are equally available and likely to be actualized at the expense of the others (authentic competition), "constitutive" exclusion participates in the *intrinsic legality of the system* in that it is directly involved in the very existence and formation of semiolinguistic objects. Indeed, what unfolds in the relations of "constitutive" exclusion is simply the articulation between a "possibility" and an "impossibility" in language, that is, semiolinguistic legality as such (as laws instituting a semiolinguistic objectivity). As has already been seen (cf. 3.2.8) and as we will see again (cf. 8.3.4.2), the impossibility to enter into a paradigm, and more generally the impossibility of applying certain variations or certain transformations to semiolinguistic data—an impossibility expressed in the form of an admissibility judgment—is an *essential* impossibility which indicates the existence of an underlying legality and which contributes just as much in determining the value of elements being considered. Thus, relations of "constitutive" exclusion are relations which participate in the institution of semiolinguistic identities precisely in their opposing semantic values.

The articulation of the two sorts of exclusions and the general schema of functioning which stems from it is the following: Consider a term C which is a member of a paradigm (A, B, C) in an F_1 context and which is excluded from it in an F_2 context. As we have seen, it is through this latter "constitutive" exclusion that term C is brought into semiolinguistic existence and that, more particularly, it establishes its value in opposition to A and to B (taken here indistinctively). But in addition to this, since this opposition is functionally "governed" by the F_2 context, the identities which stem from it are also "informed" by it, particularly in the sense that this context instructs their values. The semantic value of C therefore has two sources: its opposition to (A, B) and its anchoring to the F_2 context.

Let's now look at the (A, B, C) paradigm. The existence of this paradigm attests to a semantic trait common to its three terms (generic trait), and the competition between these terms is expressed by each of them through a specific trait. Moreover, as F_2 informs the value of C constructed on the basis of its opposition to (A, B), F_1 informs the constitution of paradigm (A, B, C), and, in particular, because it receives these three terms, it informs their generic component.

Now, the question is about the coordination of these different elements, and the logic which will lead us to it is, in line with Saussurean thought, that of the articulation of a homogeneous background. However, we will see that although, according to the preceding point of view, the two types of relations of exclusion seem hierarchized, they reveal themselves to be functionally inseparable.

Let's indeed consider term C in context F_2, hence in a relation of exclusion with respect to paradigm (A, B). It is clear that the relevance of this exclusion rests upon the existence of a background shared by A, B, and C. There is indeed no sense in instructing a difference between objects which have nothing in common. The separation which institutes C in its opposition to A and to B therefore supposes the basis of the paradigm (A, B, C), and, correlatively, a certain F_1 context which installs this paradigm. But the terms of the paradigm suppose the previous moment of their constitution, that is, their positioning with respect to linguistic legality as it is expressed through relations of exclusion. This is the case for term C, notably, which supposes the existence of a context such as F_2 which excludes it from the paradigm set by F_1. *We thus see that term C constructs its specific identity on the double basis of its competition with A and B (by virtue of F_1) and its exclusion from (A, B, C) by virtue of F_2.*

This conclusion is reinforced by the fact that the relations of competition do not suffice to qualify the value of the competing terms. Indeed, to say that A competes with B says nothing about the stakes of such competition—it says nothing regarding the establishment of specific content. As we have seen (cf. 5.2.4), value is constructed at the intersection of two dimensions, the first being relational (which is given here by a relation of competition) and the other being the dimension of exchange (which opens onto objects which are exterior to the order of the first dimension) and which the paradigm lacks when taken alone. Also, in order to go beyond the simply *formal* fact of a relation of competition, it is appropriate to enhance the paradigmatic relation between C and (A, B) with a piece of information that is external to this paradigm. This information is in part provided by the F_2 context in that it governs the exclusion of C from (A, B, C). To conclude, it should be noted that, on the one hand, this information is also *constitutive* in that, as has been emphasized, it is through constraints of exclusion that linguistic legality is developed, and, on the other hand, it should be noted that we can now understand why and how the absence of C from the (A, B) paradigm determined by F_2 allows A and B to acquire the semantic value otherwise attributed to C—the reason being that the paradigm (A, B) is constructed on the basis of the exclusion of C from (A, B, C).

This functional configuration which intricates two regimes of exclusion will be reworked and inscribed within the apparatus of the Saussurean sign where it will also receive a specific morphodynamic determination (cf. Chap. 6). But, for the time being, we propose an illustration using the example of the semantics of the French verb *couper* ('to cut'). It will be a matter of simultaneously putting into place the paradigms of competition (in this case between synonyms) in which differential significations are constructed and into place the contexts of distributional exclusions.

Concerning the synonyms of the verb *couper* ('to cut'), in the *TLF*[133] we will take the following pieces of informations:

– [the object designating a part of a whole] *séparer* ('to separate'), *détacher* ('to detach'): To cut a piece of bread, the branches of a tree; to cut an arm; to cut the grass;

---

[133]Dictionnary *Trésor de la langue française*.

- *tailler* ('to cut/trim'): To cut a suit, a dress;
- [the object designating a thing taken as a whole] *diviser* ('to divide'): To cut wood, bread; to cut into pieces, to quarter (to cut into four pieces), to cut in half;
- [the object designating a thing taken as a whole] *inciser* ('to make an incision'), *entailler* ('to make a notch in'): *Un croc coupe la lèvre supérieure* ('A fang cuts the upper lip');
- [the object designating a thing considered from the standpoint of its volume, length, or area] *diviser* ('to divide'), *séparer* ('to separate'): *Une route qui coupait la plaine par son milieu* ('A road cut the field through its middle');
- [when talking about a group of people] *diviser* ('to divide'), *scinder* ('to split'): *L'assassinat du duc d'Orléans coupa la France en deux* ('The assassination of the Duke of Orléans cut France into two');
- [followed by a complement introduced by *de* ('from')] *séparer* ('to separate'), *isoler* ('to isolate') (a person or a group of people from a larger group): *On me coupait du monde* ('I was cut off from the world');
- *retrancher* ('to cut away/off; to substract'), *supprimer* ('to remove/eliminate'), *ôter* ('to remove, to take off'), *enlever* ('to remove') [one or several parts of a whole]: *Couper la fin d'une émission, couper un passage* ('to cut the end of a program, to cut a passage');
- [interrupt a sequence] *barrer* ('to strike something out'), *faire cesser* ('to make something stop'), *arrêter* ('to stop'), *interrompre* ('to interrupt'): *Couper une voie de communication, une route* ('To cut a communications channel, a road'); *Couper le gaz, l'eau; couper la parole* ('To cut the gas; to cut somebody off [to interrupt someone who is speaking]');
- [pronominal usage] *se faire une entaille* ('to cut oneself'), *inciser* ('to make an incision'), *entailler* ('to make a notch in'): *Se couper au doigt* ('to cut one's finger').

Among the main syntagmatic schemas which determine both the paradigms of synonyms and the exclusion of other synonyms, we will retain:

S_1: *V en* ('*into*') (parts, pieces...)
S_2: *V N de* ('from') N

The S_1 and S_2 schemas allow to distribute the synonyms of the verb *couper* ('to cut') in its non-pronominal usages into three classes:

C_1: class of terms accepted by S_1 and rejected by S_2
C_2: class of terms rejected by S_1 and accepted by S_2
C_3: class of terms rejected by S_1 and rejected by S_2

We have:

$C_1 = \{$*diviser* ('to divide'), *scinder* ('to split')$\}$
$C_2 = \{$*séparer* ('to separate'), *détacher* ('to detach'), *isoler* ('to isolate'), *retrancher* ('to cut away/off; to substract'), *ôter* ('to remove, to take off'), *enlever* ('to remove')$\}$

C_3 = {*supprimer* ('to remove/eliminate'), *interrompre* ('to interrupt'), *arrêter* ('to stop'), *faire cesser* ('to make something stop')}

We can see that in the acceptations taken into consideration here, the terms belonging to C_1, C_2, and C_3 all have meanings of "rupture" or of "division" (in French: "*scission*"). We can also see that they can be distributed according to their "degree of presence", a presence which is either real or simply envisioned and which measures a figure/ground type salience of the part(s) resulting from the splitting.

Indeed, (class C_1) *couper* ('to cut'), *diviser en quatre parties* ('to divide into four parts'), is to produce four parts considered to be present, and their existence being acknowledged, they are considered on equal footing. When (class C_2) the parts resulting from division are conferred unequal degrees of importance from the standpoint of their effective presence or of the attention they are given, we are thus steered towards semantics of "detachment". The various parts produced still exist in their material reality, but either they are "put forth" by means of the relation to the whole from which they stem (*couper une tranche de pain*—'to cut a slice of bread'), or they are considered as needing to disappear from the field of attention (*couper les branches*—'to cut the branches'). Finally, in this semantic progression, the last degree is reached (class C_3) when the act of division provokes the effective disappearance of "that which has been cut": *couper la parole, le gaz, la fin du texte* ('to cut off someone who is speaking, to cut the gas, the cut end of the text').

The meanings of the term *couper* ('to cut') are therefore distributed along an axis of qualification which measures "the *degree of presence* (real or attributed)" of the products of an act of division relatively to their original totality. When schema S_1 is mobilized, this axis is then divided into two opposing parts, the one pertaining to the terms of C_1, the other to those of C_2 and of C_3. When it is schema S_2 which is summoned, then the semantic axis relating to the "presence of the parts" is categorized into two competing sub-domains, one pertaining to C_1 and C_2, the other pertaining to C_3 (cf. Fig. 5.6).

What would remain to be done would be to take into consideration the usages of *couper* ('to cut') in the sense of "*entailler*" (pronominal usage of *couper*). In this acceptation of *couper*, in contrast to the previously examined ones, the act of division does not produce parts. Also, it will be necessary to enrich the semantic space of *couper* with a second axis of qualification which evaluates the *degree of accomplishment* of the act of division *from the standpoint* of the production of parts.

Fig. 5.6 Distribution of meanings of the verb *couper* ('to cut')

	S_1		S_2
C_1		C_2	C_3
diviser *scinder*		*séparer* *détacher* *isoler* *retrancher*	*supprimer* *interrompre*

5.3.4 Admissibility

To return to our previous conclusions and considering, *on the one hand*, that the conflicts of actualization *versus* virtualization which express relations of competition between the terms of a paradigm establish the semantic values of units differentially, i.e. into reciprocal oppositions, and, *on the other hand*, that the distributional constraints, through the judgments of possibility (or of admissibility) which manifest them, involve the legality of the semiolinguistic system, we see being woven around the regime of exclusion (in its generic form)—which then takes the value of a structural germ—the interdependence of the S&P relations, the manifestation of a semiolinguistic legality (the admissibility judgments), and, finally, the structural objectivity (via the differential identities).

For the time being, we will focus on the examination of the S&P relations with linguistic legality as it is expressed through admissibility judgments and as such judgments indicate the fact of a demarcation between semiolinguistic possibilities and impossibilities.

5.3.4.1 Generalities

Regarding admissibility judgments, the touchstone of a very large span of contemporary linguistics, much has been said and discussed, and the polemics are far from being resolved, simply because each of the opposing parties proposes valid arguments at its own level of object comprehension, and the controversies could never be overcome without adopting a point of view which subsumes them.

Emphasizing the weakness of admissibility judgments, it was therefore possible to put forth their contextual variability (notably with respect to discursive types, to textual genres) as well as their situational and socio-cultural variability. In an experimental context, among the very numerous factors which influence or perturb the admissibility judgment, we will have identified, in addition to the obvious factor of the "subject" (problem of the idiolects), the "condition of trial" factor (the idea which subjects have of the aim of the trial), the conditions of presentation (written/spoken), the "method" factor (direct/indirect observation), and the "form of trial" (choice of question), etc.[134] Moreover, we know that even regarding utterances which are phrastic isolates devoid of contexts and which are produced in a laboratory setting, linguistic judgments do not always converge.

But if the principle of an "absolute" semiolinguistic legality (global and intangible), serving as source and motivation for authentic semiolinguistic knowledge (which exhausts its object by means of conceptualization) is surely excessive, we nevertheless cannot deny the fact of partial and coherent distributions of admissibility judgments, distributions that are correlative of various perspectives to which facts of language lend themselves, and which deliver so many possible planes of

---

[134]Cf. Al (1975).

semiolinguistic rationalization. In truth, as will be shown (cf. Chap. 9) by using a phenomenological approach of an existential type (which M.-P. provides, cf. Chap. 7)—that is, an approach which recognizes in the empirically given an immediately signifying nature and according to which the primordial fact is that of expression, and also according to which the phenomenal field does not pertain to an order of empirical rationality but to an order of significativity—what will appear is that just as the geometrical perspective is never but a possible representation of spontaneous vision (cf. *PW*, pp. 51–52), a representation which indeed draws from natural vision but which neither captures nor exhausts its integrality or its essence— other representations, as well justifiable, being also possible—, likewise, grammar, syntax, or semantics, as schemas of determination of an "originary speaking mass", are but possible rational manners of approach, that is, conceptual representations which emanate from it, indeed, but which grasp only a layer of functioning and of existence when the "originary speaking mass" is reflected and finalized according to such formats, a layer of which the selection and rational elaboration are quite legitimate in that they respond to particular and effective linguistic practices or projects. In other words, grammar, syntax, and semantics relate schemas of necessity and orders of objectivity which are not the essence of semiolinguistic reality, but which are enabled by virtue of a conceptualization echoing semiolinguistic practices engaged in view of such or such expressive end. Taking heed of these precautions, it is indeed legitimate to recognize in the diversity of linguistic facts various local and always modulatable orders of systematicity and of necessity, rendered through admissibility judgments, and which both interest and legitimate rational linguistic knowledge.

5.3.4.2   Intuition and Existence Judgment

As Milner insists, grammatical activity, which consists in attributing properties to language data removed from their context of production, such an attribution being made namely on the basis of a polarity (acceptable/unacceptable, grammatical/ agrammatical…), constitutes a "fact". And the appreciation of the "well-formedness" or not of the material examined is of an intuitive nature.

Indeed, any speaker has the power of *intuiting* a linguistic event with respect to a certain plane of linguistic practices and, dually, of its coherence with the order according to which it regulates its viewpoint on language. And it is indeed a matter of *intuition* because the admissibility of an utterance is "flagrant": It is immediate and certain. The admissibility or not of an utterance is always apprehended as being obvious and immediate. It does not require any demonstrative development; it is not to be proven. We may cite Husserl in this respect: "The impossibility of their combination [between words] rests on a law of essence, and is by no means merely subjective. It is not our mere factual incapacity, the compulsion of our 'mental make-up,' which puts it beyond us to realize such a unity [...] the impossibility is rather objective, ideal, rooted in the pure essence of the meaning-realm, to be grasped, therefore, with apodictic self-evidence."[135]

Quite to the contrary, metalinguistic attention, which is artificial, and to which the "intuition of receivabilty" can be submitted, can only prove to be troubling: "[T] he exercise of this faculty of judgment (admissibility judgment), naturally but unconsciously at work in the usage we make of language, is distorted from the moment it is explicitly solicited."[136] Also, when "the speaker (not the linguist) knows if 'something is sayable or not,' he or she does not [however] know why"[137]—which, moreover, is corroborated by the fact that linguistic "illegality" is fully perceived by the members of societies who do not have a reflexive view and knowledge regarding their language. "In fact, we observe that ethnological enquiry regarding languages has always been possible, even among subjects who are unable to refer to any 'grammatical' notion. Their judgment will often be raw. It will namely take the form of a judgment pertaining to one's belongingness to a community: "Never, the subject will say, would anyone among my group say that." And when asked to explain such an answer, the subject will often be incapable of doing so."[138]

An important point, already mentioned, but which must nevertheless be stressed, is the "objective" content of admissibility judgments. Without a doubt, an admissibility judgment can be taken as the expression of a more or less appropriate character of a piece of data with respect to a rule (or prescription). But in this case, the existence as such of the datum is not in question. If, in compliance with Husserl's conceptions, we define a norm (understood as a prescription) as a rule relative to objects which are essentialy independent of this rule,[139] then the transgression of the norm is without regarding the existence of the objects it regulates. Now, admissibility judgments are not of this kind, they pertain to the very existence of linguistic objects. To alter the structures which institute linguistic objectivity is not simply to contravene a norm regarding objects configured elsewhere and to stipulate the "correct usage" without any sanction other than a favorable or hesitant appreciation. It is to fundamentally touch upon the very principles of their existence.

We may recall in this respect the distinction established by Husserl between "nonsense" and "absurdity"—a distinction which Chomsky will revisit from another angle in terms of grammaticality and of semanticity.

---

[135]*RL4/F,* p. 62.

[136]Soutet (1997, p. 182).

[137]Samain (2000), personal communication.

[138]Milner (1989, p. 54).

[139]"[...] every normative [...] discipline rests on one or more theoretical disciplines, inasmuch as its rules must have a theoretical content separable from the notion of normativity [...]" (Husserl 2001, p. 33).

We have seen (cf. 3.2.8) that Husserl distinguishes between "nonsense" (that which is "senseless"[140]) and the "absurd" ("or counter-sensical"[141]): "true meaningless [must not be confounded] with another quite different meaninglessness, i.e. *the* a priori impossibility of a fulfilling sense."[142]

Whereas the first (nonsense) concerns the forms proper of linguistic objectivity, the second (absurdity) concerns the intuitive or imaginative correlates (the fulfillments) by which the targeted meaning takes the form of an actual representation.

In the first case, the only one which interests us here, what is in question is the very *existence* of an object of meaning. Thus, considering an incoherent assortment of words: "[I]t is apodictically clear that no such meaning can exist, that significant parts of theses sorts, thus combined, cannot consist with each other in a unified meaning",[143] and the apodictic consciousness of the impossibility of such an assortment attests to essential laws of meaning, in other words, "[to] laws governing the existence or non-existence of meanings in the semantic sphere."[144] Also, the complex expression which is "nonsensical", in that it contravenes the format of semiolinguistic objectivity, is deprived of any semiolinguistic existence.

5.3.5   Admissibility, Differentiality, and S&P Relations

What should now be noted is that the relations of negative difference with respect to the plane of content condition the *very existence of signs*: The disappearance of a boundary in the substance of content has for consequence to establish continuity between, i.e. to homogenize, the two subdomains (the signifieds) which it institutes following relations of reciprocal delimitation. Such a structural "collapse" thus affects the existence of the signifieds, and at the same time it affects the existence of the signs which imply them.

Hence, the arrow "⇒" (which governs the installation of boundaries in the substance of content) is functionally linked with linguistic existence *versus* non-existence.

Moreover, with respect to forms and principles, we have seen that (i) the S&P relations, as well as (ii) the relations of "actualizing" exclusion instituting differential identities of meaning, and (iii) relations of "constitutive" exclusion (distributional constraints) which underlie linguistic existence and non-existence, proceed from a shared "generic" principle of exclusion (the exclusion which founds the paradigmatic and which intersects it with the syntagmatic). Thus, the relations of

---

[140]*RL4/F*, p. 67.
[141]Ibid.
[142]*RL1/F*, p. 202.
[143]*RL4/F*, p. 67.
[144]Ibid., p. 68.

"constitutive" exclusion, which are at the systemic source of admissibility judgments, proceed from the same rationality, the same internal principle, as the S&P relations.

This is to say that difference and S&P relations are "functionally linked".

Finally, regarding the analytical methods and practices which consist in varying the items up to the limits of admissibility (of a certain semiolinguistic possibility), it is indeed in the syntagmatic and paradigmatic as variational axes that the modalities of the *existing* and *non-existing* are put into play and encountered in language. More specifically, in this practical perspective, but also in their functional sense, the S&P relations, as they administer the variations of a given syntagm, thus constitute an operational structure which takes hold of the possible and of the impossible in language. Such is the case, for example, with differential pairs which linguistic analysis constantly employs, and which precisely and methodologically put into play the departure from linguistic legality—in other words, the departure from the sphere of existence in language.

We thus observe a functional correlation between the S&P relations and existence *versus* non existence in language, an opposition itself paired with the order of semiolinguistic possibility and impossibility.

In the end, the S&P relations, the articulations of (i) possible *versus* impossible and (ii) existing *versus* non-existing in language, the "$\Rightarrow$" relation, as well as the differential relations categorizing the content's substance, are seen to participate in a same unitary functional logic: The ones are the reflection of the others at their own and different levels of semiolinguistic functioning and apprehension.

In particular, in what concerns the progression of our demonstration, we will retain here that the relation of determination "$\Rightarrow$" which at its output regulates linguistic existence and non-existence, structurally refers to the order of the S&P relations.

We will now look at the functioning of the S&P relations as they regulate the actualization *versus* the effacement of boundaries in the substance of content. In order to do this, we will have recourse to the morphodynamic apparatus as a generic model of the emergence of differentiating structures (negative differences) in a substrate space. We will then arrive at a model of the sign, one which is quite close to that established in Piotrowski (1997) and reworked in Piotrowski (2009a, b, 2010).

References

Al, B. (1975). *La notion de grammaticalité en grammaire générative-transformationnelle : étude générale et application à la syntaxe de l'interrogation directe en français parlé*. Leiden: Presse universitaire de Leyde.
Amacker, R. (1975). *Linguistique saussurienne*. Genève–Paris: Droz.
Benveniste, E. (1966). *Problèmes de linguistique générale I*. Paris: Gallimard, coll. *Tel*.
Benveniste, E. (1971). *Problems in general linguistics* (M. E. Meek, Trans.). Coral Gables: University of Miami Press.

CLG/B: Saussure, F. de (1959). *Course in general linguistics*. (Baskin, W., Trans.).

Culioli, A. (1999). *Pour une linguistique de l'énonciation: Formalisation et opérations de repérage, T. 2*. Paris: Ophrys.

Ducrot, O. (1968). *Le structuralisme en linguistique*. Paris: Le Seuil, coll. Points.

Godel, R. (1969). *Les sources manuscrites du Cours de Linguistique Générale de F. de Saussure*. Genève: Droz, coll. *Publications Romanes et Françaises*, 61.

Hjelmslev, L. (1971). *Essais linguistiques* (p. 47). Paris: Éditions de Minuit, coll. Arguments.

Husserl, E. (2001). *Logical investigations: Prolegomena, Investigations I & II*. (Findlay, J. N., Trans.). London & New-York: Routledge.

Itkonen, E. (1991). *Universal history of linguistics*. Amsterdam: John Benjamins Publishing Company.

Jakobson, R., & Halle, M. (2002). *Fundamentals of Language*. Berlin: Mouton de Gruyter.

Kant, E. (1944). *Critique de la raison pure*. Paris: PUF, coll. Bibliothèque de Philosophie Contemporaine.

Lévi-Strauss, C. (1983). *Structural anthropology* (Vol. 2) (M. Layton, Trans.). Chicago: University of Chicago Press.

Merleau-Ponty, M. (1993). Indirect language and the voices of silence. In Smith, M. B. (Ed. & Trans.). (1993) *The Merleau-Ponty aesthetics reader: Philosophy and painting*. Evanston, Illinois: Northwestern university Press.

Milner, J.-C. (1989). *Introduction à une science du langage*. Paris: Le Seuil, coll. Des Travaux.

Petitot, J. (1985). *Les catastrophes de la parole*. Paris: Maloine, coll. Recherches Interdisciplinaires.

Piotrowski, D. (1997). *Dynamiques et structures en langue*. Paris: CNRS Éditions, coll. Sciences du Langage.

Piotrowski, D. (2009a). *Phénoménalité et Objectivité Linguistiques*. Paris: Champion, Collection Bibliothèque de Grammaire et de Linguistique.

Piotrowski, D. (2009b). Place et raison de l'analyse phénoménologique en linguistique. *L'Information Grammaticale, 124*, 3–15.

Piotrowski, D. (2010). Morphodynamique du signe; I – L'architecture fonctionnelle. *Cahiers Ferdinand de Saussure, 63*, 185–203.

Rastier, F. (1982). Paradigmes et isotopies. *Actes Sémiotiques, 5*(24), 8–16.

Saussure, F. de (1993). *Saussure's third course of lectures on general linguistics (1910–1911)* (Harris, R. & E. Komatsu, Ed. & Trans.) Pergamon: Elsevier.

Saussure, F. de (2006). *Writings in general linguistics*. In Bouquet, S., Engler, R., Sanders, C., & Pires,M. (Eds.). Oxford: Oxford University Press.

Soutet, O. (1997). *Linguistique*. Paris: PUF, coll. Premier Cycle.

Chapter 6
The Morphodynamics of the Sign

In this chapter, we will first present (Sects. 6.1 and 6.2) the principles as well as the functional architecture of the morphodynamic apparatus. Adopting a utilitarian logic, we will limit our exposition to the sole elements necessary to our demonstrative progression. For reasons of necessity, it is therefore a minimalist and summary version[1] of the morphodynamic approach which we will deliver here. During a second stage (Sect. 6.3), we will merge the morphodynamic model with the Saussurean structural apparatus. It will be a matter, through a single gesture, of overcoming some shortcomings in Saussure's construction and of producing an adequate mathematical determination of its functional components. Following this (Sect. 6.4), having at our disposal a morphodynamics of the Saussurean sign, we will establish and discuss its phenomenological (Husserlian) signification. Finally, and to prevent any misguidance, we will expose the specificity of the morphodynamic perspective in contrast with the functionalist approaches.

6.1 General Schema

Introducing the question of morphodynamics, we propose to present it as a solution to the problem of the categorization of a homogeneous substrate space.

Let's consider in this respect a "substrate" space W and see by virtue of which principles the elements of W (the *tokens*) can be grouped and distributed under class identities (the *types*). Of course, the problem arises only if W and its elements are not specifically governed by laws, as is the case for example with a physical

---

[1]For a complete presentation of the model, its foundations, and its scientific significance, we refer to the works by J. Petitot, mainly Petitot (1985a, b).

© Springer International Publishing AG, part of Springer Nature 2018
D. Piotrowski, *Morphogenesis of the Sign*, Lecture Notes in Morphogenesis,
https://doi.org/10.1007/978-3-319-89848-3_6

substrate. In this case, the categorization of *W* would derive from such laws and would thus be intrinsic to it. Conversely, we exclude the principle of an external, therefore abstract form which would be projected upon a shapeless substrate (matter) in order to institute a "substance" (organized matter). The point of view which will be retained here is the intermediary standpoint of a hylemorphic (or morphogenetic) conception which "internalizes" within *W* a certain formative principle, but without relating it to the "concrete" nature and order of its elements.

The case which interests us is therefore that of a "homogeneous" substrate, which, in a "technical" sense, signifies that the categorizations proceeding from the action of a group on *W* produce only a single equivalent class. As *W* does not possess in itself the principle of its internal differentiations and organization, the level of complexity required by its categorization must therefore be sought "elsewhere", but without however "leaving" the plane of *W*. To do this, the "classical" solution is that of an "embedding", which consists in inscribing *W* (thus said to be an "external" space) within a "richer" space of forms *F* ("internal" space) the organization of which *W* will inherit.

In the "elementary" version of the morphodynamic approach, the "internal" space is a space of dynamical processes defined by potential functions $f_w: M \to R$ (*M* being a compact variety, R the set of real numbers), and the connection between *W* and *F* takes the form of a control (σ-field: $W \to F$) which associates f_w elements from the *F* space of dynamics to the *w* units of *W*.

We may recall that the states which the dynamics determine, i.e. the states towards which the system is likely to converge, are given by the minima (or "attractors") of the functions which represent these dynamics. In the case where the dynamics possess several attractors, they would *compete* for the system's actualization. In order to decide between them, the recourse to a device of selection is therefore necessary, that is, a rule which stipulates which among these attractors determines the "actual" state versus the other states which are then "virtualized." We will retain here "Maxwell's convention" which gives priority to the attractor having the weakest value (on graphs, it is the deepest potential well).

It will now be necessary to specify the modalities of an organization intrinsic to *F*, and to account for the "return" to *W* of the schemas of structure accomplished in *F* as controlled by *W*.

In what concerns the first point, we will pay attention to the f_w dynamics of F from a *qualitative* point of view, that is, independently from the coordinates systems where these potential functions are defined. To do so, we will apply a transformation group on the coordinates systems relatively to which the f_w functions are expressed. In practice, this amounts to retaining, as identifying elements intrinsic to these dynamics, only the number and relative position of their respective "attractors" (the minima of these functions[2]). These characteristics (number and relative position of the attractors), which are invariable due to the actions of the transformation group, thus deliver the qualitative type of the f_w dynamics of *F*.

---

[2]We ignore here the maxima.

In complement to the transformation group operating on F, and which instructs qualitative types in it, we enhance the functional apparatus with a topology T (neighborhood structure) so as, on the one hand, to geometrize (spatialize) the dynamics space and, correlatively, to introduce a qualification in terms of stability/instability. Specifically, a function $f(x)$ is said to be *instable* if all neighborhoods of $f(x)$ comprise elements (other functions) which are not of the same qualitative type as f (x). Which amounts to saying that any variation of $f(x)$, be it infinitesimal, may provoke its qualitative modification (we would then deem it a "catastrophe").

At this point, it would be fitting to examine two major results in the field of morphodynamics.

The first concerns the internal characterization of stable and unstable forms. It is *Morse's theorem* which establishes that (under specific conditions which we will skip) a function is structurally stable if and only if (i) all of its critical points are "nondegenerate" and (ii) all of its "critical values" are distinct.

We may recall that a critical point of a function $f(x)$ is a point in which all first derivatives are annulled (it is a point in which the tangent to the curve of $f(x)$ is horizontal), and that the value of $f(x)$ at a critical point is called its *critical value*. Moreover, a critical point is said to be degenerate if the determinant of the matrix of second derivatives in this point is null (intuitively, this means that this point "superimposes" several critical points—a case of "collapse").

What is reciprocally asserted by Morse theorem is that a dynamics has only two possible sources of instability, that is, (i) the degeneration of the critical points, or (ii) the equality of its critical values. Furthermore, any stabilization of an instable dynamics will consist either in (i) "separating" or "dissolving" superimposed critical points, or (ii) in introducing an inequality among the critical values.

The geometrical and dynamic configuration which stems from this is the following.

Let's call *orbit* of $f(x)$ the set of functions which are of a same qualitative type as $f(x)$. In cases where $f(x)$ is unstable, the orbit of the $f(x)$ function defines a boundary separating the possible stabilizations of $f(x)$. Indeed, and intuitively: Any variation of $f(x)$ which remains in its orbit leads to a $g(x)$ function having the same qualitative type as $f(x)$ (by definition of the orbit), and therefore to a $g(x)$ function which is unstable similarly to $f(x)$ and thus subject to the same stabilizations, that is, subject to be stabilized into functions of the same qualitative types as those stemming from the stabilizations of $f(x)$. Given that all stabilized classes obviously do not overlap, the orbit of $f(x)$ thus appears as a set which acts as a separation (boundary) between the various types of stabilization classes of $f(x)$.

In the case where the instability is of type (ii) (equality between the critical values of attractors m_1 and m_2), the stabilization consists in favoring one or the other of the attractors. If we then consider the forms thus stabilized from the standpoint of the relations between their attractors, the "scenario" is that of a competition with respect to actualization and in which the advantage given to one of the states (correlatively to its attractor) is at the expense of the other. For rather obvious semantic reasons, this configuration (which corresponds to a *qualitative opposition*) is said to be a "conflict configuration"—cf. Fig. 6.1.

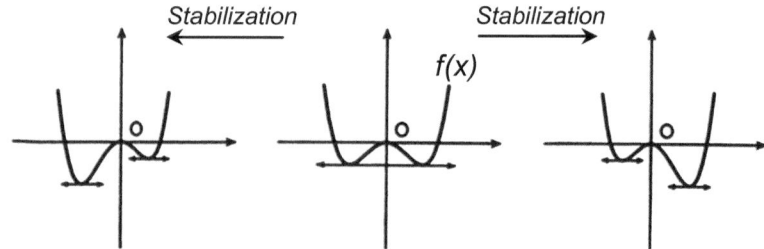

Fig. 6.1 Instability: configuration of *conflict*

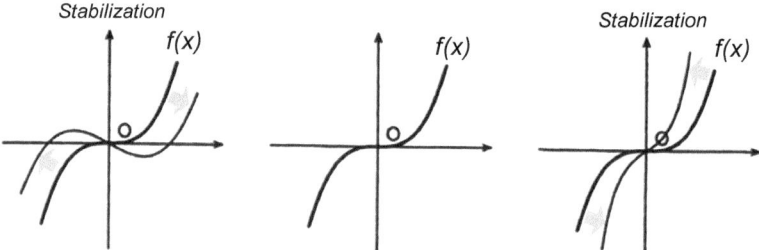

Fig. 6.2 Instability: configuration of *bifurcation*

In cases where the instability is of type (i), the stabilization will consist either in making an attractor appear or in making the degenerate critical point disappear (and therefore also all attractors). If we then consider these two stabilized forms from the point of view of the relations between their respective attractors (or absence of attractors), we can see that what is at stake is not of the order of "preponderance" (as in cases of "conflict"), but rather of "existence" (attractor vs. absence of attractor). This configuration (which corresponds to a *privative opposition*) is called "bifurcation" inasmuch as through the introduction of an attractor, it provides the system with an opening towards (hence the possibility for) a new state—cf. Fig.6.2.

Paraphrasing Morse's theorem, we can then say that "any instability is a 'mixture' of instabilities of the 'bifurcation' type or of the 'conflict' type."[3]

If we now consider the preceding functional configurations (which specify in their mutual relations the states of a certain system) in terms of evolution, therefore, from the standpoint of their progressions in a space of possible states, in the first case ("conflict" catastrophe), during the crossing of the boundary which constitutes the orbit of the instable state $f(x)$, we "brutally" pass from a state determined by an actual attractor "dominating" a virtual attractor to a state where the virtual attractor is actualized at the expense of its competitor. In the second scenario (the "bifurcation" catastrophe), the overcoming of the instability of $f(x)$, according to its orientation, either opens the possibility for a new state, or causes a previously actualized state to disappear.

---

[3]Petitot (1985a, p. 163).

The second major point of the morphodynamic perspective concerns the key role of singularities (instable forms).

Indeed, instable forms carry within themselves all the information regarding the stabilized forms which surround them, this being true for both their geometry (spatial distribution) and their qualitative types. Already, as we have seen, the instable forms constitute boundaries separating classes of more stable forms, and they determine the distribution of the subdomains which they delimit. Moreover, and via Morse's theorem, they determine the qualitative types of the forms which derive from them by stabilization. Also, a singularity constitutes a sort of "germ" inducing a geometric distribution of the more stable forms. In other words, an instable form intrinsically comprises all the information relatively to the spatial distribution of the stabilized forms inscribed within it: "as organizing centers, the singularities are, in a way, *morphogenetic principles*."[4]

Taken in all of its generality, the problem of the spatial distribution of the stabilizations of an unstable dynamics is too complex. But under certain conditions, which then give way to a "good situation", it is possible to simplify this problem, generally of an infinite dimension, as a problem within a finite-dimensional space. Now, these conditions are satisfied in the case where the dynamics are potential functions like those retained by the morphodynamic approach. More specifically:

Let's write as $f^\sim$ the orbit of f (that is, its equivalent class, which groups the functions of a same qualitative type as f), and as K_F the set of unstable forms of F (called the catastrophe set of F), which is a system of boundaries categorizing F (in that the unstable forms interface with their various stabilizations).

In the "good cases", and locally on f, the global space F can be described as the "sum" of $f^\sim$ and of another subspace which, for reasons that will appear later, is noted W as the substrate space. The "good situation" is therefore the one in which "(i) locally, in f, the (F, K_F) pair is isomorphic with the direct product of $f^\sim$ by the (W, K_W) pair, where K_W is the intersection of K_F with W. In such a case, (W, K_W) is said to be a *transverse* model of K_F on f."[5] Complementarily, the "good situation" requires transverse models to be "equivalent for a natural equivalent relation."[6]

As we have said, the mathematical configuration studied by the morphodynamic approach satisfies these two conditions and, hence, any function of F neighboring f is qualitatively equivalent to a function of the transverse space W of a finite dimension "c." Also, and given B = $\{g_i(x)\}$ (i = 1, ... c) a basis of W, "around" f, the elements g of W are expressed as sums $f + \Sigma u_i.g_i(x)$.

Thus, the study of the spatial distribution of the stabilizations of f can be achieved in the context of the transverse space W by the examination of the deformations of f obtained by varying the u_i parameters which weight, in the expression $f + \Sigma u_i.g_i(x)$, variations oriented following the c directions of the base B.

---

[4]Ibid., p. 155.
[5]Ibid., p. 158.
[6]Ibid., p. 153.

From this angle, also, the u_i parameters appear as the factors which control the stabilization of the instable form f.

We therefore see that the transverse model W enables to find and to functionally situate the component of the substrate space in its role as a control space of the dynamics. Indeed, "W being of a finite-dimension, it is isomorphic to a neighborhood W' of the origin of R^k and we can therefore interpret W as a σ-field: $W' \to F$ which associates to $w \in W'$ the element $f_w \in W$."[7] In other words, we can consider W as the *embedding* of a neighborhood W' of the origin of R^k within F. This is how the elements of the substrate space (which corresponds to W') pertain to two points of view: On the one hand, as units of the source-space of the σ-field, and on the other hand, as an image of their embedding in the functional space which they control. In this way, "to control forms is to embed the control space within their functional space."[8] Furthermore, and considered as an "embedded space", the substrate space, intrinsically homogeneous, receives the categorical determinations (according to a qualitative geometry) of the dynamics which it controls: "[the functional space] is endowed with an inherent catastrophe set whose trace will manifest, through an embedding, in the control space."[9]

In other words, and to conclude: The categorization K_W of the "substrate space" W is "the trace it bears of the instabilities of the [dynamic] forms which it controls"[10] and of the relations of conflict between internal states it determines.

Also to be noted is the major result according to which the geometry of the stabilizations (or dually, of the interface system K_W), which accounts for the interactions (between attractors) and for the distribution of the various dynamics, is conditioned by the dimension of the control space. Under conditions of stability, the complexity of the morphologies is thus highly constrained. Furthermore, morphodynamics was able to establish the exhaustive and complete typology of the catastrophe schemes (interactions between attractors) under conditions of dimensions inferior to 4 (cf. Thom 1977).

6.2 Illustration and Precisions: *Cusp* Geometry

We will begin by a short presentation of the unfolding (construction of the space of the stabilizations) of the *cusp* singularity, specifically $f(x) = x^4$. Then we will examine its "semantic productivity", that is, the various sorts of relations of signification which are accomplished within it in function of the various possible paths of stabilization.

---

[7]Ibid., p. 154.
[8]Ibid.
[9]Ibid.
[10]Ibid., p. 189.

6.2.1   Cusp Mathematics

The control space W of the singularity $f(x) = x^4$ is of dimension 2. In other words, the stabilizations of the cusp singularity are accomplished following two directions of variation, i.e. are controlled by two parameters (u and v). The normalized expression of its universal unfolding is $F(x, u, v) = x^4/4 + ux^2/2 + vx$. It is then a matter of studying the geometric distribution of the stabilizations of $f(x)$ in function of the parameters u and v controlling the two stabilization vectors $g_1(x) = x^2$ and $g_2(x) = x$.

To do this, we examine the conditions for which the first derivative in x of $F(x, u, v)$ is annulled—that is: $F'(x, u, v) = x^3 + ux + v = 0$. We know that such an equation is annulled either in a single point, in three distinct points, or in three points among which two are coincident. The values of u and v for which the equation $F'(x, u, v) = 0$ admits a double root (in addition to the simple root) are those which annul the second derivative of $F(x, u, v)$, that is, $F''(x, u, v) = 3x^2 + u = 0$. From this, the condition between u and v is deduced: $4u^3 + 27v^2 = 0$.

This condition defines, within the space W, a region noted K_b (K for "catastrophe space" and b for "bifurcation") grouping the functions $F(x, u, v)$ which present a simple minimum and an inflection point. K_b thus constitutes a domain of partial stabilizations of $f(x)$, a domain which serves as a boundary between the stabilizations resulting from a deformation of the inflection point of its elements (bifurcation catastrophe), that is, either functions presenting a unique minimum or functions presenting two minima separated by a maximum. Specifically, K_b describes on the plane (O, u, v) a semicubical parabola (a cusp), with a turning point at the origin (cf. Fig. 6.3).

The catastrophe set K also has a boundary noted K_c which divides into two subregions the "interior" domain of the cusp, i.e. the domain where the forms having three distinct critical points find themselves. The boundary K_c groups the

Fig. 6.3 *Cusp* singularity: the 2-dimensional space of stabilization

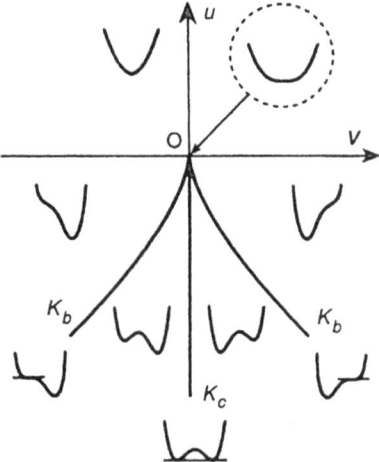

Fig. 6.4 Names of the
domains of *W*

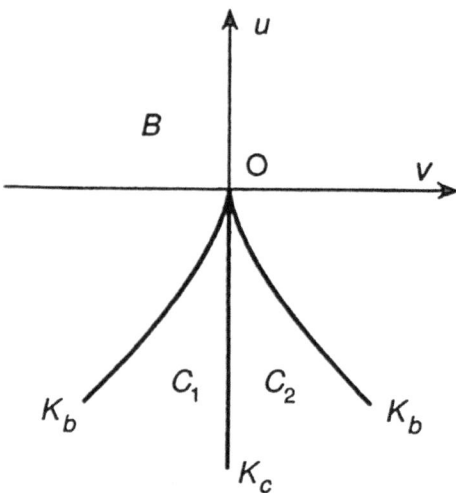

functions whose two minima have an equal critical value: It is therefore a matter of instabilities of the "conflict" type which are resolved on one side or the other of K_c by "favoring" one minima or the other. K_c is precisely the negative semi-axis of the parameter *u*.

The geometry of the stabilization of the cusp singularity therefore has the following general aspect (we represent in the different domains of *W* the qualitative type of the forms which the parameters *u* and *v* determine in it):

In order to facilitate the discussions, we will attribute to the various domains of *W* the following titles (cf. Fig. 6.4).

Let's now look at the relational contents which the cusp schema schematizes.

We have already signaled the dynamic nature of a universal unfolding: The space of the stable forms is derived from a "germ", i.e. from a singularity, by means of its successive deformations which produce stabilized forms from it. Such series of deformations can be defined as paths in the control space *W*. We will then consider the differential organization of *F* (the system of thresholds spatially distributing the qualitative types) not *statically*, like a global and arrested differential configuration, but *dynamically*, like a production, an act of local generation of differential boundaries. We thus pass from a *semantics of* structural *configuration* to a *semantics of* structural *production*, in which the attractors (the dynamics) will receive the relational values in function of the trajectories which lead to them.

G/S Trajectory. Let's consider the stabilization trajectory starting with the germ *f* located at *O* and leading to a point *β* of C_1. We have thus produced a stable form f_1 which presents two minima: The one being absolute and the other being relative (Fig. 6.5).

But the attractor defined by the absolute minimum M_1 does not come into opposition (competition for actualization) with M_2 which is in a virtual position, inasmuch as the actualization of M_2 is inscribed as a possibility for the structural

Fig. 6.5 Stabilization path
without generation of a
qualitative opposition

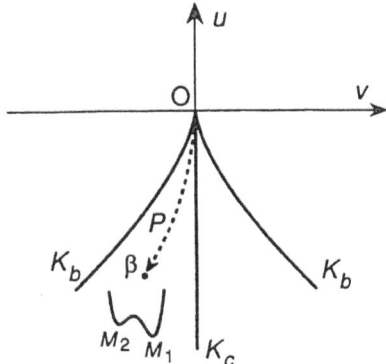

configuration considered here, very specifically in that the actualization of M_2 *will have been taken into account* during the process (of stabilization) which culminated in the actualization of M_1. To do this, it is therefore necessary to consider a stabilization path where, previously to the obtention of the stable form f_1, an *authentic situation of competition* for actualization between M_1 and M_2 is accomplished, a situation of competition in which these two attractors are "on equal footing" and where the actualization of the one or the other is just as foreseeable. Such a path, illustrated in the following figure (Fig. 6.6), consists in "leaving" the origin O while remaining on K_c (up to one of its points α), and then in "leaving" K_c and "entering" C_1 (up to point β).

This stabilization trajectory, noted G/S, accounts for the generic/specific relationship. Indeed, this trajectory controls a series of deformations of an initial instable form $f(x)$ presenting a degenerate minimum of three (coalesced) critical points so as to produce, firstly, in α, a form with two minima having a same critical value (conflict instability), and in β, via a conflict catastrophe, a form endowed with an absolute minimum (determining an actual state) prevailing over the relative minimum (virtual state).

From the standpoint of actualizations, we therefore pass from a state to which the unique minimum of $f(x)$ refers, that is, a state considered independently from any opposing connection, firstly (in α), to two states of equal opposing "weights", then, secondly (in β), to a unique actual state M_1 acquired by virtualization of a competing state M_2. Thus, we have passed (i) from an instable and relationally undetermined form (the degenerate minimum of $f(x)$ does not enter into relation with other "attractive" states likely to oppose it) which relates the undifferentiation of a substrate W, to (ii) a dynamics of the "conflict" type which establishes (in α) an authentic competition between two attractors (having equal chances to be actualized), and, finally, (iii) to the resolution of the conflict by the actualization of one attractor at the expense of the other.

The G/S path thus appears as a process through which an originally undifferentiated state is articulated at its end following the principle of a qualitative opposition. With the G/S trajectory we therefore pass from a unity of genus,

Fig. 6.6 *The Generic/*
Specific stabilization path

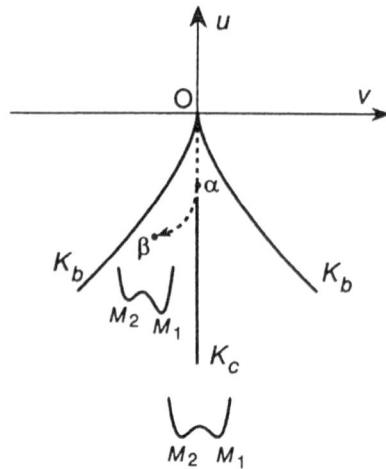

i.e. from a generic term, to two species identities: Two specific states in polar opposition. That is to say that the *G/S* trajectory schematizes the generic/specific relation.

E/I   Trajectory.   The   *E/I*   trajectory   relates   the   construction   of   the extensive/intensive relation—a relation of an eminently topological nature which we know to designate the contrast "between a precise term and a vague term."[11] The "precise" term, also called "intensive", is a term which tends to "concentrate" its meaning within a semantic region, whereas the "vague" or extensive term is characterized by the "fact of being able to occupy any part of the zone."[12] The pair of lexemes *day/night* may serve to illustrate this: *Day* is an extensive term in that it can just as well designate the diurnal part of the day (in which case it is in opposition to *night*) as it can designate the day in its entirety, whereas the intensive term, which focalizes its meaning on the nocturnal fraction of the day, is *night*.

The *E/I* relationship is schematized by a trajectory which begins in *O*, follows K_c, and successively invests the C_1 field and the *B* field after crossing the boundary of K_b (cf. Fig. 6.7).

As in the *G/S* trajectory, the *E/I* trajectory, in its first phase, determines in β an actual state M_1 in polar opposition to a virtual state M_2. Following this, in a second phase (of β towards γ) during which the K_b boundary is crossed, the relative minimum M_2 disappears to the sole benefit of M_1. In γ, the dynamics only presents a single attractor, and thus does not institute an opposing articulation in the substrate space. But this undifferentiation of the substrate space is not the same as that delivered by the instable germ $f(x)$ (located at *O*. Cf. Fig. 6.3). Indeed, the undifferentiation in γ is not *originary*: It results from a fusion of opposing attractors and therefore from an overlapping of terms that were previously constituted in their

---

[11]Hjelmslev (1985, p. 34).

[12]Ibid., p. 41.

Fig. 6.7 The *Extensive/ Intensive* stabilization path

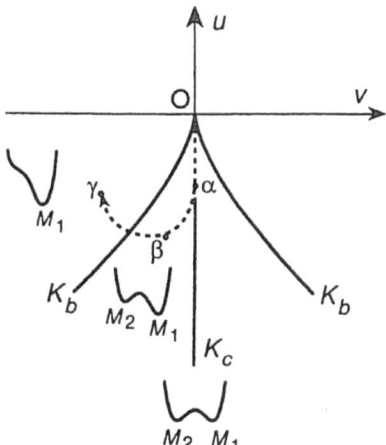

relational identities. The dynamic form present in γ therefore relates the broadening of the field of determination of M_1 to the point of overlapping M_2.

We therefore see that the E/I path reconstructs the mode of the extensive/ intensive relation. The "extensive" term, that is, the term which tends to "broaden" its field so as to designate the totality of the category, is the term M_1. The "intensive" term, which focalizes its meaning in its opposition to the extensive term, is the term M_2.

6.2.2   *Illustration*

The cusp catastrophe enables, as we have seen, to reconstruct these fundamental opposing connections which are the generic/specific and intensive/extensive relations—and, correlatively, to uncover the internal reasons which are of a dynamic and topological nature. The fact remains that these opposing schemas are rarely to be found in a "pure state", and that the examination of the terms belonging to the same lexical fields generally reveals far more complex configurations—though, following Morse's theorem, they are factorized into conflict and bifurcation types of connections.

What the morphodynamic approach teaches us is that the differential configurations (conjugating privative and qualitative oppositions) rapidly become highly complex, and this, as soon as we increase, be it only by a single "notch", the dimension of the substrate space—and we will be required to do this in order to account, albeit very partially, for the usages (which we previously examined—cf. 5.3.3.4) of the French verb *couper* ('to cut').

Let's therefore return to the example of the verb *couper*. Firstly, by considering only a single variational schema, for example V *into* (*parts, pieces...*). In this

Fig. 6.8 Stratification of the substance of content for synonyms of *couper* ('to cut')

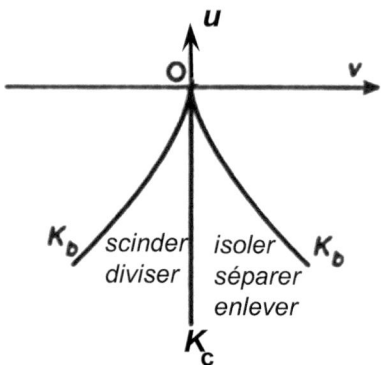

simplified example, the S&P (syntagmatic and paradigmatic) modules of expression and the substance of content are specified as follows:

– S&P—Syntactic Schema:

S$_1$: *V en* ('into') (*parties, morceaux...*) ('parts,' 'pieces'...)

– Expression:

F$_1$ = {*diviser* ('to divide'), *scinder* ('to split')};
F$_2$ = {*séparer* ('to separate'), *détacher* ('to detach'), *isoler* ('to isolate'), *ôter* ('to remove, to take off'), *enlever* ('to remove'), *supprimer* ('to remove/eliminate'), *arrêter* ('to stop')...}

– Content: A boundary[13] K$_c$ corresponding to a qualitative opposition delimits the subregions of the substance of content specific to classes F$_1$ and F$_2$ (cf. Fig. 6.8).

Let's now consider the case where the structuration of content proceeds from taking into account the two variational schemas (S$_1$ and S$_2$). The various modules are then defined as follows:

– S&P—Syntactic Schemas:

(i) S$_1$: V *en* ('into') (*parties, morceaux...*) ('parts,' 'pieces'...)
(ii) S$_2$: V N *de* ('from') N

– Expression:

(i) F$_1$ = {*diviser* ('to divide'), *scinder* ('to split')}; F2 = {*séparer* ('to separate'), *détacher* ('to detach'), *isoler* ('to isolate'), *ôter* ('to remove, to take off'), *enlever* ('to remove')...}

---

[13]To simplify the presentation, we are not taking into consideration here the K$_b$ boundaries corresponding to privative oppositions.

(ii) $F_2 = \{$*séparer* ('to separate'), *détacher* ('to detach'), *isoler* ('to isolate')...$\}$; $F_3 = \{$*supprimer* ('to remove/eliminate'), *interrompre* ('to interrupt'), *arrêter* ('to stop')...$\}$

– CONTENT: In the case examined here, the substance of content is categorized following the principle of two syntactic schemas (S_1, S_2) which determine three paradigmatic classes (F_1, F_2, F_3). In order to generate these three classes, it is necessary to have recourse to an "original" singularity of which the successive stabilizations lead to dynamic forms (potential functions) comprising two maxima and three minima—each of these minima therefore corresponding to a paradigmatic class, or, this being equivalent, to a differential acceptation of *couper* ('to cut').

The control space W of this singularity, called "butterfly", has 4 dimensions and its (W, K) stratification is rather complex. We will not lend ourselves here to a detailed and laborious examination of the opposing interactions engaged by the attractors of the various stabilizations of the "butterfly" singularity. We will limit ourselves, for modestly illustrative purposes, to the presentation of a single sectioning of the (W, K) stratification so as to reveal, on the one hand, the structural complexity developed by a syntagmatic configuration which is nevertheless quite simple with respect to linguistic data (schemas S_1 and S_2), and, on the other hand, the richness of the differential relations which the terms composing classes F_1, F_2, and F_3 are likely to establish; in other words: the richness of the plays of meaning which such terms are likely to accomplish. In Fig. 6.9, taken from (Petitot 1992, p. 171), we associate a paradigmatic class (F_1 or F_2 or F_3) to each attractor, and we observe the relative distributions of the three attractors over adjacent "compartments". Thus, in Zone A, the attractor associated with F_2 determines the actual state (absolute minimum) of the internal system, and the acceptations of the terms of F_2 are established following the relations of "competition" which this attractor engages inside of A with the states which are virtualized (relative minima) but which are capable of actualization in such or such adjacent region of A. For example, from the standpoint of the A/B boundary, F_2 of A is in competition with F_3; and F_1 of C is in competition with F_3 of C from the standpoint of C/D.

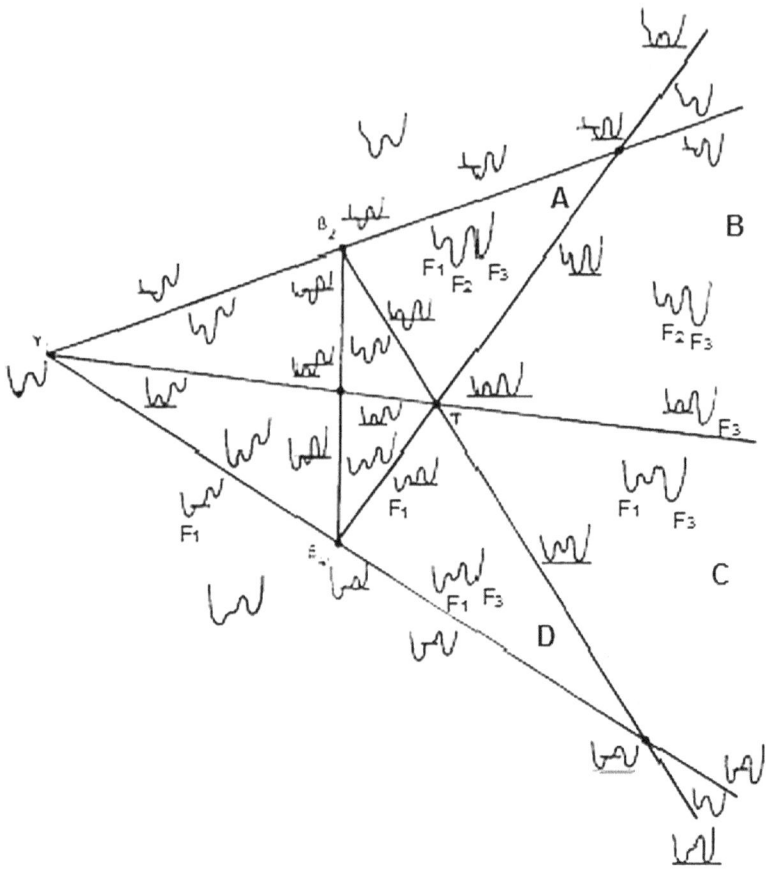

Fig. 6.9 Section of the stabilization space of the singularity "butterfly" $f(x) = x^6$

6.3 The Morphodynamics of the Sign

6.3.1 Construction

Considering the previously established schema of the sign (cf. 5.3.2; Fig. 5.5), it is now necessary to instruct (to describe and explain) the principle and modalities of emergence of the differential forms which categorize the substance of content with respect to the vertical double arrow which relates the S&P (syntagmatic and paradigmatic) determinations.

The explanation will be most straightforward: As the morphodynamic apparatus accounts specifically for the "controlled" emergence of boundaries in a substrate space, it will suffice to adjoin it to the preceding schema which, incidentally, lacks such a principle of morphogenetic functioning. This is what is provided by

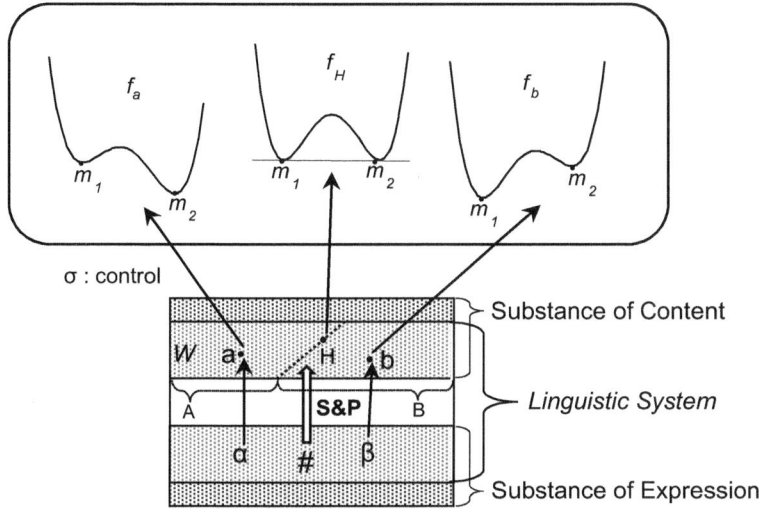

Fig. 6.10 The morphodynamical solution to the semiolinguistic categorization of a substance of content

Fig. 6.10, where for illustrative considerations we have only retained the "conflict" catastrophe configuration.

In such a morphodynamic "extension", the relation of exchange → is promoted to a functional role of control which we will say to be "primary." Specifically, the relation of exchange between a unit of expression, for example α, and the unit "a" of the substance of content towards which it points, is indeed prolonged by the σ-field of W towards F. It happens that by the effect of the functional composition "σ o →" and *with respect* to the processes of categorization which constitute their structural outcome, the → relation receives, by way of status feedback, the functional role of control, hence one which is "primary." The expression term α, for example, through its relation of exchange with a unit "a" of the substance of content, determines, via the secondary control σ of W towards F, a dynamics f_a of which the m_2 state is actualized at the expense of m_1, which would be actualized if the control had been commanded by the unit of expression β.

In order to account for the *modus operandi* of the S&P relations, as determining the differential categorization of the substance of content, and dually as a functional framework having a hold on linguistic legality, we will return to the "fundamental" equation "Opposition = Difference + S&P relations" (cf. 5.2.1). As the relations of "difference" are of two sorts, the one being "negative" and operating in the substance of content so as to install signifieds within it, and the other being "distinctive" and appertaining to the signifiers, we must therefore consider the S&P relations as a functional module engaging connections, on the one hand, with the plane of signifiers (F_E connection), and, on the other hand, with that of the signifieds (F_C connection).

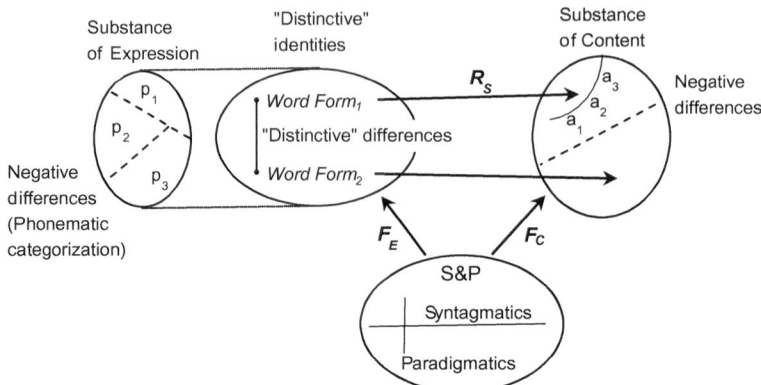

Fig. 6.11 The functional architecture of sign

Figure 6.11, which takes the previously acquired schema of the sign and "externalizes" its S&P relations in a position of functional module, presents this new architecture (to simplify, we have omitted here the morphodynamic module which "dominates" the substance of content).

In this schema, a sort of functional redundancy can be observed between the composition of F_E and of R_S (the "exchange" relation, also called "relation of signification") on the one hand, and, on the other hand, the F_C connection. Indeed, the "direct" connection F_C is superimposed over an "indirect" connection obtained by the functional composition of F_E and R_S. Also, to avoid having to maintain an exceeding functional articulation, it is natural to "remove" the F_C relation. We obtain the following final architecture, in which, focusing on the essential, we have omitted the phonematic component which "underlies" the word forms, and in which we have introduced the morphogenetic component through the cusp format (Fig. 6.12).

But in order to perfect the unity of this functional schema, and to establish its full linguistic signification, there remains to instruct more specifically (i) the status in language of the *σ-field*, (ii) the question of the *actualization* of differential contents, and (iii) the question of (grammatical, semantic…) *admissibility*.

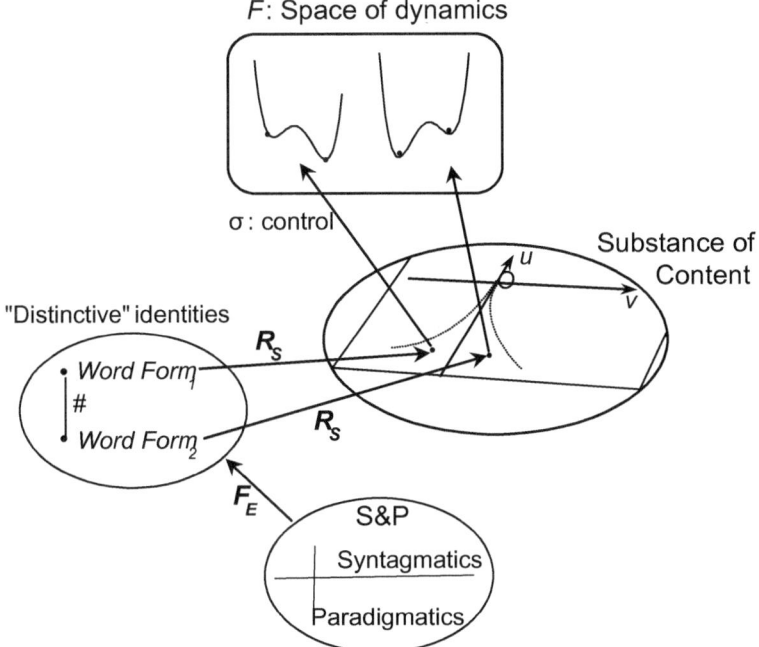

Fig. 6.12 The morphodynamical and functional architecture of sign

6.3.2   Complementary Determinations

To address these various points, we will start with the more general question of the linguistic status of the processes of instantiation of boundaries into substance.

6.3.2.1   S&P Connection ↔ Differential Forms

A morphology of interfaces instantiated within a certain subregion S' of the substance of content produces within it opposing identities of meaning, that is, signifieds. The question of the actualization of the networks of boundaries is therefore attached to that of the elaboration of signifieds, hence, more generally, to that of *their existence*: Quite generally, we know that the presence or absence of a system of boundaries in the substance of content translates the existence or non existence of signifieds. It follows that the *modalities of instantiation of boundaries are functionally correlated to the module of S&P constraints*. We may recall that the S&P relations administer the access to the values of possibility in language, that is, they administer the observations of existence or of non-existence in language—because, tying in with Husserlian views (cf. 4.8.3.3), *admissibility judgments are the predicative expression of a consciousness of the success or failure of a process of constitution of linguistic identities of content.* We therefore foresee the principle of the functional unity between the S&P forms and the regimes of categorization of the substance of content: *This unity resides in their relation to the modalities of existence in language of values of meaning.*

More specifically: We know that the predicates of admissibility carry a differential sense. In other words, inasmuch as they are relevant as regards language, the predicates of admissibility are intrinsically correlated to variational procedures. We may recall that, as has been well established, it is not the impossibility[14] of a syntagmatic construction, for example **AB'*, which is significant, but the fact of an *AB* and **AB'* pair signaling the encounter of a constraint in language over the course of a process of paradigmatic variation *B–B'* conducted over the syntagm *AB*. Existence or non-existence in language, i.e. linguistic possibility or impossibility, are thus inseparable from the variational procedures which, it must be emphasized, are not reducible to paradigmatic alternations of syntagmatic components (cf. the procedures of pronominalization, of passivization, etc.). And in this respect, the S&P module, of which we will nevertheless preserve the denomination, must indeed be considered as the *general module of variational (or transformational) procedures.* We thus see the motive from which proceeds the correspondence between the S&P structures and the systems of boundaries instituting differential identities within a substance of content: *Both directly touch the forms of existence and of objectivity in language.*

6.3.2.2 σ-Field ↔ S&P Connection

The functional connection between the S&P module and the system of differential forms, which we now seek to characterize rigorously, is provided to us with the model of the universal unfolding of a singularity (instable form). We may recall summarily that a universal unfolding constitutes a "universal" geometry of the forms obtained *by stabilization* (following different paths) of a singularity. The (external) control space *W* of these forms "receives" in return a stratification expressing configurations of instability in the functional space of forms (for reasons of simplicity, we will limit ourselves here to the forms of the universal unfolding and of the stratification of the cusp singularity).

Let's begin with a paradigmatic variational schema *B–B'* operating upon the syntagm *AB* and with the admissibility clauses *AB* and **AB'*. We will acknowledge that the opposition of terms *B* and *B'* is not an actual and intangible fact of language. Indeed, this opposition which is encountered in the event of the transformation of *AB* into *AB'* would have been ignored if the speaker had his or her eyes set upon the opposition of *B* with, let's say, *B''* through the *AB–*AB''* variation. Also, the differentiating forms in language, the forms which institute the signifieds

---

[14]A linguistic impossibility (syntactic distortion, semantic inconsistencies, agrammaticality...) noted using an asterisk "*".

in their opposing identities, must be considered not as established and definitely distributed forms, but as *forms produced and renewed in the event of linguistic activities*. In this perspective, we will characterize AB and $*AB'$ clauses by means of (i) a singularity and (ii) a stabilization path. Let's examine this.

The terms B and B' delimit a certain subregion of the substance of content which, by way of the σ field, serves as a control space. This substrate space is not, as such, invested with boundaries characterizing the opposition contracted by B and B': It lends itself indeed to a multitude of characterizations accomplishing the most variegated oppositions. Also, the opposition between B and B', which will be instantiated from the instructions provided by clauses AB and $*AB'$, must be conceived not as a *present* opposition, but as a *potential* opposition: as a differential germ to be actualized by linguistic activity. In other words, *clauses AB and $*AB'$ comprise the information of a singularity*, that is, a potential for the actualization of the opposition between B and B'. Let's now address the second point: the stabilization of this singularity.

It is now a matter of characterizing the dynamical processes determining (i) the actualization of B in opposition to B' when AB is produced and (ii) the "suspension" of the boundaries when it is $*AB'$ which is produced.

We know that the accomplishment of oppositions is achieved following singularity-origin stabilization trajectories which, in our example, characterizes the AB and $*AB'$ clauses. Also, the actualization of B versus B' can be directly qualified by means of a *stabilization path*. For example, if the term B, by means of the exchange relation $\rightarrow$ and in its (qualitative) opposition to B', points towards units of the zone presented in grey in the Fig. 6.13 (*remark*: in this figure, for reasons of simplification, we "crushed" the functional levels S', W, and F), then the actualization trajectory of B triggered by the production of the AB syntagm, on the basis of the AB and $*AB'$ clauses, will be given by a "prescribed" path C. We may recall that this path "handles" an instable qualitative opposition between attractors (also noted B and B') and resolves

Fig. 6.13 The prescribed path of stabilization

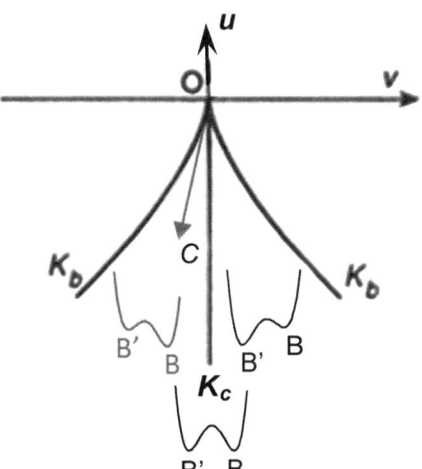

the instability to the benefit of *B* (which becomes actual in opposition to virtual *B'*): Most exactly, what is actualized is a dynamic form determined (via σ) by the units of *S'* towards which *B* points.

Also, the functional connection between the S&P module and the module of the signifieds (substance of the content categorized by differential regimes) is founded upon the following principle: *The predicates of admissibility attached to variational data (for example AB and *AB') determine a singularity of the content and pre-scribe a path of stabilization oriented towards the actualization of a dynamic form associated (by the σ control) to the term which is the object of a variational processing and which is present in the admissible construction (that is, B).*

We will recall that the pieces of information necessary to the localization of the "organizing center", i.e. of the singularity at the source of the morphology of interfaces, and that the pieces of information determining the direction of the sta-bilization trajectory are given by the S&P module via → the exchange relation.

6.3.2.3 Inadmissibility in Language

Let's now examine what happens if a speaker produces a "deviant" syntagm, for instance *AB'*. In such a case, the production of *AB'* determines a stabilization trajectory; we will say that it "forces" a stabilization trajectory, which tends towards the actualization of *B'* at the expense of *B* which is then virtualized. Indeed, the utterance of *B'* confers it an effective presence and forces the linguistic system to take the direction of its actualization. Also, the dynamic configuration is the fol-lowing (Fig. 6.14).

Fig. 6.14 Conflict between prescribed and forced paths

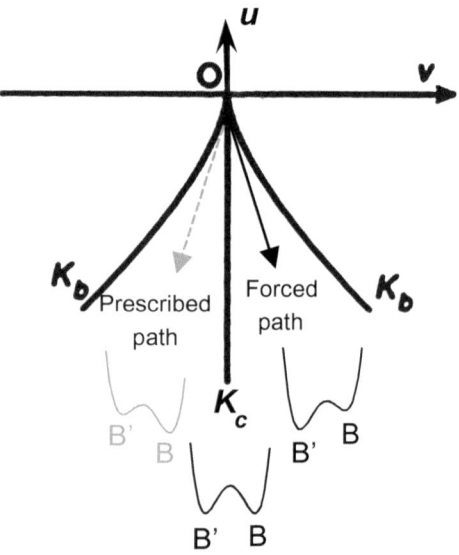

Two incompatible stabilization paths affront one another, each aiming to actualize an oppositional pole. In principle, this conflict has no solution: Taken between two tendencies, the path will remain at the boundary K_c ($v = 0$), either in a position of instability ($u < 0$), or eliminating any form of opposition ($u \geq 0$). In both cases, no stable opposing value is actualized, and no signified is promoted into existence. Also, linguistic impossibility (S&P level) indeed relates inexistence in language (level of the forms of content).

In order to illustrate this functional configuration, let's return to the variational schema *V en trois* ('in three') (*morceaux, parties*...) ('pieces,' 'parts'...). The admissibility data (*enlever ('to remove'), *ôter* ('to remove, to take off'), *détacher* ('to detach'), *scinder* ('to split'), *couper*, ('to cut'), *diviser* ('to divide')...) *en trois* ('in three') (*morceaux, parties*...) ('pieces,' 'parts'...) will be qualified in the linguistic system by a certain original singularity and by certain stabilization trajectories oriented towards the actualization of the lexemes *couper* ('to cut'), *scinder* ('to split'), *diviser* ('to divide')... in opposition to the lexemes *ôter* ('to remove, to take off'), *détacher* ('to detach'), *enlever* ('to remove')... The various locations of the substance of content towards which these terms point carry the positional information required to situate the instantiation loci of the differential thresholds; in such a way that the (qualitative) opposition specified in the syntagmatic schema *V en trois* ('in three') (*morceaux, parties*...) ('pieces,' 'parts'...) is accomplished by a boundary separating the zones grouping on the one hand *scinder* ('to split'), *diviser* ('to divide'), and *couper* ('to cut'), and on the other hand *ôter* ('to remove, to take off'), *enlever* ('to remove'), and *détacher* ('to detach'). This amounts to positioning the reference of the axes (u and v) of the deployment of the cusp singularity and to prescribing a stabilization trajectory "towards" the region towards which *couper* ('to cut'), *scinder* ('to split')... points. That is (Fig. 6.15).

Fig. 6.15 The prescribed path of stabilization for the syntagmatic schema *V en trois* (*morceaux, parties*...)

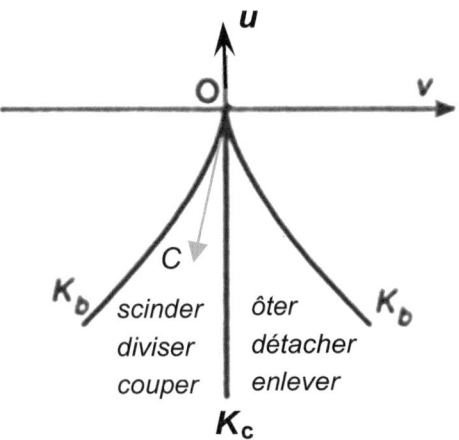

When the syntagm *enlever en deux* ('to remove in two') is uttered, the stabilization trajectory oriented towards *enlever* ('to remove') is "forced" and opposes the prescribed trajectory. We then observe a conflict in stabilization dynamics: The originary instable "germ" finds itself "in language" oriented towards the actualization of [*scinder* ('to split'), *diviser* ('to divide')...] (in opposition to [*ôter* ('to remove, to take off'), *détacher* ('to detach')]), whereas the accomplishment of *enlever en deux* ('to remove in two') "compels" the actualization of [*ôter* ('to remove, to take off'), *détacher* ('to detach')] (in opposition to [*scinder*('to split'), *diviser* ('to divide')...]). This conflict between "prescribed" and "forced" stabilizations resulting in nothing, the achievement of a differentiating boundary is prevented and at the same time the linguistic object is annulled. No opposition in speech is accomplished, and *enlever en deux* ('to remove in two'), producing no signified, is indeed an impossible construction in language.

6.3.3 Results

Before addressing the phenomenological aspect, we will observe that in addition to producing a regulated characterization of the problematic connection between relations of difference and S&P relations, which institutes the play of semiolinguistic oppositions (cf. 5.2.1.1: the equation Opposition = Differences + S&P relations), the functional device developed here, on the one hand, accounts for the assimilation of the relations of exchange and of comparison, and, on the other hand, establishes and explains both the duality of signifiers and the indivisible and asymmetrical unity of the sign.

First, concerning the last point: The control relation, primary then secondary, which relates signifiers to dynamics (f_i) determining differential structures in the substance of content, is a relation which, from one standpoint, has recourse to units (control parameters) pertaining to a space (a control space) endowed with its own regimes of constitution (for example, a metrics, or, with respect to what interests us, a phonological legislation), and, from another standpoint, functionally establishes these units as being "pure" factors in the constitution of differential identities pertaining to a distinct sphere.

If the focus is put on this second aspect, therefore on the morphogenetic function of the signifier as assigned to it by the morphodynamic device, it is the integrated and dissymmetrical character of the sign as an association of a signifier and of a signified which is emphasized.

Indeed, and first of all approximately: The existence of signifiers, as terms of a process of emergence of differential values, being conditioned by signifiers, the latter are therefore by construction "consubstantial' to the signifieds. It is a quasi-analytical truth.

But if the signifieds cannot be conceived separately from the signifiers, such is not reciprocally the case: The fact of signifiers (as control factors) is, always by construction, a functional prerequisite to the establishment of signifieds, but not reciprocally. Signifiers and signifieds are therefore not the asymmetrical poles of an integrated unit. This internal dissymmetry reveals its functional sense the moment we examine in greater detail the principle of the integration from the signified to the signifier.

Indeed, as has been seen, it is precisely as they participate in the contrasting S&P relations, which prescribe trajectories of stabilization from a "structural germ" towards some differential distributions in the substance of content, that signifiers determine the actualization of signifieds, and that the entities of expression are invested with differences in meaning which they therefore control and institute. But to the "prescribed" trajectories dually respond "forced" trajectories which are like the structural reverse of the possible meanings in language. The differential signification assigned to a signifier in discourse indeed proceeds, in a fully oppositional logic, from the "constitutive" exclusion (cf. 5.3.3) of other signifiers in the syntagmatic place it occupies within. We may recall for example (cf. 5.3.3.1) that the differential value of *couper* ('to cut') (vs. *detacher* 'to detach') in the syntagm *couper en deux* ('to cut in half') bases itself on the impossibility of the construction **détacher en deux* ('to detach in half'). Therefore, the possibility for a signification carried by a signifier rests, through paradigmatic variations, on the possibility of syntagmatic assemblages which are impossible in language. We therefore understand the functional sense of the dissymmetry of the sign. Because if the signifier and the signified shared a same status and a same function, in sum, if they played equivalent roles as constituting elements of the sign, the annihilation of the one would entail the annihilation of the other and, reciprocally, it would then be impossible to involve, in language, for purposes of semantic construction, syntagmatic configurations overstepping linguistic legality. But such is not the case, as indicated by the "maintenance" of the signifier despite that no signified is actualized: When the process of differentiation of the content fails in response to a violation of linguistic legality, thus annihilating any semantic existence in language, the signifier however subsists nonetheless in linguistic consciousness as phonematic or graphematic complexes opening onto a significatory void. The dissymmetry of the sign is therefore in part the functional correlate of a system which, *via* the S&P relations and as these functionally put linguistic impossibility into play, *incorporates the modalities of its own transgression*.

But it is not only the sign which presents a dual nature. As we have been able to derive from Saussure's writings (cf. 5.2.2.2), the signifier is not without ambivalence itself, being in part material and in part turned towards the semiolinguistic system. This equivocal constitution of the signifier which is seen to preserve, in its semiotic identity, the presence of the substances which carry it, and which phenomenological analysis recognizes according to its own terms (cf. 4.4 and 4.5), is configured precisely by the morphodynamic apparatus.

Indeed, in the morphodynamic device, the signifiers belong to an "external" space of expression endowed with its own order (phonological or graphemic). Also, albeit they may be oriented towards to substance of content in view of differential categorization, and although they are considered from the point of view of their formal position in the overall device, that is, as control parameters, the signifiers preserve a "concrete" quality, accessible as such in an act of specific perception and, therefore, they present a dual nature. The duality of the signifiers is thus structurally rendered and explained by their ambivalent position as control parameters: in part anchored to an expression substance, in part turned towards the signifiers.

In what concerns the first point, that of the problematic assimilation of the two constitutive dimensions of a value (exchange and comparison relationships), we will observe that the morphodynamic device performs the overlapping of the exchange and of the comparison by a conversion of the status of the exchange relation.

This point deserves particular attention. In this respect, we may contrastively consider Prieto's point of view concerning the question of the unity between exchange and comparison. According to Prieto,[15] among the regulating regimes of the semiotic realm is that of *relevance*, by virtue of which "classes" constructed according to the principle of a "causal" relation between two object planes (for example, on the one hand, objects of an empirical world, on the other, symbols, the causal relation being one of denotation between symbols and objects) are requalified as extensions of "concepts", that is, as families of objects presenting the same characteristics (descriptive concepts) which therefore differentiate them from other objects of a same discursive universe. The principle of relevance is therefore a principle of cognitive practicability of classes, in the sense that, given certain "classes", it is important in order to use them to "distinguish and differentiate the objects which belong to the class from those which do not", and, to this end, to elaborate concepts "of which the extension coincides with the class." As the classes proceed from relations of exchange, and the concepts (or their extensions) from relations of comparison (generic/specific), we therefore see that Prieto's semiology assembles within a logical and functional equilibrium these two relational modalities. But what should be noted is that there is not here an authentic integration of these two relational dimensions: The unity between exchange and comparison is, as we have seen, suspended to a higher principle of relevance which fundamentally proceeds from an epistemic *praxis*. Furthermore, there is clearly a subordination of comparison to exchange: "The class logically precedes the concept."

In the morphodynamic device, conversely, the relations of exchange and of comparison are not assembled, while preserving their specificities, to form a practicable cognitive equilibrium, but assimilate reciprocally within an integrated functional unit, and this, at the necessary cost of a *modification of their nature*: Therefore, *exchange* (between objects of various universes) is promoted to the rank of *control* (of differentiating dynamics). On the other hand, *comparison*, which for Prieto operates through predicative determinations, is rendered by regimes of *boundary emergence*, the differential structuration of a substance of content actualizing within it opposing identities of meaning (signifieds).

---

[15]Prieto (1990, pp. 62–63).

6.4 Phenomenological Signification

We observe and quite directly so that the Saussurean morphodynamics of the sign, which therefore expose the forms of linguistic objectivity, coincide with the complex structure of semiolinguistic intentionality such as described by Husserl (cf. 5.4).

Firstly, indeed, we observe that, similarly to semiolinguistic intentionality which conjugates two orders of "directedness"—one being of a perceptual nature and the other of a signifying orientation—the morphodynamic model of the sign articulates two object planes which are in part unlinked although they are functionally conjugated: On the one hand, there is the plane of signifiers, taken as phonematic arrangements, and which therefore stem from a simply "perceptual" grasp, and, on the other hand, the plane of signifieds as differential identities of meaning.

Second, and more essentially, we observe that in the infrastructure of the Saussurean sign, the signifiers and signifieds hold, by their functional positions, structural significations which are by all means similar to those of the primary and thematic objects of the attentional field, respectively.

Indeed, the signifiers, as being "simply perceived", are involved in the morphodynamic structure of the sign as control parameters for the constitution of signifieds.

Now, it is clear that from the standpoint of "structural economy", what is significant in the morphodynamic performance is the process of differentiation which unfolds in a substance of content; because the system as a whole, as in its final reason, presides over the genesis of signifying morphologies and thereby constitutes only the machinery in which is outlined, at the forefront, linguistic existence and non-existence. Which amounts to saying that the configurational moments which prevail in the internal logic of the dynamic architecture of the sign, those which Husserl calls "themes" in the sense that they occupy a higher position on the scale of consciousness investment, are precisely the signifieds, as differential values.

It follows that, correlatively, and with respect to the horizon of functioning of the system which mobilizes them, the signifiers appear to be somewhat incidental: They are but "intermediaries", in all likelihood required in functional terms, but secondary with respect to the stakes. The signifiers indeed find themselves to be engaged in the control of emergent forms, but as these occupy the forefront of the "morphodynamic scene", they are met with "disinterest" from the very moment they are mobilized, inasmuch as, intrinsically, in their functional signification, they orient towards the signifiers to which they are, so to speak, devoted.

It must also be noted that in the morphodynamic apparatus, the necessary connection between the signifiers and signifieds is a dissymmetrical and dynamical relation, in which the signifiers therefore have a functional role *at the service* of the emergence of differential identities of meaning, which *then count in priority* for consciousness.

But the phenomenological signification of the morphodynamic apparatus extends well beyond this first correspondence between, on the one hand, functional and structural positions (that is, the control parameters and the differential values) and, on the other hand, phenomenological determinations (respectively, primary and thematic objects). However, in order to establish the full phenomenological scope of the morphodynamic apparatus, it should first be strongly emphasized that the morphodynamic architecture is to be understood as an act (in the Husserlian sense) rather than as a process—in other words, and more clearly, the morphodynamic schema does not have a "functional" quality in the sense of classical cognitivism, functionalist and computational. It does not exhibit a hierarchy of levels or of object planes, which would be successively reached over the course of an operational progression. To the contrary, in the morphodynamic schema, the various planes coexist within an organic complex where the various elements mobilized establish their identities following reciprocal functional connections. Thus, the morphodynamic complex operates following the logic of an act, that is, here, following the mode of the apprehension of a simply perceptual material (the sign taken as concrete) which thereby finds itself to be functionally invested (and promoted to the rank of signifier) in a global and unitary aim towards the constitution of signifieds. And if a form of hierarchy can nevertheless be discerned within it, it will be a matter of a hierarchy in terms of *thickness* and not in terms of *sequencing*. Indeed, if it is therefore inconsequent, due to its functional signification, to dismember the morphodynamic apparatus into a succession of object planes, it is however quite legitimate to distinguish different "phases" which, in their logic of reciprocal overlapping, participate in the texture of the sign in its natural unfolding towards meaning and its fulfillments.

More specifically, these various phases (or strata) are coextensive to the various states of functional engagement which the morphodynamic apparatus organizes. Each of these strata proceeds from the selection and the highlighting of certain structural features involved in the system and, correlatively, from the neutralization or from the "passage to the background" of those which are not retained—and each of these strata also produce specific objects of verbal consciousness. On the basis of this, the main strata of verbal consciousness, that is, the main sorts of "semiotically engaged" objects of which consciousness can configure the presence and of which it can take hold are described as follows.

We will first only retain the simple position of the control parameter, mainly attributed to a phonic complex in that it falls under an act of semiolinguistic intentionality. The verbal consciousness which corresponds to it is a simple *consciousness of the availability* for meaning: The signifier is only grasped as likely to participate in an upcoming verbal configuration, and in total ignorance of the role which it will play within. The consciousness of availability is nothing more than a consciousness of the singular moment of an "opening towards…" without any determination whatsoever regarding the orientation of such "opening". This represents a first state of semiotization, where the concrete object, at first limited to itself, abandons so the speak what is of "concern to its own self" and presents itself as a "window onto" something beyond, but without regards to the field to which it potentially gives access nor to the function it will receive in a global semiotic configuration for which it declares itself to be available. For example, it is to this stratum of verbal consciousness that the syllabic portions pertain, such as they are primarily perceived in the progress of a discourse, that is, as they are still in the uncertainty of the semiotic function that will be incumbent upon them (thus, as a morpheme or as a simple part of a broader term). The notion of "word-sound", which is more than a "sound" but not yet a "signifier" (Husserl, cf. 4.4) covers this stratum of verbal consciousness and the subsequent one.

At a higher functional degree, and supported by an underlying consciousness of availability, we will take into account the connection of control, but from the sole point of view of its existence (abstraction made of its own identity, that is, its reference to such or such region of content). The object of consciousness thus retained proceeds from a simple *consciousness of involvement* (in meaning). We find here Benveniste's plane of "semiotic signification",[16] a plane solicited in the trials of "lexical decision" in which it is a matter of recognizing a stimulus in its simple quality as a word or as a logatome (pseudo-word). We may indeed recall that, from Benveniste's point of view, natural languages combine two regimes of signification: the signification stemming from the linguistic system and the signification such as is accomplished through discourse. What *essentially* distinguishes them are the modalities, serving as criteria, according to which these two regimes of meaning let themselves be apprehended. Whereas the signification of a semiotic unity appears only under the prism of the presence/absence opposition, that which emanates from discourse is suitable to being "understood", hence grasped in its specific identity. In other words, since we are considering the sign as an entity of the linguistic system, "it is not question of defining the meaning […]. On the plane of the signified, the criterion is: Does it signify or not? To signify is to have a meaning, no more"[17] and "in semiology, it is not a matter of defining what the sign means."[18] On the plane of discourse, on the other hand, when it is a matter of "language in use and in action", meaning resides in "what was intended" in the act: in what to speaker means to say. In other words, the sense of the sentence is in "the linguistic actualization of [the speaker's] thought"[19] or in the "idea it expresses."[20] That is to say that in this case, it does not suffice for the sign to simply be "recognized",[21] hence, to be grasped as having a signified, without any mention other than this signified's existence. Discourse calls to be "understood", and this involves a semantic apprehension having hold, beyond the simple presence of meaning, over a specific identity of meaning. It is precisely the stratum of "semiotic" signification (in the sense of Benveniste) that the consciousness of engagement accounts for: A form of expression is recognized as being an authentic

---

[16]Benveniste (1974).

[17]Benveniste (1974, p. 222).

[18]Ibid.

[19]Ibid., p. 225.

[20]Ibid.

[21]Ibid., p. 64.

signifier with respect to the existence or not of a functional connection of control, which therefore attests to its own involvement in a world of meanings.

The next stratum solicits the functional control connection in its specific identity (reference to a particular subdomain in the substance of content) but without there already being an available consciousness of the signified, which pertains to the next stratum.

Now this stratum of consciousness constitutes an opaque locus of the approach exposed here. Because if the stratum of the signified which follows it does not entail particular theoretical problems and finds itself to be comforted by the possibility to perform tests of semantic categorization soliciting an implicit differential contrast (tests consisting in judging the subsumption of a term under a given category of meaning), the possibility for a linguistic consciousness of meaning which is not a consciousness of a signified demands theoretical broadening, and in view of this, the concept of "motif", introduced by Cadiot & Visetti (2001), will be key.

If the concept of motif is suitable for overcoming our theoretical shortcomings, it is because it refers to "a dynamic of constitution of a *relation to...*, of an access towards... of a mode of apprehension, of presentation, of construction...".[22] The motif as concept thus pertains to a phenomenological approach and therefore coincides, at least as far as the broad principles are concerned, with the morphodynamic assembling which we have also seen to carry a phenomenological meaning. Also, there is nothing unnatural at first in turning towards the concept of motif in order to shed some light on the most obviously obscure part of the sign, taken in its whole thickness.

In what concerns the theoretical content of the "motif", in what we can retain from it for our own purposes, we may simply say that it enables to overcome the difficulties which can be found within the simplest form of a functional connection (relation of exchange) thrown between a term and some element of the substance of content. Because since it is a matter of recognizing the semantic identity of a form/ meaning functional connection as defined in the morphodynamic approach, the only support which we have at hand is in the element of the substance of content to which the control function points. It would then be necessary to reduce the semantic identity of the control to the terminal value of an "indicative" type of relation (cf. 2.2), that is, a relation established between a sign and an object, be it empirical or categorical, constituted for its own sake. This option is untenable for at least two reasons. On the one hand, this denies the essence of the semiotic fact (an essence which reveals itself in the "meaningful" sign), and, on the other hand, it contravenes to a logic and a dynamic of stratification (of the strata of verbal consciousness) which is that of a gradient of semantic determination, and of which the ultimate stage finds itself in the act of fulfillment. Indeed, between the simple consciousnesses of availability to that of engagement, a first level of semantic determination has been crossed. Further on, going from the consciousness of the signified to that of fulfillment, we will pass from an identity of negative signification

---

[22]Cadiot and Visetti (2001, p. 114).

to a positive and accomplished identity, as delivered by the indicative sign. We therefore conceive that between the consciousness of engagement and that of the signified, we cannot situate a verbal consciousness in which meaning would be fully accomplished (such a consciousness is located at the ultimate stratum of fulfillment) and, reciprocally, that we need to conceive of a "lesser" level of semantic qualification than that of the signified but one which is sufficient in order to recognize the specificity of a connection of control versus another.

The notion of motif, in that it captures this intermediate level between the consciousness of engagement and that of the signified, precisely meets these expectations. Indeed, the content of a motif, although it is intelligible and identifiable, escapes any arrested qualification precisely inasmuch as it pertains to a praxical intentionality and not to an act of knowledge: A motif relates the particular style of a relation which a subject establishes with his or her milieu, a relation which installs a world of signifying qualities and of forms in response to the primordial fabric of vague solicitations which solicits him or her. We find here the themes of an existential phenomenology (cf. 3.3 and Chap. 7). The motifs thus place themselves "at a level lower than that of the distinction between mono- or polysemy, inasmuch as such distinction presupposes more stabilized regimes of meaning",[23] and, likewise, their order is situated "quite previous to any logic regarding the classification of referents or regarding categorization."[24] Finally, the concept of motif guarantees the existence and ensures the semantic unity and identity of a consciousness of involvement in meaning carried by the terms. This is because through the motif, it is indeed a matter notably of "preserving the intuition of a central principle of unification [...] by virtue of which some words at least specifically have a [sense understood as a] *motif*."[25]

But there is more. Because the motif, in that it does not constitute a potential generating its own acceptations for usage, in that it does not comprise the principles of its own various semantic actualizations, and in that it nevertheless participates in an actual production of meaning, the motif, therefore, calls for complementary operations in order to adjust and stabilize its matter. It is at first a matter of *profiling* with respect to which differentiation plays a central part: "[motifs] are insufficient when it a question of organizing, and *a fortiori* of explaining, the diversity of the usages which in reality have recourse to processes of profiling and of thematization."[26] That is, that the consciousness of motifs lends itself to ulterior elaborations leading to signifieds—and this is precisely what the apparatus of the strata of verbal consciousness exposes.

Thus, prolonging the conscious phases of *availability* and *involvement*, and proceeding from the identity (as *motif*) of the functional connection between the units of expression and the substance of content, we will recognize a new stratum

---

[23]Ibid., p. 96.
[24]Ibid., p. 97.
[25]Ibid., p. 96.
[26]Ibid., p. 104.

specific to verbal consciousness: *motif* consciousness. This kind of consciousness naturally fuels, through schemas of dynamic differentiation which take hold of it, an ulterior verbal consciousness, that of the signified.

The following stratum of verbal consciousness is, so to speak, the focal point of the morphodynamic apparatus, in that it restitutes an act of signifying directedness. At this level, a consciousness of the *signified* is elaborated as a consciousness of a differential structuration instituting negative identities of signification.

Let's finish our journey through the depths of verbal consciousness by addressing the consciousness of *fulfillment* (or, in Merleau-Pontian terms, *consummation*—cf. 7.6) which is not explicitly situated in the morphodynamic schema, but which nevertheless constitutes the logical though unnecessary continuation of the consciential thickening of the sign: It is a question, in the act of fulfillment (cf. 3.2.7), of carrying a negative and simply intentional object (the signified) to a higher degree of positivity and of effectivity, through, for example, the actualization of a mental representation, through a categorical determination, or yet through the reference to a referent. This extends beyond the semiolinguistic field.

Complement: Our description would not be complete if we were to neglect a plane of factualities which, without pertaining as such to the strata of verbal consciousness, nevertheless participates in semiolinguistic objectivity, at least in its metabolism. This is the plane of the substance of expression, that is, the plane of the phonic or graphical facts constituted as "concrete", hence independently from any semiolinguistic engagement, and that are, as such enclosed upon themselves. As shown by Husserlian analysis and as confirmed by Saussure's writings, the signifier presents an ambivalent nature. According to phenomenological analysis (cf. 5.4), this stems from the fact that the matter (for example, graphical) of the signifier, despite its relegation to a secondary status in the attentional field relatively to thematic consciousness (of the signified), nevertheless maintains its character as a simple object of perception. According to structural analysis (cf. 5.2.2.2–5.2.2.3), the ambivalence of the signifier stems from the involvement of its concrete character in the operation of delimitation which determines it in part. Finally, in the morphodynamic apparatus, the structural position of the signifier as control parameter, as we have seen, is ambivalent, in that it puts into coincidence the material character of expression and its functional engagement.

As the signified proceeds from the signifier, the signifier itself mounts itself upon an element which is exterior (as endowed with an intrinsic constitution) to the linguistic system; an element which "persists" in its "concrete" characters when it is apprehended through a intentional semiolinguistic act, and which will return to an inert state, after the sign has been "consumed" (cf. 7.6). The substance of expression, if it has no place in the stratification of verbal consciousness in the strict sense, can nevertheless claim to find one if we broaden its sense: if we consider, at a level lower than that of the structural regimes which institute semiolinguistic phenomenality and objectivity, the necessary and persistent fact of a perceptive materiality of which the semiolinguistic consciousness makes use but without ever assimilating it.

6.5   Contrast: Morphodynamism Versus Functionalism

We wish to return here to some of the fundamental aspects of the morphodynamic model (*MD*) which essentially distinguishes it from the functionalist (*F*) approaches. In all likelihood, the various characteristics which we propose to highlight have already been exposed and discussed, but it is also true that the diagrams which, in the end, schematically restitute the structural and functional configurations specific to the *MD* and *F* approaches present apparent similarities, and that these similarities find themselves moreover to be reinforced by the use of a descriptive vocabulary which appears to relate common architectural characteristics on both sides. There is therefore a risk of misunderstanding: that of a functionalist assimilation of the *MD* model, which would consist in considering the *MD* apparatus, through the "surface" configuration which its diagrammatics delivers, as the simple reconstruction of a functionalist system over other conceptual a priori. In the *MD* model, the descriptive principles of the functionalist conception would then find themselves to be preserved, and this model would never be but an interpretive variant of the functionalist representation of semiolinguistic processes. It is this misunderstanding which we would promptly like to explore and lift.

To do this expeditiously, but without conceding anything of essence, the core of the misunderstanding could be presented as follows: On the surface, the *MD* and *F* models have for common and fundamental schema that of a two-leveled hierarchy, respectively made of the levels of expression and of content, and with dynamic relations between these levels. According to the functionalist perspective, the levels of expression and of content respectively comprise atoms of expression (lexemes) and of content (concepts). According to the *MD* approach, the components of expression and of content respectively carry the names of signifier and of signified. In both cases, the units of expression engage functional dynamic connections with units of content: inference or causality connections in one case, connections "of control" in the other. These connections govern joint actualizations of units of expression and of content: by the propagation of activation in one case, by the determination of dynamics in the other. Of course, as previously emphasized, the functional architectures are notably more complex (cf. illustration in Sect. 6.2.2, Fig. 6.9). The levels of expression and of content can comprise subdivisions which are themselves involved in internal and external interactions, whereas the connections can operate positively (activation) or negatively (inhibition) with possible feedback—but these are only complexifications which by no means modify the conceptual a priori and the functional principle of the two-leveled elementary schema and which we will therefore retain as basis for comparison.

Under this minimalist, albeit essentialist light—but, as we will see, at the cost of a serious alteration of the concepts of the *MD* model—the *MD* and *F* models seem to effectively overlap, and the specificities of the ones resemble the specificities of the other. Indeed, the strata of verbal consciousness, stemming from Husserlian phenomenological analysis, of which the *MD* model intends to deliver a structural interpretation and provide a mathematical characterization, indeed appear assimilable to the levels put forth by functional analysis: To the strata of signifiers and signifieds respectively correspond the levels of lexemes and concepts, and the strata of consciousness which come into shape through the various reflexive graspings of the "control" connection seem as much to be founded upon the inferential connections as upon the activation links between the lexemic and conceptual planes. Indeed, the consciousness of the existence or of the identity of a control function between a signifier and the plane of signifieds would be directly transposed unto the consciousness of the existence or of the identity of an inferential or activation link between the lexeme and the concept. Thus, the functionalist framework appears to adequately receive and articulate the set of descriptive specifications set up by the *MD* model. This would lead to conclude there is a homology between the functional and morphodynamic characterizations.

But such a conclusion is faulty, and the error which leads to it is that of an abusive assimilation of the categories of causality (logical or material) and of control. In a way, the source of the confusion denounced here resides in the graphical fact of a sole and same arrow symbol which illustrates, in the diagrammatics of the *F* and *MD* models, functional connections which have radically differing contents and significances. Because if the category of implication (logical or material) and the category of control have shared roots in the Kantian schema of causality—as specific realizations of this schema within distinct ontologies, respectively material (and logical) and structural—they exist in two mutually foreign and irreducible instances, and their assimilation therefore constitutes an epistemological error. In order to properly reveal the conceptual and descriptive specificities of the *F* and *MD* conceptions, it will be necessary to direct attention towards the functional and structural significations towards which the arrow symbol directs in the diagrams pertaining to the two approaches examined.

The fundamental difference between the functional link (*F* model) and the control function (*MD* model) resides in the fact that the first has a content of successiveness whereas the second has a constituting effect. Indeed, the functional connection regulates a cause to effect sequence, hence a sequence between two entities or two states each likely to have an autonomous determination. Thus, the relation of material implication associates, in the order of a necessary successiveness, entities (for instance lexemes and concepts) which are constituted for themselves, at their own level, and bearing their own determinations. In this sense, the functional link is second relatively to the states it associates in a relation of necessary implication. Conversely, the control function is first with respect to the states which it assembles in that they derive from it: The control function, which responds to a structural and holistic ontology, has a constituting scope. Its inputs and outputs are not data or states that are established independently from the functional roles they receive, and which institute them as much in their qualities as in their existence.

A more meticulous examination of the two-leveled functional architecture provides a good illustration of this difference. Indeed, we directly see that the lexemic and conceptual levels are not interdependent: The conceptual atoms belong, in terms of number, quality, and relations, to a universe of objects untied from eventual lexemes, and reciprocally. The concept in its system is therefore entirely defined at its own level, and the connection which links it to lexemes can no longer have any other sense than activation. In fact, and we are still reducing our discussion to the most elementary scenario, the dynamic connection between a Lexeme and a Concept (that is $L \rightarrow C$) is nothing more than an operational writing of a set-theoretic identity, that is, a pair (L, C). The $L \rightarrow C$ link is nothing more than the (L, C) pair enhanced with a dimension of effectuation: It defines itself as a specific pairing, and its identity is fully given with that of its poles. The arrow of the functional link $L \rightarrow C$ is therefore without its own content otherwise than in terms of effectuation, because it is devoid of other content than the pairing which it relates and which therefore is, we must reiterate, fully qualified by the terms it joins. We therefore see that the connection is second with respect to its terms. Its sole content is that of the dynamicity which it introduces, for example in the form of the propagation of activation.

Consequently, and conversely to what is articulated by the *MD* model, no consciousness of connection is conceivable here—in other words, a functionalist approach does not allow to respond to a lexical decision test on the basis of a specific consciousness of an $L \rightarrow C$ connection, simply because such a connection is reducible to a pairing (L, C) and that the consciousness of the existence or identity of this connection is to be correlatively reduced to the existence or identity of a concept C paired with L. In sum, from a functionalist perspective, a test of lexical decision regarding a term L will consist in taking note of the possible existence of a (L, C) pair, but always passing by C.

If the "auxiliary" status of the functional connection between the lexemic and conceptual levels cannot be doubted, what remains to be shown is how and in what respect the control connection at work in the *MD* apparatus is first relatively to its terms.

In order to do this, we will first observe that in the *MD* approach, the "word-forms" (henceforth *WF*) refer to units of substance of content (*USC*) which themselves determine dynamics (*D*) and, at their outputs, attractors which define particular identities of content. One may object that, in the end, by introducing in this manner additional levels of processing, we remain in a functionalist approach. Simply, instead of a direct (L, C) pairing, we would have a triple pairing (WF, USC) and (USC, D) and (D, C). In such case, the control functions would not essentially distinguish themselves from the material or logical implications, and the *MD* system would indeed be irreducible to a functional device. But this objection does not hold, for two essential reasons.

First, the connection between the *WF*s and the *USC* is not a connection between *WF*s and content identities. Second, the dynamics mediately associated to the *WF*s are not retained from the standpoint of the identities which their attractors would determine in a given coordinate system (for example, of signification), but rather from the point of view of their qualitative types and of the oppositional relations between these qualitative types (cf. 6.1). These two points are essential and, although they have been vastly developed in the preceding pages, they would benefit from being succinctly returned to.

First, concerning the connection of the *WF*s to the *SC*. Here, the usage of the term "matter", in the sense of an indeterminate manifold, would have been preferable to that of "substance". Indeed, following Hjelmslevian terminology, substance designates semiologically formed matter, that is, the shaping of a manifold of singularities that are heterogeneous and instable, without configured states, so as to produce individuation and identity. Thus, the connection of the *WF*s to the *SC* is not a connection to specific values of a global space of meanings, but a connection to a dustiness of indeterminate singularities which, precisely, the linguistic system will grasp in order to institute homogeneity, differentiation, and, *in fine*, values of meaning. The *SC* thus understood corresponds more or less to the subjective and labile space of Fregean representations: It is a matter of the manifold of representational occurrences, specific to experience and to the momentary states of each subject, of which the reproducibility is uncertain and the contours poorly defined and unstable. It follows that the connection of the *WF*s to the *SC* is not a precisely defined relation as such—though we must recognize it has some content (because this relation establishes a bridge with a particular diversity of content occurrences), its identity is uncertain and gives itself as "to be determined". This shapeless and private psychic mass will be seized by language in order to institute public content values—and we shall see how, this being our second point.

The structural operation performed by the linguistic system will consist in taking hold of this significatory matter, in homogenizing it, and, correlatively, in categorizing it by instructing differences within it. This double operation institutes in the *SC* a system of differentiating thresholds which express, within it, opposing relations (relations of competition with respect to actualization) between the qualitative types of dynamics associated to the units of substance. If we sacrifice to a quasi-functional writing (as a succession of planes), the *MD* apparatus distributes its levels in a loop which accounts for its holistic nature as well as for the constituting character of the control connection which is implemented within. Firstly, the *WF*s thus enter into a "shifting" connection with a manifold of occurrences of content. This connection is prolonged by a field which associates to each token of the *SC* a certain dynamic, a dynamic which is to be considered from the standpoint of its qualitative type (number and relative positionings of the attractors) as well as on the basis of the relations of opposition with the dynamics associated with other competing *WF*s. In return, then, the oppositions established between these dynamics are expressed in the *SC* by a differential categorization in the form of boundaries which institute within it negative identities of content; in short, these are signifieds, in the sense of Saussure. This "retroactive" moment therefore assigns to the uncertain units of the *SC* a well-defined differential identity, and in a retroactive prolongation, it determines the initial link between the *WF*s and the *SC*.

What we will retain here, for what interests us in the *MD* approach, is that unlike the functionalist model, we do not start with two planes endowed with constituted units which would be associated by a link's "secondary" effect without any other capacity or content than its being operational (passage from *L* to *C*), but that there are *WF*s which are invested with a dynamic power of homogenization and of structuration with respect to representational matter (the *SC*) which starts as formless, and which produces within it opposing identities *in fine*. That is to say, the connection here is instituting: It is at the source of the units which it generates and connects in a same gesture, and, on another hand, as we have seen, this connection, in contrast to what occurs in the functionalist model, has content which it receives from the identity of the meanings of which it governs the existence and the value.

Whereas the functionalist framework only articulates two levels of verbal consciousness (we may recall that the identity of the connection between planes is reducible here to that of its poles), it is a whole other matter in the *MD* model, which has at least five. This is because it is not a matter of determined identities of content which are at first "associated" to the *WFs*. It is only by the effect of dynamics, which these *WFs* control, that negative identities find themselves within the *SC*. The control connection is thereby first relatively to the identities which it establishes at its poles, and this primacy of the functional connection can be translated at the level of the mind by the possibility of a specific consciousness referring to it—a consciousness which is likely to present at least two aspects: that of the existence of such a connection and that of its identity. Thus, the *MD* model distributes at least four planes of verbal consciousness: that of the *WFs* in a simple position of control parameters (consciousness of availability), those of the existence and identity of the control function (respectively the consciousness of engagement and of motifs), finally, the consciousness of the negative identity of meaning (consciousness of the signified) which the connection establishes at its term. To this, it is necessary to add a fifth level of consciousness, complementary from the standpoint of the intentional logic at work within the *MD* system: that of a consciousness of fulfillment acquired when the simply negative values which are the signifieds are carried to an actual and accomplished level of determination, for example in the form of a mental image or concept.

But, if we were to concede to all of this, if were to proceed to the numbering of strata of verbal consciousness by reference to the functionalist model, it would be necessary to record an additional stratum: that which corresponds directly to the level of lexemes. Indeed, the stratum of the *WFs* (word-forms), having a control function (stratum which therefore founds a consciousness of availability), is not the exact analogue of that of the lexemes—simply because the lexemes are identities which are defined and constituted independently from their connections with the stratum of concepts, whereas the *WFs* in a position of control receive their semiolinguistic identity from their relations with the *SC*. Thus, in all rigor, it is necessary to distinguish the *WF* "in itself", as individuated and qualified outside of the semiolinguistic system, for example as a simple phonematic composition, and the *WF* as it is incorporated in the semiolinguistic device and mobilized to its ends (consciousness of availability and of engagement), and which receives from the functional position it occupies a new identity (stratified) with respect to both the phenomenological and structural planes.

In the end, therefore, the "pheno-morpho-dynamic" approach highlights six strata of verbal consciousness, comparatively to the two levels of the functionalist approach. To conclude, we may add that if, as has been mentioned in the beginning, the terms "strata" and "level" seem substitutable, care must be taken to not confound them in that they pertain to distinct structural rationalities. Whereas the "levels" have identities conceived independently from those of the other levels, the "strata" participate in an organic schema in the sense that the units of each of them are configured through relations established with the units of the others. The phenomenological profile of the units pertaining, for example, to a consciousness of availability is thus modeled by the function which is devoted to them within the system which grasps them and which indissolubly associates them with values of content—precisely in that these units participate in the existence of such values.

Complement: After having thus distinguished the morphodynamic approaches from the functionalist approaches, it is difficult to not do the same with the non-sequential approaches: the interactionist or connectionist approaches.

Generally speaking, and conversely to the morphodynamic approach, the connectionist approaches have input and output planes which *directly encode* semiolinguistic units or features (for instance syntactic or semantic) that have been *previously identified*. The matter then consists in making emerge, in the framework of a more or less supervised task of prediction or categorization, micro-values carried by other units and connections (said to be internal). Thus, *we always initially endow ourselves* with all the syntactico-semantic features, and the modelization effort concerns the learning of a connective weighting which aims to restitute the linguistic functionings retained according to their afforded latitudes (issue of admissibility). *There is nothing of the like in the morphodynamic approach* where, as we have seen in detail, *the semantic space is at first endowed with no referential* (of which each basic vector will be given by a syntactic, semantic or other kind of feature). The substance of content is a homogeneous substrate and it is in a logic and process of *emergence* that the differential structure is constituted (catastrophe boundaries) which categorizes the semantic continuum into opposing identities of meaning, correlatively instituting the local references.

References

Benveniste, E. (1974). *Problèmes de linguistique générale II*. Paris: Gallimard, coll. *Tel*.

Cadiot, P., & Visetti, Y.-M. (2001). *Pour une théorie des formes sé-man-tiques: motifs, profils, thèmes*. Paris: PUF, coll. Formes Sémiotiques.

Hjelmslev, L. (1985). *Nouveaux essais*. Paris: PUF, coll. Formes sémiotiques.

Petitot, J. (1985a). *Les catastrophes de la parole*. Paris: Maloine, coll. Recherches Interdisciplinaires.

Petitot, J. (1985b). *Morphogenèse du sens: 1, Pour un schématisme de la structure*. Paris: PUF, coll. Formes Sémiotiques.

Petitot, J. (1992). *Physique du sens: de la théorie des singularités aux structures sémio-narratives*. Paris: Editions du CNRS.

Prieto, L. (1990). Classe et concept. In R. Amacker & R. Engler (Eds.), *Présence de Saussure*. Genève: Droz.

Thom, R. (1977). *Stabilité structurelle et morphogénèse*. Paris: InterÉditions.

Chapter 7
The Merleau-Pontian Perspective

7.1 Introduction

We now propose to resume, in its general lines and main articulations, the work aiming to elucidate the being of speech conducted by M.-P. in Chap. 6 of his *Phenomenology of Perception* (henceforth *PhP*), and which, as we will see, leads to an impasse. Nevertheless, regardless of its shortcomings, this sixth chapter uncovers certain phenomenological truths concerning the act of speech and also lays the bases of approaches which, as furthered in ulterior studies such as *The Prose of the World*, *Signs*, and *The Sensible World and The World of Expression*, will provide a solution.

The angle of exposition we have retained here is therefore not forlorn from the start, and, contrastingly, it has the virtue of putting into play the very difficult and delicate progression towards the recognition of the phenomenological constitution of speech. Therefore, the key concept of "diacriticity", ultimately introduced by M.-P. in response to the issues regarding expressivity and perception, will be located in a framework which reveals its full rationality and explanatory power.

Adopting an approach leading to conceive of diacriticity as a fundamental structure of semiolinguistic phenomenon, it will also be a matter of revealing the forms and principle of this ensouling interiority by which speech is a living presence and by which it acquires experiential density.

But it will also be a matter, in conclusion, of showing that the differential and dynamic forms which the Saussurean model of the sign expose, and which pertain to an epistemic plane far removed from that of existential phenomenology, can nevertheless be put into relation with the order of living speech, by means of a kind of analytical conversion. We will indeed see that the morphodynamics of the Saussurean sign has an "existential rationality" in that its functional architecture can be seen as the projection of an originary diacritical constitution onto an objectal plane of determination. By positing

© Springer International Publishing AG, part of Springer Nature 2018   163
D. Piotrowski, *Morphogenesis of the Sign*, Lecture Notes in Morphogenesis,
https://doi.org/10.1007/978-3-319-89848-3_7

the forms of linguistic knowledge in this manner with respect to the forms of living speech, on the mode of a radically reductive but precisely qualifiable conversion, we will gain two things: First, we will attest to a certain empirical validity of the formal apparatus owing to its anchoring to the order of living speech, then, but in counterpart and specifically evaluating the losses entailed by such a conversion, we will situate the limits and circumscribe the perimeter of legitimacy of "objective" knowledge as pertains to language.

Chapter 6 of *PhP*, devoted to speech, has immediately and to its own ends recourse to the controversies and concepts instructed in the chapters preceding it. Thus, the study of speech, on the one hand, revives the critique of the empirical and intellectualist positions as well as it defends an expressivist conception, and, on the other hand, it reinvests the distinction between *abstract* and *concrete* movements. We may first recall its main elements before continuing along the direction induced by the examination of these questions.

7.2 Against Empiricism and Intellectualism

First, *against empiricism*—according to which the world is originarily delivered as a manifold of sensations, in which the perceptive faculty spontaneously distinguishes contiguities, regularities, resemblances, etc., and, by means of associations, produces in the mind a world of things, and, by means of abstractions, a world of ideas—and also *against intellectualism*—according to which the immediate matter of sensations is likewise a mosaic of qualities, but where thought installs, by means of constitutive syntheses governed by sovereign concepts, a universe of determinate objects—so, against empiricism and intellectualism, according to which the meaning of the perceived world does not reside within it, but, rather, extracts itself from it or projects itself onto it, M.-P. defends an expressivist position arguing that the sensible qualities relate along a perceptual mode the vital significations which constitute the originary framework of experience: "[In experience] we are not given 'dead' qualities, but rather active properties",[1] and in this "layer of living experience through which other people and things are first given to us",[2] it is qualities inhabited by an existential value, by a "meaning for us" that "sensing"[3] apprehends. What is encountered in an immediate manner is therefore not a mosaic of mute sensations, to be explored or informed, but indeed a fully signifying presence: "[the roots of perception] do not consist in the 'elements' of sensuous impression, but in originary and immediate expressive characters. Concrete perception [...] is never resolved into a simple complex of sensuous qualities [...] but each time accords itself with a determined and specific tonality of expression."[4]

---

[1] *PhP/L*, p. 52.
[2] Ibid., p. 57.
[3] Ibid., p. 52.
[4] Cassirer *in* Rosenthal and Visetti (2008, p. 185).

To elucidate the expressive fact, to found in law and in reason the fact of a tangible presence of meaning, in sum, to explain how "the expressive sense [...] adheres to perception itself, in which it is immediately grasped and 'experienced,'"[5] we know that the Merleau-Pontian solution consists in placing oneself before the moment when sensible qualities are constituted as signifying, that is, this originary moment of a face-to-face between the life force of "one's own body" and an environment of uncertain solicitations, of "poorly formulated question[s]"[6] to which one's own body attempts to respond in search of syntony: "Thus, a sensible that is about to be sensed poses to my body a sort of confused problem. I must find the attitude that *will* provide it with the means to become [some] determinate [quality]; I must find the response to a poorly formulated question. And yet, I only do this in response to its solicitation. My attitude is never sufficient to make me truly see blue or truly touch a hard surface. The sensible gives back to me what I had lent to it, but I received it from the sensible in the first place."[7]

Thus, a world is constituted as a "background", that is, as a world where the forms and qualities signify the acts and engagements of the subject who installed them as such, at the term of a sort of successful "coupling" of one's own body with its pre-sensible and pre-objectal environment; forms and qualities which, in turn, draw something of a geometry (the "background") whose "geodesics" (level lines or lines indicating a greater slope) imperiously solicit the actions of the perceiving subject. We then call "concrete" the movement which "adheres" to its background in the sense that it accomplishes the program of action which this background expresses.

But one's own body also carries this power to install a world. To do this, through an operation of thematization, one's own body posits itself as an object to its own gaze, and, also, the background is placed at a distance and its directing power is neutralized. The body, then withdrawn from the "background" which canalizes its engagements, and being free to invent and to deploy new schemas of action, then installs around itself a new fabric of meanings. Thus, the "abstract" movement projects its own background.

In what concerns language, the fundamental error of empiricism and of intellectualism, and which M.-P. relentlessly denounces, is that of approaching language from the outside, from a "third person" perspective. The error is to cast upon the facts and activity of speech the gaze of a "detached" observer who, following the descriptive grid he or she projects (a grid which is likely to comprise several levels of abstraction and a varying number of dimensions), only records states of speech in their correlations or in their sequencings, and, in continuity with this, relates the intelligibility of languages to an order of a system or, in a more operational perspective, to a functionalist format (neuro-physiological, computational, representationalist...). For M.-P., such an error is just as likely to be found with the

---

[5]Cassirer *in* Rosenthal and Visetti (2008, p. 185).
[6]*PhP/L*, p. 222.
[7]Ibid.

empiricists as with the intellectualists for whom linguistic functioning is comparable to a mechanism or to a computation: "Whether stimuli trigger, according to the laws of the nervous system, the stimulations capable of provoking the articulation of the word, or whether states of consciousness bring about the appearance of the appropriate verbal image by virtue of acquired associations, in both cases speech takes place in a circuit of third person phenomena."[8] In sum, and following this optic, "Man can speak in the way an electric lamp can become incandescent",[9] and, in the end, if there is a linguistic rationality, it does not reside within the speaking subject. Speech is triggered by stimuli; it is not carried by a subject's intention: "There is no one who speaks, there is but a flow of words that occurs without any intention to speak governing it."[10] Which amounts to saying that, from this angle, speech is a *process* and not an *action*, in the sense that "it does not manifest the inner possibilities of the subject."[11]

7.3   Concrete/Abstract

And M.-P. nuances these considerations on the basis of clinical observations which attest to an implication of the subject in the activity of speech. Indeed—and it is at this moment that the concrete/abstract opposition is addressed again, albeit laterally —certain linguistic troubles affect the capacity of subjects to employ words outside of their "concrete" context of usage. Thus, just as patients limited in their capacity for "spontaneous" movement can only move in some manners if the environment invites them to do so (the movement is then "concrete" because it "adheres to its background"), likewise, some patients find themselves incapable of speaking other words than those which constitute a verbal reaction to the situation: "The same word that remains available to the patient on the level of automatic language escapes him on the level of spontaneous language."[12]

This form of aphasia attests to the existence of a "speaking subject" or of an "attitude" behind speech, in other words, to the existence of an "intentional language"[13]: "Thus, behind the word we discover an attitude or a function of speech that conditions it."[14] But if speech thus demarcates itself from a strictly mechanistic schema of functioning, its triggering motive still remains external to its own order, in that it is on the plane of thought that the trouble may have been located: "If 'concrete' language remained a third person process, then spontaneous language or

---

[8]Ibid., pp. 179–180.
[9]Ibid., p. 180.
[10]Ibid.
[11]Ibid.
[12]Ibid.
[13]Ibid.
[14]Ibid.

authentic denomination became a phenomenon of thought, and so the origin of certain types of aphasia had to be sought in some mental disorder."[15] And this is this how intellectualism recovers some of its legitimacy.

Be it as it may, and before seeing how M.-P. places on a same plane of inadequacy and exceeds by a single consideration the empirist and intellectualist conceptions, let's note that the analogy which is nevertheless manifest between abstract/concrete movement and spontaneous/concrete speech is not readily exploited by M.-P. to defend the gestural conception of speech which he will nevertheless introduce a few pages further. And this for the sole reason that at this stage, the analogy does not hold.

Indeed, for spontaneous speech to be assimilable to spontaneous movement, it would be necessary for it to be the matrix of meaning; it would need to be capable of emanating and installing a world of significations, as the spontaneous gesture installs its own. But at this stage of the discussion, this is not so, because, as M.-P. stresses, and despite that intellectualism posits a subject behind the word, for the intellectualist as well as for the empiricist speech remains *exterior* to meaning: "These two theories, however, concur in the claim that the word *has* no signification."[16] The demonstration is direct. For empiricism, the matter is settled; for intellectualism, it is in thought and in its categorical operations that meaning resides, and not in speech: "Thought has a sense and the word remains an empty envelope."[17] And as a conclusion: "In the first [account: empirism], there is no one who speaks; in the second [intellectualism], there is certainly a subject, but it is the thinking subject, not the speaking subject."[18] And, on the whole, the one and the other are equivalent: "With regard to speech itself, intellectualism hardly differs from empiricism, and it is no more able to do without an explanation through automatic reflexes. Once the categorial operation has been accomplished, the appearance of the word that accomplishes it remains to be explained, and again an explanation is found through a physiological or psychological mechanism."[19]

But what is fundamental in order to establish an accurate comprehension of the activity of speech, in order to grasp its authentic essence and precise nature, is exactly what the criticized approaches deny, that *speech is meaning-laden*: "Thus, we move beyond intellectualism as much as empiricism through the simple observation that *the word has a sense*."[20]

---

[15]Ibid.
[16]Ibid., p. 182.
[17]Ibid.
[18]Ibid.
[19]Ibid.
[20]Ibid.

7.4 The Word has a Sense

This observation, that "the word has a sense", fully places us within a problematics of expressivity and, concomitantly, steers towards questions regarding the logic and possibility of a reciprocal incorporation of "form" (understood here as the perceptible face of the sign) and meaning. We refer to the paragraphs which have been devoted to this (cf. 2.4), and we will limit ourselves to recalling the arguments which M.-P. advances in *PhP* to defend his views.

First argument: If speech were only the encoding of previously elaborated thought, "then we could not understand why thought tends toward expression as if toward its completion."[21] *Second argument*: Thought only truly exists during the moment of speech, as attested by the fact that "the thinking subject [...] is in a sort of ignorance of his thoughts so long as he has not formulated them."[22]

It must therefore be recognized that "for the speaker [...] speech does not translate a ready-made thought; rather, speech accomplishes thought",[23] and this despite such an observation being counter to any naïve or spontaneous observation. Indeed, at a first glance, "one might believe that [the listener himself] gives the words and the phrases their sense, and even the combination of words and phrases is not some external contribution."[24] In other words, if we retain from the activity of speech only its most "apparent" forms in that they constitute its resulting and stabilized states, that is, the "word-forms" and their associated meanings, then we are indeed led to think, following a logic of communication, that all speech is but a transmission of signals encoding the thought of a speaker—signals from which the listener, having at his or her disposal a table of corresponding meanings, reconstructs the supposedly transmitted thought. But in truth, as emphasized by M.-P., speech can transmit nothing in this manner which the listener does not already possess: "Here as everywhere, it seems true at first glance that consciousness can only find in its experience what it had itself put there."[25] Then, "the experience of communication would be an illusion",[26] this running counter to all which experience seems to indicate—and this is a new argument which M.-P. advances in favor of authentically signifying thought, because "the fact is that we have the power to understand over and above what we may have spontaneously thought."[27]

In the end, it is necessary to admit the presence of meaning in speech itself. And since meaning is produced during the moment of speech, to understand speech is to relive the utterance and to make the speech of another speak within oneself: "Through speech, then, there is a taking up of the other person's thought, a

---

[21]Ibid.

[22]Ibid., p. 183.

[23]Ibid.

[24]Ibid., p. 184.

[25]Ibid.

[26]Ibid.

[27]Ibid.

reflection in others, a power of thinking *according to others*, which enriches our own thoughts."[28]

This theme of "speaking" and of "understanding" as "moments in the unified system of self-other"[29] will be largely employed in ulterior works. Thus, "successful communication occurs only if the listener, instead of following the verbal chain link by link, on his own account resumes the other's linguistic gesticulation and carries it further",[30] and also: "It summons me and grips me; it envelops and inhabits me to the point that I cannot tell what comes from me and what from it. "Whether speaking or listening, I project myself into the other person, I introduce him into my own self" (*PW*, p. 19). Such is also notably the case with the activity of reading, which similarly gives life to the author's thought when the reader takes hold of the system of original meanings to which the text progressively initiates him or her: "I get closer and closer to [Stendhal], until in the end I read his words with the very same intention that he gave to them [...]. In the same way, the author's voice results in my assuming his thoughts [...]. It is only then that the reader [...] can say [...], 'In this light at least, I have been you.' [...] I am Stendhal while reading him. But that is because first he knew how to bring me to dwell within him."[31]

7.5   The Verbal Gesture—Introduction

It is at this point which, to put it laterally, M.-P. introduces the gestural dimension. Thus, after having recapitulated what was previously established, that "the sense of words must ultimately be induced by the words themselves",[32] he continues stating that "more precisely their conceptual signification must be formed by drawing from a gestural signification, which itself is immanent in speech."[33]

At this stage of the argument, if speech can claim to have a gestural nature, it is because it principally pertains to an existential order. To speak is not simply to assemble words into signifying combinations, but to act in the world instituted by a language, as one's body acts within its environment. And to know how to speak is not to possess a system of rules and conventions, it is to know one's possibilities for action within the universe of experience of language: "[...] I begin to understand the sense of words by their place in a context of action and by participating in everyday life, [so too] I begin to understand a philosophy by slipping into this

---

[28]Ibid.

[29]*PW*, p. 18.

[30]Ibid., pp. 28–29.

[31]Ibid., p. 12.

[32]*PhP/L*, p. 184.

[33]Ibid.

thought's particular manner of existing."[34] Moreover, and also more fundamentally, this verbal gesturality, like the gesturality of one's own body, generates its meaning —from which conceptual significations may therefore be retrieved. Thus, "we are clearly led to recognize a gestural or existential signification of speech, as we said above."[35] In short, "speech is a gesture, and its signification is a world",[36] and furthermore: "[language is] a means of [...] constructing a linguistic universe of which we later say—once it is precise enough to crystallize a significative intention and to have it reborn in another—that it expresses a world of thought, as it gives it its existence in the world."[37] But if we accept without too much difficulty an existentialist conception of speech, as a form of an "I can" in a world which responds to it and which it instructs, therefore a conception which treats language as a "fundamental activity by which man projects himself toward a 'world,'"[38] it is, however, more difficult to clearly conceive the specificity of the signifying power which inhabits the verbal gesture: In what and by means of what is such a gesture capable of inducing meanings which must be acknowledged to not have the same nature and scope as those stemming from the transcendental power of a body as an embodied vector of a life force?

Moreover, and paradoxically, if, in all evidence, phonetic gesticulation does not cover the set of signifying resources of one's own body, it exceeds them. Indeed, in the authentic act of expression ("speaking speech"), "existence is polarized into a certain "sense" that cannot be defined by any natural object; existence seeks to meet up with itself beyond being, and this is why it creates speech as the empirical support of its own non-being. Speech is the excess of our existence beyond natural being."[39]

At this stage, therefore, and abstraction being made of the existentialist argument, the question fully holds: In what specifically is speech similar to gesture, and how are immanent meanings woven within it?

These questions having been posed—questions which M.-P. does not address head-on at this stage of his exposition—we can return to the argumentative progression of Chap. 6 which, after having recognized that "the word bears the sense",[40] interrogates that which occults a nevertheless blatant truth: "From where does the mistake come from and how is it that we apprehend meaning and the word as exterior and foreign from one another?"

The sources of this mistake are multiple and intersecting. But there is, first of all, the fact of "already constituted [...] thoughts that we can silently recall to ourselves and by which we give ourselves the illusion of an inner life",[41] an illusion which is

---

[34]Ibid., pp. 184–185.

[35]Ibid., p. 199.

[36]Ibid., p. 190.

[37]*PW*, p. 31.

[38]*PhP/L*, p. 197.

[39]Ibid., pp. 202–203.

[40]Ibid., p. 183.

[41]Ibid., pp. 188–189.

doubled by "the illusion of already possessing within ourselves, with the common meaning of words, what will be necessary for understanding any text whatsoever."[42] These are "illusions" which then lead to seeing the word as nothing but the dressing of a content of thought that was available before any speech.

But these illusions do carry some truth in that, despite that they distort the vision and comprehension of the fact of speech, they are induced by the power of language proper and they manifest dispositions and capacities which are authentic and intersecting, and of which the main ones are called *consummation* and *sedimentation*.

7.6   Consummation and Sedimentation

The term *consummation* serves to designate this propensity of speech to project us into a world of pure thought, the inclination to "self-consummate" being in the sense where the words, as soon as they are pronounced, disappear from our mind by installing within it a scene of ideas or referring to things and, correlatively, the words thus left aside fall laterally as would the inert residue of spent vitality.

The theme of consummation widely traverses the Merleau-Pontian reflection regarding speech, and it would be easy to multiply the citations regarding it. The following excerpts will suffice to clarify this particular aspect of the phenomenon of speech.

– "When someone [...] succeeds in expressing himself, the signs are immediately forgotten; all that remains is the meaning. The perfection of language lies in its capacity to pass unnoticed. *But therein lies the virtue of language:* it is language which propels us toward the things it signifies. In the way it works, language hides itself from us. Its triumph is to efface itself and to take us beyond the words to the author's very thoughts, so that we imagine we are engaged with him in a wordless meeting of minds. Once the words have cooled and been reaffixed to the page as signs, their very power to project us far away from themselves makes it impossible for us to believe they are the source of so many thoughts."[43]
– "There is [...], in the exercise of language, the consciousness of saying something, the presumption of a fulfillment of language."[44]
– "The power of language [gives] us the illusion of going beyond all speech to things themselves."[45]
– "This [...] marvel that linguistic meaning directs toward something beyond language, is the very prodigy of speech, and anyone who tries to explain it in

---

[42]Ibid., p. 185.
[43]*PW*, p. 10.
[44]Ibid., p. 40.
[45]Ibid., p. 41.

terms of its 'beginning' [the inherited language] or its 'end' [the world of knowledge] would lose sight of its 'doing'."[46]

And if speech consummates itself by installing a thought and by devitalizing the words which lead to it, the reverse operation may be considered as an abstract exercise, that is, of bridging the gap between the sign and meaning, not absolutely, at the risk of falling into the absurdity of a consubstantiality of sound and of meaning, but by reducing the distance "to one's liking": "Even what I call signification appears to me as thought without any admixture of language [as pure thought] only through the power of language to carry me toward what is expressed; and what I call a sign, reducing it to an inanimate envelope […] approaches as closely as possible to signification as soon as I consider the way it functions in living language […] language carries the significations which hide as much as reveal its operations, and which, once born, will appear to be simply coordinate with the inert signs."[47]

In order to finish with the theme of consummation, let's nevertheless stress that it is never anything more than a "limit-idea." Certainly, by consummating itself, speech installs thoughts in our mind, but albeit these thoughts may be accessible in themselves, far removed from the words which induced them, the rupture with speech is still not absolute: "Speech forgets itself as being a contingent fact […] and […] this gives us the ideal of a thought without speech […]. Even if 'thought without speech' is no more than a limit-idea and a bit of an absurdity [*counter-sensical*], even if the sense of a speech act can never in fact be delivered from its inherence in some speech."[48] In truth, "no speech completely effaces itself before the meaning toward which it points",[49] and: "There is thus an opaqueness of language. Nowhere does it stop and leave a place for pure meaning […] meaning appears within it only set in a context of words."[50]

In truth, thoughts remain inhabited by the power which generated them, not in the mode of an explicit and distinct presence, but in the mode of a solicitation which always calls for other acts of speech. It will be likewise for "spoken" language which we will see to proceed from a *sedimentation* of consummated words. It will be necessary, later on, to introduce the concept of *diacriticity* in order to attain and unveil the reason and principle of this perpetual reference of speech towards itself as much as to the "background of silence" from which it proceeds and with respect to which it configures itself. We will return specifically to this.

In what concerns *sedimentation*, as a specific property of natural languages relatively to the other signifying systems (music, painting…), it is question of this aptitude of language for setting the moments resulting from the act of consummation (meaning and form) and of arranging them into a world of shared signs and

---

[46]Ibid.

[47]Ibid., pp. 30–31.

[48]*PhP/L, p*, 196.

[49]*PW*, p. 40.

[50]*IL&VS*, p. 79.

significations: "What is simply true [...] is that, of all the expressive operations, speech alone is capable of sedimenting and of constituting an intersubjective acquisition. [Whereas] each artist takes up the task from the beginning, he has a new world to deliver, while in the order of speech, each writer is aware of intending the same world with which other writers were already concerned."[51]

The successive sedimentations then arrange and configure the "conceptual and final sense of words"[52] as well as the phonic forms which will have carried them, into a complex landscape which is nevertheless traversed by regularities which reflect the adjustments and equilibriums which the speaking subjects, in the living exercise of their speech, will have attained. This is "spoken" language: A set of verbal significations and connections that are arrested and available, and which weave the space of linguistic agreement and of shared speech: "[I]t might be said that *languages* [*langages*], that is, constituted systems of vocabulary and syntax, or the various empirically existing 'means of expression,' are the depository and the sedimentation of acts of *speech* [*parole*], in which the unformulated sense not only finds the means of expressing itself on the outside, but moreover acquires existence for itself, and is truly created as sense."[53]

7.7   Sedimented Language/Speech[54]

To "sedimented" language, "speech" responds in a dual manner: "We may say that there are two languages. First, there is language after the fact, or language as an institution, which effaces itself in order to yield the meaning which it conveys. Second, there is the language which creates itself in its expressive acts, which sweeps me on from the signs toward meaning—sedimented language and speech",[55] or furthermore "sedimented language is the language the reader brings with him, the stock of accepted relations between signs and familiar significations without which he could never have begun to read. It constitutes the language and the literature of the language. [...] But speech is the book's call to the [...] reader. Speech is the operation through which a certain arrangement of already available signs and significations alters and then transfigures each of them, so that in the end a new signification is secreted."[56]

But, likewise as the sedimented significations always remain quivering with new speech, likewise sedimented language is not an immutable theater of words and

---

[51]*PhP/L*, pp. 195–196.

[52]Ibid., p. 193.

[53]Ibid., p. 202.

[54]"*Parole parlante et parole parlée*" is translated in *PhP/L* as "the speaking and the spoken word", but "*langage parlé et langage parlant*" is translated as "sedimented language and speech" in *PW*.

[55]*PW*, p. 10.

[56]Ibid., p. 13.

syntax, of which the elements and their connections would only be able to function mechanically, in sum, as a language in the third person: Sedimented language is equally also in wait for speech, because having the status of a norm, it offers itself by essence to verbal distortions, inductive of living meaning, in order to open onto a beyond of simply available significations: "[T]he act of expression [the *speaking speech*] constitutes a linguistic and cultural world, it makes that which stretched beyond fall back into being. This results in spoken speech, which enjoys the use of available significations like that of an acquired fortune."[57] "From these acquisitions, other authentic acts of expression [...] become possible. [...] Such is the function revealed through language, which reiterates itself, depends upon itself, or that like a wave gathers itself together and steadies itself in order to once again throw itself beyond itself."[58]

7.8 Intermezzo

By means of consummation and of sedimentation, the successive acts of expression end up weaving the system of an acquired language: from a vocabulary as a table of correspondences between form and meaning, from a syntax and from other rules exposing temporary regularities but always to be revived by means of decenterings and distortions which only the speaking word can accomplish. This is where there is a risk of error, in considering acquired language as the objective being of language, as the principle of order (cf. Saussure or Hjelmslev) and of intelligibility of the facts of speech, then taken to be singular and labile. Acquired language has the virtue of clarity: It univocally posits signs with respect to meaning, having taken prior care to unlink them and to qualify them, and it also reveals systematicity: "If it still seems to us that language is [...] transparent [...], this is because we remain for the most part within constituted language, we provide ourselves with available significations [...]. The sense of a phrase appears intelligible to us throughout, even detachable from this phrase and defined in an intelligible world, because we presuppose as given all of the participations that it owes to the history of the language and that contribute to determining its sense."[59]

Steering away from such a misapprehension, it is towards speaking speech, where "the meaningful intention is in a nascent state",[60] that it would be suitable to look so as to posit it as a reference and source of any form of observable linguistic functioning and structuration. It is indeed from speech that sedimented language is

---

[57]*PhP/L*, p. 203.
[58]Ibid.
[59]Ibid., p. 194.
[60]Ibid., p. 202.

born, and this generation calls to be explicated in its various phases and in its internal regimes. Here, the issue of semiogenesis arises: As a morphogenetic process over the course of which the expressive fact, resulting from speech, polarizes itself into a signifier and a signified, and, in breaking the unity of the sign, it accomplishes what M.-P. calls a consummation, that is, it projects the subject towards a set of thoughts in rupture with their signifying medium, i.e. their signifier, which then becomes residual.

We will also and already note a few correspondences with the Husserlian and Saussurean approaches discussed in the preceding chapters. The morphodynamics of the Saussurean sign which, through an analytic of the field of verbal consciousness, delivers a phenomenological description of the sign, coincides in part with the problematic articulations of M.-P. who, in turn, brings new clarity to it.

Thus, in what first concerns consummation: We have seen that the device of the Saussurean sign is fully oriented towards the production of signifieds, which, themselves, as negative identities, open onto their fulfillment in terms of positive identities (of substance). The act of fulfillment therefore has the effect of installing thought within a world of ideas or representations that are foreign to its order within the field of consciousness (attentional) where the signifiers and signifieds are configured. In this movement, where the signified is converted into an idea, consciousness therefore finds itself to be projected outside of the sphere of language, and this, at the same time as the signifier falls back into a state of being a simple acoustic or graphical phenomenon.

Concerning sedimented language, we find it in the morphodynamic device in that the configurations which it defines and posits as a scaffold for the activity of speech are grasped and even raised to the status of laws. Indeed, the distributional rules of "acquired" language, or, as we might call them, its system, involving the syntagmatic and paradigmatic dimensions, are inscribed within the morphodynamic device in the form of prescriptions regarding the emergence of boundaries (via the stabilization paths) in a substance of content; in other words, in the form of prescriptions conditioning the formation of signs and therefore existence in language. Hence, the morphodynamic device contains the order of a sedimented language, precisely in that the schemas of linguistic regularity which it comprises are functionally paired with the process of formation of the signs and more broadly with the conditions of existence in language.

For the time being, however, it is fitting to move forward regarding the clarification of speech. And continuing in the direction set forth in Chap. 6, there will be two paths to explore: Firstly, it is necessary to see how M.-P. makes possible and argues the conception of speech as gesture. Then, and almost correlatively, it will be necessary to pay heed to the principle of an immanent gestural sense, most specifically through a more detailed analysis of the relations between thought and speech.

7.9 The Verbal Gesture—Continued

In what concerns the principle of speech as verbal gesticulation, we have seen that it stems from an expressivist and existentialist position. By and within speech, meaning acquires a tangible existence and installs itself as if in another world: "The operation of expression, when successful, [...] makes the signification exist as a thing at the very heart of the text"[61]: "We discover here, beneath the conceptual signification of words, an existential signification that is not simply translated by them, but that inhabits them and is inseparable from them."[62] Speech thus produces a world of significations of which words are the actual forms which configure and mark out its topography; a world which speech invests and, if required, makes compliant with its own ends. That is, speech, "the operation of expression", "opens a new field or a new dimension to our experience."[63] The act of speech "brings [the signification] to life in an organism of words, it installs this signification [...] like a new sense organ".[64] Thus, "expression confers an existence in itself upon what it expresses, installs it in nature as a perceived thing accessible to everyone, or inversely rips the signs themselves [their concrete characters] from their empirical existence and steals them away to another world."[65]

This principle being established, let's see in what respect speech is similar to a gesture.

We will first observe that bodily movement and the act of speech both present a holistic and finalized character. Regarding speech, it is established that the utterance is not a summative succession of words, but indeed an integrated totality accomplishing a certain intent to signify. In this respect, which is central in linguistics, we may simply cite Benveniste: "A sentence constitutes a whole which is not reducible to the sum of its parts; the meaning inherent in this whole is distributed over the ensemble of the constituents",[66] and, moreover: "rather than contributing to it, words accomplish the meaning of the sentence." Thus, the act of speech carries its own end and therefore its totality from the moment the first word is uttered. Likewise for bodily movement: "[T]he originality of movements that I execute with my body: my movements anticipate directly their final position [...] I do not find [my body] at one objective point in space [like an object] in order to lead it to another, [...] I have no need of directing it toward the goal of the movement, in a sense it touches the goal from the very beginning and it throws itself toward it."[67]

But it is from the angle of their practice that the parallel between gestures and speech is the most flagrant. Indeed, similarly to when the empirical world has and

---

[61]Ibid., p. 188.

[62]Ibid.

[63]Ibid.

[64]Ibid.

[65]Ibid.

[66]Benveniste (1971, p. 105).

[67]*PhP/L*, pp. 96–97.

delivers things following a geometry and a play of qualifications which express their immediate relations to a certain capacity of action, likewise language delivers a world of words and of constructions as they "constitute a certain field of action held around me."[68]

To speak therefore amounts to moving through speech within a world of words: "I relate to the word just as my hand reaches for the place on my body being stung. The word has a certain place in my linguistic world [...]. The only means I have of representing it to myself is by pronouncing it."[69] And likewise that the body knows its world on the mode of a "power to do", speech knows words on the mode of a "power to say", which is therefore "power" *by virtue of* words: "[K]nowing a word or a language [*langue*] does not consist in having available some preestablished neural arrangements [or some verbal representations] [...] the words that I know [...] are behind me, like the objects behind my back or like the horizon of the village surrounding my house; I reckon with them or I count upon them, but I have no 'verbal image' of them",[70] or: "Likewise [for movement], I have no need of representing to myself the word in order to know it and to pronounce it."[71]

The gestural nature of speech then appears clearly. In the same manner as bodily movement installs a sensible world with respect to a subject which invests it, language "is the subject's taking up of a position in the world of his significations."[72] And it is not a matter here of metaphors: The principle of constitutive interactions which simultaneously install a subject and his or her world just as much concerns the gesture of speech with respect to the world of significations. "The term 'world' is here not just a manner of speaking: it means that 'mental' or cultural life borrows its structures from natural life and that the thinking subject must be grounded upon the embodied subject."[73] Furthermore, "the phonetic gesture produces a certain structuring of experience [...] just as a behavior of my body invests —for me and for others—the objects that surround me with a certain signification."[74]

7.10   The Verbal Gesture—Difficulties

It is at this stage that certain difficulties arise, difficulties of which M.-P. was aware. Indeed, if the expressivity of the gesture is effective, if it is endowed with an immanent meaning, it is not because the sense of the gesture would be as such

---

[68]Ibid., p. 186.

[69]Ibid.

[70]Ibid.

[71]Ibid.

[72]Ibid., p. 199.

[73]Ibid.

[74]Ibid.

"given" in the manner that a sensible quality would be (as significant as it may be). M.-P. uses the example of the angry gesture, of which the meaning is intrinsic: "The gesture does not *make me think* of anger, it is the anger itself."[75] However, M.-P. continues, "the sense of the gesture is not perceived like, for example, the color of the rug."[76] For it to be possible to understand the gesture, for its perception to not be limited to a biomechanical action, it is necessary to have access to its internal motive, to the bodily register from which it emanates, hence, to a certain proximity in terms of vital mode and existential engagement: "I do not 'understand' the sexual gesture of the dog, and even less that of the beetle",[77] and this scene has meaning only if "this behavior becomes a possibility for [the spectator, in other words] if it [is] found among [its] internal possibilities."[78] Which amounts to saying that "[t]he sense of the gestures is not given but rather understood, which is to say taken up by an act of the spectator. The entire difficulty is to conceive of this act properly and not to confuse it with an epistemic operation."[79]

To understand a gesture, to grasp its immanent meaning, is therefore to seize for oneself, so far as this may be possible, the internal dynamics of this gesture which accomplishes a particular engagement of the body and which installs at its horizon a meaning which is then accessible: "The gesture I witness sketches out the first signs of an intentional object. This object becomes present and is fully understood when the powers of my body adjust to it and fit over it",[80] or: "The sense of the gesture thus 'understood' is not behind the gesture, it merges with the structure of the world that the gesture sketches out and that I take up for myself."[81] Likewise for the verbal gesture, as we have seen that through it, the listener takes on and reanimates the significatory aim of his or her interlocutor.

For this to be possible, it is necessary for the interlocutors to share the resources which instruct a same universe of existence. But if this condition may be satisfied given that it is a matter of one's own body and of the sensible world, this is by no means the case for verbal gesticulation.

Indeed, to begin with, verbal significations do not overlap with those which emanate from one's own body—and M.-P. openly acknowledges this: "At first, it seems impossible to allow either words or gestures an immanent signification, because the gesture is limited to indicating a certain relation between man and the perceptible world, because this world is given to the spectator through natural perception, and because the intentional object is hence offered to the observer at the same time as the gesture itself."[82] To the contrary, "the verbal gesture [...] intends a

---

[75]Ibid., p. 190.
[76]Ibid.
[77]Ibid.
[78]Ibid.
[79]Ibid.
[80]Ibid., p. 191.
[81]Ibid., p. 192.
[82]Ibid.

mental landscape that is not straightaway given to everyone, and it is precisely its function to communicate this landscape."[83] Also, the verbal gesture must be situated on another plane of experience than that of the body with respect to a concrete world. If speech is a gesture, this gesture is therefore executed within the world which is its own: not a world of sensible objects and qualities, but a world of signs and meanings. And this world "[that] nature does not provide, [...] culture here offers [it]",[84] such a world being a space for symbolization, a landscape of values and a framework of usages, which original meaning-intentions will invest to their own ends. This shared space in which the verbal gestures accomplish themselves is therefore that of spoken (also called sedimented) language: It is "[the] available significations, namely, previous acts of expression, [which] establish a common world [...] to which current and new speech refers, just as the gesture refers to the sensible world"[85]—and, moreover "the sense of the speech is nothing other than the manner in which it handles this linguistic world, or in which it modulates upon this keyboard of acquired significations."[86]

But a question then arises: that of the "primordial speech". Because if the verbal gesture takes place within a world of significations which constitute the mental landscape of a community of speakers, it is necessary to explain how, starting from nothing, language was able to progressively install such a world. "For the miracle to happen, the phonetic gesticulation must make use of an alphabet of already acquired significations, and the verbal gesture must be performed in a certain panorama that is shared by the interlocutors, just as the comprehension of other gestures presupposes a perceived world shared by everyone in which the sense of the gesture unfolds and is displayed."[87] We should insist: "[Since] speech is a genuine gesture and, [as] just like all gestures, speech too contains its own sense",[88] it is indeed necessary for this gesture, in the manner of the bodily gesture, to have an environment which will instruct new meanings by virtue of its own power. Now, this environment, we have seen, is sedimented language, resulting from the accretions of speech. A "primordial speech" is therefore necessary in order to install a first world of significations. And, by analogy with this originary moment where a vital power responding to the questioning of an unsure environment simultaneously establishes a body and its world, it will be necessary to consider the possibility of a soliciting halo of which the verbal gesture would be similarly the mode of its encounter and the principle of a response. The conception of the diacritical sign will lead to this.

To tell the truth, this problem of "primordial speech" does not find a conclusive outcome in *PhP*, nor is it deepened. At most in the lines where, interrogating the

---

[83]Ibid.

[84]Ibid.

[85]Ibid.

[86]Ibid.

[87]Ibid., p. 200.

[88]Ibid., p. 189.

apparently fortuitous link" between the verbal sign and its signification",[89] and confining the arbitrariness of the relations between forms and meanings to the sole products of consummated speech ("If we consider only the conceptual and final sense of words, [...] the verbal form [...] seems arbitrary"[90]), M.-P. relates the immanence of gestural meaning to its emotional texture: "We must, then, seek the first hints of language in the emotional gesticulation by which man superimposes upon the given world the world according to man."[91]

The sense attached as such to the verbal gesture will therefore be of an emotional nature; Speech would be the very existence of the emotion it expresses, and, in this originary moment, nothing would separate it from its meaning: "[I]f we took the emotional sense of the word into account, what we have [...] called its gestural sense [...] we would then find that words, vowels, and phonemes are so many ways of singing the world, and that they are destined to represent objects, not through an objective resemblance [naïve theory of onomatopoeia] but because they are extracted from them, and literally express their emotional essence."[92] And then: "[The choice of vowels or consonants...] would not represent so many arbitrary conventions for expressing the same thought, but rather several ways for the human body to celebrate the world and to finally live it."[93]

We will agree that this conception of emotional content immanent to the verbal gesture is far too reductionist. And it is, moreover, following another path, that of diacriticity, that M.-P. will examine this question in greater depth. But for the time being, we will return to the previously introduced question concerning the relations between thought and speech.

7.11 Thought/Speech

We therefore accept the idea that meaning resides in speech itself, and that "thought [...] does [not] exist outside the world and outside of words."[94]

But in fact, if thought is indeed present within the verbal body, it does not find itself there in the manner of thematized content. The thought which is accomplished through verbal gesticulation is thought *in action*, and not thought of which the contours are determined, and which was posited for its own sake so as to offer itself to a reflexive gaze. This sort of "thematized" thought arises *after* speech. It is neither contiguous nor preexisting to the linguistic gesture which, as a matter of fact, carries it into existence: "The orator does not think prior to speaking, nor even

---

[89]Ibid., p. 192.
[90]Ibid., p. 193.
[91]Ibid., p. 194.
[92]Ibid., p. 193.
[93]Ibid.
[94]Ibid., p. 188.

while speaking; his speech is his thought. The listener similarly does not think about the signs."[95] It remains that thought which follows words does not stem from naught: It is the conceptual outcome of a meaning which weaves itself through the verbal connections and which responds to a certain intent to signify. During the moment of speech, thought exists indeed but according to a verbal being, in grasp of words. If, by "thought", we mean a content of defined intellection, then "the 'thought' of the orator is empty while he speaks",[96] and "the speaking subject does not conceive of the sense of what he says, he no more represents to himself the words he employs."[97] This is because in the act of speech, the sense is not a focal point of the mind, but is distributed over the totality of the gesture which accomplishes it: "[T]he sense [is] present everywhere, but nowhere [is] it posited for itself."[98] In fact, the sense which diffuses itself through speech is perceived in its verbal being: "[W]hen a text is read [...] we do not have a thought on the margins of the text itself. The words occupy our entire mind."[99] It is indeed the verbal gesture which carries its meaning. The content uttered or heard is nothing other than an articulation of words. And it is at the end of the text or of the discourse, while leaving the plane of verbal existence, and as if signaling "the lifting of a spell [...], that thoughts [...] will be able to arise."[100]

The meaning possibly delivered by speech, and which therefore configures itself in this speech along the mode of verbal existence, is not present within it as such: "[T]he spoken word [...] is pregnant with a meaning which can be read in the very texture of the linguistic gesture [...] and yet is never contained in that gesture."[101]

But, then, of what sort is this "verbal significance" allowing a thought which does not reside within it to result from it? This question presents two narrowly interwoven aspects. The one pertains to this "power that speaking subjects have of going beyond signs toward their meaning"[102]; and it should be reiterated that this is, under the name of consummation, a "fundamental fact of expression [that is] *a surpassing of the signifying by the signified which it is the very virtue of the signifying to make possible.*"[103] The other touches the nature and principle of the meaning carried by a verbal gesticulation which therefore has the vocation of converting itself into thought.

---

[95]Ibid., p. 185.

[96]Ibid.

[97]Ibid., p. 186.

[98]Ibid., pp. 185–186.

[99]Ibid., p. 185.

[100]Ibid.

[101]*Signs*, p. 89.

[102]*IL&VS*, p. 118.

[103]*Signs*, p. 90.

7.12   The Verbal Gesture—The System

This second question has already been approached. We have seen that the answer which M.-P. provides in the first stages of his reflection, that the verbal gesture, in the same measure as one's own body of which it shares the resources, has the power to modulate the world of acquired significations and to bring forth new meanings within it. We have also seen the limits and difficulties of this conception: on the one hand, the enigma of primordial speech bestowing meaning into existence, and on the other hand, the irreducibility of the verbal significations to the sole powers of one's own body.

Surely, therefore, the idea of a verbal gesticulation emanating meaning in an equal measure as one's own body would lead to an impasse. There would, however, be some advantage in pursuing its examination, most specifically in the detail of its internal forms and of its *modus operandi*, which, at this stage of the presentation, remains undefined. It will then be a matter, abstraction being made of the very problematic principle of speech as drawing from one's own body, of finding in the articulatory forms of the verbal gesture a new formula of its immanent sense.

In essence, the verbal gesticulation proceeds on the one hand by way of interlacings and overlappings, and on the other hand, by way of convergence and condensation. To speak is to put into sequence, to superimpose and to progressively integrate a series of elementary verbal gestures which *in fine* install at their fore, as the focal point of the tensions which animate them, a certain signification: "The clarity of language is not behind it in a universal grammar we may carry upon our person; it is before language, in what the infinitesimal gestures of any [...] vocal inflection reveals to the horizon as their meaning."[104] In the exercise of authentic speech, it is therefore not a matter of serving again verbally encoded meanings, but of using words in a manner such as "The cross references multiply [and] more and more arrows point in the direction of a thought I have never encountered before."[105] In short, the meanings of speech are like "ideas in the Kantian sense[:] the poles of a certain number of convergent acts of expression which magnetize discourse without being in the strict sense given for their own account."[106] Then a moment arises when this coherent accumulation of punctual gestures, which are so many semantic adumbrations, ends up being crystallized and causing the rise of an object of meaning in consciousness: "Once a certain point in discourse has been passed, [the sketches (*Abschattungen*)] suddenly contract into a single signification. And then we feel that *something has been said*."[107] We can then say that thought will have been expressed "when the converging words intending it are numerous and eloquent enough to designate it unequivocally."[108] Language is therefore an "oriented

---

[104]*PW*, p. 28.
[105]Ibid., p. 12.
[106]*Signs*, p. 89.
[107]Ibid., p. 91.
[108]Ibid.

system."[109] This must be understood not as a system which carries internal dynamics which would make it tend towards certain of its possible states, but a system which tends to produce an "exterior" where it installs the subject as if in a new dimension of experience: "A language is […] a methodical means of […] constructing a linguistic universe of which we later say—once it is precise enough to crystallize a significative intention and to have it reborn in another—that it expresses a world of thought, as it gives it its existence in the world."[110] But if language installs the speaker in a world of significations which is external to the system it constitutes, it is in its interior that it configures this world: "Before language carries the significations […] it must secrete through its internal organization a certain originary sense upon which the significations will be outlined."[111] This returns us to the previously raised questions concerning the principle and nature of immanent gestural meaning.

It is by two means which converge towards the principle of diacriticity that M.-P. returns to these questions in his works ulterior to *PhP*.

The first line of questioning resumes the already explored path, that is, the one where one's own body is considered as the originary source of all significations. But it does so while introducing inflexions likely to be rather inconclusive albeit important in that they will contribute to the intelligibility of the solution retained by the second line of questioning: the solution of differentiality.

We will therefore return a few moments to the body and to the conception of speech as gesture, as M.-P. addresses them in *The Prose of the World*.

7.13 The Verbal Gesture—Animating Interiority

We first find a few arguments comparable to those developed in *PhP*. Thus, speech is a gesture inasmuch as the relation of the speaker to his or her verbal repertoire is similar to that of one's own body with the world surrounding it: "When I am actually speaking I do not first *figure* the *movements* involved. My whole bodily system concentrates on finding and saying the word, in the same way that my hand moves toward what is offered to me."[112] Furthermore, this "action at a distance by language" which installs meanings at the exterior point of words is "[an] eminent case of corporeal intentionality"[113]: "my consciousness of my body immediately signifies a certain landscape about me, that of my fingers a certain fibrous or grainy style of the object. It is in the same fashion that the spoken word […] is pregnant with a meaning which can be read in the very texture of the linguistic gesture […]

---

[109]Ibid., p. 88.

[110]*PW*, p. 31.

[111]Ibid.

[112]Ibid., p. 19.

[113]*Signs*, p. 89.

and yet is never contained in that gesture."[114] Also, "speech is comparable to a gesture because what it is charged with expressing will be in the same relation to it as the goal is to the gesture which intends it."[115]

But a few lines later, M.-P. goes beyond the first parallel he had established between speech and gesture following the principle of one's own body instituting a world of meanings echoing its own power, that is, the principle of one's own body as a "strange signifying machine"[116] which "itself sets up its own background"[117] in a world of which it was able to reflexively neutralize the first signifying stratum. He then considers not only a world of objects of meaning as though configured but a world of objects of which the constitution is never complete, in that these objects, as the backgrounds against which they become figures, are always sources of solicitations, of requests which demand complementary explorations, a new engagement of the embodied subject, and without the possibility of completion.

Let's recall, in this respect and in case it were necessary, that to "see" an object is not to acquire a representation of it on the basis of the visual impressions it produces, it is "either to have it in the margins of the visual field and to be able to focus on it, or actually to respond to this solicitation by part one focusing on it."[118] Thus, to see is to respond to a world's call to existence. And furthermore, to "look fixedly" at an object is to delve into it: It is to explore its interior horizon. It is in this manner that speech is embodied: As the world traverses me with its solicitations to which I respond by engaging the active power of my body in order to install object forms and qualities within it, in equilibriums which are always transitory, likewise, I am traversed in my subjective being by a quivering and vague tension of meaning, a tension which my verbal gesture will invest by its forms so as to shape it into meanings.

But, as the sensible world only takes shape and acquires value in regard to an active being, likewise meanings only exist as counterparts of the gestures produced in response to the signifying tension which motivates it: "Signification arouses speech as the world arouses my body: by a mute presence which awakens my intentions without deploying itself before them."[119] More specifically: "In me [...] the significative intention [...] is at the moment no more than a *determinate gap* to be filled by words."[120] And furthermore, situating oneself this time within the conception of a world of acquired meanings in which speech brings forth new meanings, "[In me, the significative intention is] the excess of what I intend to say over what is being said or has already been said."[121] What we will emphasize here

---

[114]Ibid.

[115]Ibid.

[116]*PhP/L*, p. 114.

[117]Ibid.

[118]Ibid., pp. 69–70.

[119]*Signs*, p. 89.

[120]Ibid.

[121]Ibid.

is that meaning is in relation with the verbal gesture in an "intentional" mode, and that the meanings towards which speech is directed "[is] our own taking possession or acquisition of significations which otherwise are present to us only in a muffled way."[122]

From these considerations, we will retain two things.

On the one hand, the issue of the body as originary source of meanings is maintained but broadened to all of its components: The verbal gesture is no longer the sole protagonist of the operation of expression; It is the active principle opening onto a world of significations (in an intentional mode) but not by virtue of its own one and only power. The verbal gesture is the embodied response to the "mute" tensions which inhabit the speaker which he or she transfigures into objects of meaning: "[T]he thematization of the signified does not precede speech [...] it is the result of it."[123]

On the other hand, the theme of the "emptiness of thought" as a tension seeking resolution and seeking existence through a verbal gesture finds itself to be specified here. The "speaking silence" which nurtures the act of speech is precisely this "muffled" and still undefined meaning-intention: "[S]peech [...] seeks to embody a significative intention which is only *a certain gap*",[124] or only a "speechless want."

7.14   Synthesis

In sum, at this stage of M.-P.'s reflection, the main elements of questioning surrounding speech are the following: *On the one hand*, there is the originary matter of a "speaking silence", which is in a way the internalized counterpart of the primordial surroundings of solicitations addressed to one's own body, and, *on the other hand*—and also similarly here to one's own body which runs counter to its still diffuse environment in a way that, by engaging active forms within it, and seeking a response which will validate them, one's own body institutes a world of sensible qualities and of forms which signify its engagement—there is the gesture of speech responding to a muffled significatory tension, engaging it according to its own forms and power in order, when the operation succeeds, when the involvement of the verbal body and the silence in quest of a voice validate each other reciprocally, to bring a world of meanings into existence.

We thus understand that meanings are not given *in* the verbal gesture, in the sense that they would be effectively present, but *by* the verbal gesture, inasmuch as this verbal gesture is but the appropriate response which a specific bodily power brings to a silence which pressingly interpellates it, and of which it exudes at its "extremities" objects of meaning, which are therefore the content of the gestural

---

[122]Ibid.
[123]*Signs*, p. 90.
[124]Ibid.

engagement giving voice to the silence which motivates it. We therefore also understand why, in a sense, "language never says anything; it invents a series of gestures, which between them present differences clear enough for the conduct of language, to the degree that it repeats itself, recovers and affirms itself, and purveys to us the palpable flow and contours of a universe of meaning."[125] According to this conception, meaning, as an intentional object of a gestural power responding to the solicitations of a silence "rustling with speech" by engaging its active forms within it, draws its existence and determinations from the signifying power of a verbal body echoing a meaning-intention in search of a world. One will note that the issue of corporeality has been shifted: It is no longer a matter of one's own body and of its power to install a concrete world, a power which we have seen to be unable to fuel the world of speech-embodied meanings, but rather a matter of verbal gesturality and specific correlative power.

However, this shift does not really bring a solution: The enigmatic significatory resource of one's own body is still always required here, and the problematic framework of an encounter between a gestural power and a soliciting environment is maintained. But, by narrowing the question around the specific forms of verbal gesturality, we open the possibility for more specific investigations.

7.15   The Verbal Gesture—Revisited

As an introduction, we will first carry further the consequence of what precedes, that is, regarding meaning as being "induced" by gesture: Meaning is not itself present within gesture. Which amounts to saying that the constituents of the verbal gesture do not contribute to the sense of a discourse by each conveying their own part of signification, despite their being reshapable for purposes of integration, but because all together they produce a convergence towards a certain meaning.

Also, the essence of speech is to be approached on a plane of functioning where its components are in rupture from meaning: A language is not a sum of words, "[it] is the configuration that all these words and phrases draw according to their use in the French language. This would be strikingly apparent if we did not yet know the words' meaning and were limited [...] to repeating their coming and going, their recurrence, the way they associate with one another, evoke or repel one another, and together make up a melody with a definite style."[126] The intersection with the problem of primordial speech is manifest here. The principle of primordial speech is indeed that of a power to signify by means of words prior to any conventional association between form and meaning, such as registered by sedimented language, therefore of a power to signify by means of words before their acquisition of meaning, and even before being polarized into signifiers and signifieds.

---

[125]*PW*, p. 32.
[126]Ibid.

But if the problems are better posed and specified, they are not however resolved. Indeed, at this stage, we can always do without an internal significatory resource: Regularity, overlaps, etc., are not sufficient in themselves to induce a focal point which is *external* to the system. All such plays of repetitions and of systematicity suppose, in order to produce an effect of convergence, an internal animation remaining to be elucidated. In all likelihood, the cumulated effects of the intersections and overlaps "suggest even more that the whole process obeys an internal order, the power of revealing [...] what [is] in mind."[127] But it does not suffice to *suggest*, because the suggestion that "all of this" is animated from within does not give access to the principle of animation which alone installs a world of meanings. At most, from such a "suggestion", we may derive a system of semantic representations accordingly to the relational distributions observed and following a modelizing approach. There is therefore an insurmountable gap between a convergence of systematic overlaps, which never exceeds its own order, and the convergence towards an exteriority, as an installation of a new dimension of experience.

It will therefore be necessary to examine in greater detail the texture of this systematicity and to attempt to detect an internal principle of animation. In short, it will be a matter of uncovering the principle which gives to the linguistic gesture the power of "secret[ing] through its internal organization a certain originary sense upon which the significations will be outlined."[128]

7.16   Towards Diacriticity

It is at this point that M.-P. refers to Saussure and to the mode of negative identity. Indeed, and thus prolonging Saussure, it is in *difference* that M.-P. fully situates the power to signify: Difference is the mode of existence of signification, or, in other words, to signify is to express differences. The elements of spoken language are such as "each of them signifies only its difference in respect to the others."[129] Thus, "as Saussure says, signs are essentially 'diacritical.'"[130]

Thenceforth, and as a direct consequence of what precedes, it is therefore in these linguistic elements devoid of signification and which function differentially that it is appropriate to locate the originary form and the essence of the verbal gesture. It will be a matter of *phonemes*, identities which are "oppositive, relative and negative", as "components of language which do not for their part have any assignable meaning and whose sole function is to make possible the discrimination

[127]Ibid., p. 33.
[128]Ibid., p. 31.
[129]*Signs*, p. 88.
[130]Ibid.

of signs."[131] Phonemes, which deliver "an inexhaustible power of differentiating one linguistic gesture from another",[132] constitute "the real foundations of speech, since they [...] by themselves *mean nothing* one can specify. But for this very reason, they represent the originary form of signifying. They bring us into the presence of that primary operation, beneath institutionalized language, that creates the simultaneous possibility of significations and discrete signs."[133] Beneath constituted language, it is therefore in the phonological systems that we will be able to locate "this primordial level of language [...] by defining signs, as Saussure does, not as the representations of certain significations but as the means of differentiation in the verbal chain and [...] in speech."[134]

This being, it is not assured that the matter of the opening of the linguistic system towards a signifying exteriority will be clearly resolved, the linguistic system being now understood as a fabric of differential relation. In fact, it has only shifted again, now taking the form of the inextricable problem of Saussurean value (cf. 5.2.4), that is, the problem of the overlapping of "horizontal" relations between signs and "vertical" relations between the signifier and signified within a same sign. The problem encountered by "diacritical" phenomenology is by all means similar: How can the "lateral liaison of sign to sign [operate] as the foundation of an ultimate relation of sign to meaning [?]"[135] What enables to assert that "it is because the sign is diacritical from the outset, because it is composed and organized in terms of itself, that it has an interior and ends up laying claim to a meaning [?]"[136]

It would be an error to think that all this progression could have been avoided, although in the end, we encounter problems which have already been clearly formulated and to which much effort has already been devoted. This is because if the terminal problem is practically the same, the problematic framework where it resurfaces is entirely different. And whereas within the perimeter of Saussurean thought, this problem appeared insurmountable (although the morphodynamical approach gives a partial solution), we can hope that in the framework of Merleau-Pontian thinking and by embracing its perspectives, it may be surmounted.

7.17   Diacriticity and Differentiality

In order to do this, it will suffice to observe that Saussurean differentiality does not overlap with Merleau-Pontian differentiality which is called "diacritical". The relation of difference which Saussure views between phonemes or between

---

[131]*IL&VS*, p. 77.

[132]*PW*, p. 33.

[133]Ibid.

[134]Ibid., p. 31.

[135]*IL&VS*, p. 77.

[136]Ibid., p. 78.

signifiers (cf. 5.2.1.1) is a relation which operates within substances, respectively substances of expression or of content. We are therefore faced with a scenario where a form instructs a homogeneous diversity and which, though being differential, accomplishes syntheses: The boundary which establishes two negative identities in a substantial continuum accomplishes a "negative" synthesis in that it stipulates an identity which is "everything that the others are not." More specifically, whereas in positive synthesis, the conceptual unity subsumes a certain manifold of concrete occurrences, in negative synthesis, the differential unit subsumes "contrastively" the manifold of occurrences from which it distinguishes itself. The Saussurean approach therefore situates itself on a plane which recognizes sets of homogeneous occurrences (the substances), that is, the partially constituted experiential dimensions which are therefore subject to conceptual or differential syntheses. Correlatively, the relation of difference, if it authentically institutes negative identities, comprises a positive element in that the syntheses which it commands concern occurrences of substance which are accessible alternately. That is, the relation of difference is not absolutely first here: It is first relatively to the identities which it institutes, but it supposes the prerequisite and the support of a substance which it does not configure in itself.

It is here that Merleau-Pontian diacriticity radically distinguishes itself from Saussurean differentiality. This is because diacriticity does not fit into a logic of synthesis, even negative, and, correlatively, does not operate upon homogeneous substances which symmetrically share, as if facingly, the primacy of form.

The diacritical gesture, according to a first approach, creates a gap not between occurrences that were previously available and adjacent (in an appropriate geometry), but a gap which, put in approximate terms, institutes at its poles the analog of a figure and ground as given by Gestalt theory. As is the case for Gestalt, diacriticity is a regime of originary constitution, the principle of a spontaneous organization of a perceptual consciousness, and not a regime of synthesis governing, under any kind of conceptual, memory-based, or deductive authority, the direct passage from the sensorial manifold to a picture of things perceived. But, moving further ahead than does Gestalt, and introducing here the very first terms of its existential conception, the figure/background structure installed by the diacritical gesture is not reducible to the principle of a constitutive creation of salience upon an undifferentiated and inert background. This is because diacriticity is linked with the "originary rustling silence", with the "emptiness of thought" prerequisite to any speech and awaiting the words which may fill it. Without, however and of course, the diacritical gesture being under the same relation of subsumption with respect to the "lack of meaning" as is form with respect to substance.

In order to better approach this fundamental aspect of diacriticity, we may first note that what is immediately salient in this first evocation is that diacriticity exceeds the strict domain of speech. In fact, in the last stages of his reflection, restituted in the 1953 *Lecture at the Collège de France* (Merleau-Ponty 2011), we see a radical extension of the concept of *diacriticity*, which henceforth operates at all levels, with respect to all regimes of feeling and of body-motion *inasmuch as meanings are woven within.* Therefore, the meanings which inhabit sensible things, meanings which express the adjustments of one's own body with respect to its surroundings, and those which emanate from the verbal gestures are drawn closer together under a shared principle. The idea of a perception which would be diacritical through and through, does not therefore distinguish itself, in *The Sensible World and The World of Expression*, from that of a perception which is immediately an expression.

The diacritical is therefore to be understood intuitively and in a first stage as a "distance with respect to a certain level", the grasping of a salience against a certain background. But it is, so to speak, an "existential grasping of salience"—in that this general structure of perception/expression articulates in its own order a background of matter which is indecisive and soliciting, an originary environment no longer posited and thought as such as being anterior to its grasp by the vital power of a bodily schema, but relatively to the "promotion" of a thing thematically perceived or expressed, then taking position as a figure. The notion of *background,* a gestaltist legacy, is thus within this new device radically reworked and shifted. Beyond and beneath the sensitive surface of the field, it opens onto a very broad notion of a background of solicitations, motivations, horizons, and of praxis...

This conception of *diacriticity* goes hand in hand with a dynamic aspect of manifestation, as a "modality of upsurge" of the figure against its background. M.-P. speaks here of *modulation:* "[A] perception has a sense, not as being subsumed under an essence or signification participating in an idea or in a category—but as a *modulation* of a certain *dimension.*" This is the case with the example of the circle, of which the circular physiognomy can be apprehended, among many other ways, against the "background" of a constant deviation of the curve relatively to its tangent in each of its points, or, alternatively, on the basis of a radiance, of an even deployment from a central point.

M.-P. insists on the profound dissymmetry and complementarity of the two moments of this structure: "Any consciousness of a figure is thus a consciousness without knowledge of the ground." But "the ground is not simply a stage of confused perception, a context of ulterior perceptions. It is of another order, that by which the world is present through our action".[137] We deduce the principle of an underlying lack of knowledge, of an ambiguity or an incompleteness traversing the apprehension of any figure: "[C]onsciousness peers", says M.-P.—meaning that the figure is seen only because it remains carried by other latent or virtual forms, which constitute something akin to unseen points of reference or correlates, susceptible to manifest in turn.

Diacriticity therefore appears as the primary form of *differentiation*: Not as the opposition of two *terms* pertaining to a same order of individuation or of formality (on the logico-algebraic model of A vs. B), nor as the differential demarcation of two sets of occurrences participating in a shared substance, but a *gap* between a figure and its ground, a dialectic relation between fundamentally heterogeneous entities in that the one carries its part of plenitude, albeit always incomplete, in

---

[137]Merleau-Ponty (2011, p. 141).

contrast to the other which is this void "of reference" remaining to be filled, this "invisible background which composes all possible senses and makes possible the genesis of meaning from the diacritical relations between differential signs."[138]

7.18 The Diacritical Solution

It is thus that the principle and possibility of primordial speech reveals itself. The sign, in its diacriticity, indeed carries within itself the fundamental "ingredients" for coming into sensible and signifying existence: The sign articulates in its own body, which is a diacritic body, a deployment through a figure which serves as a response to a "poorly formulated question", that is, a response to the "muffled wish" of a "rustling silence" awaiting existence and qualification—but without this figure ever being separated from the background from which it emanates and towards which it always orients consciousness as if towards the resource of its expressivity: "Meaning is always the ground of a certain figure offered to us."[139]

That is to say that the diacritical sign comprises an interior: a fabric of signifying shortcomings, a reservoir of possible actualizations, of which the figures will be the manifest solutions, albeit temporary and open.

That is also to say that diacriticity carries its meaning, albeit laterally. What it delivers at the forefront, the figure, is inhabited by that from which it proceeds and incessantly solicits in search of new figures, that is, by an originary background.

Diacriticity thus appears as a local model of the body/world relation, and, more generally, of any expression and perception. And it is in this mimetic faculty that its enigmatic principle of exteriority resides. Indeed, as we have seen, the mystery of the verbal gesture is that of lateral relations, between signs, which lead them outside of themselves; it is also the mystery of primordial speech which will have seen meaning "exude" from a simple verbal gesticulation.

Already, we have seen, signs dwell amidst an existential theater, within which the advent of a world is played out. But there is also the fact that diacriticity is a structure which, beyond expression, serves as principle for all perception. Diacriticity thus appears as the shared and living fabric where signifying forms and perceived forms, and more generally, experienced forms, can superimpose one another, interpenetrate one another, and even be equivalent.

This is because verbal gestures, in that they deliver complexions of figures and backgrounds, of presences and of horizons, of determinations and of vague solicitations, are likely to "mimic" in their very forms of existence the objects of perception or of subjectivity, which therefore respond to the same internal regimes of animation.

---

[138]Kearney (2013, p. 189).
[139]Merleau-Ponty (2011, p. 179).

Since we are dealing with the question of primordial speech, we should recall that it "was not established in a world without communication, since it emerged from forms of conduct that were already common and took root in a sensible world which had already ceased to be a private world"[140] and that it may have restituted in its body the living range of the primitive conducts which progressively took place. Thus, at the origin of sedimented language, at the source of the correspondences established between a few memorized gestures and their external designations, we will foresee diacriticity as a general principle of the advent of various dimensions of experience. Moreover, it is this view of an interweaving of the fluxes of experience, an interweaving which is supported by a shared diacritical form, which constitutes the natural horizon of the elucidation of the relations between perception, action, and language—which, as developed by Visetti,[141] are therefore to be approached in terms of gears and interpenetration, in sum, as a *coupling*, between a diversity of regimes of semiotization simultaneously inhabiting the flux of experience.

7.19   Conclusion and Synthesis

At the term of this chapter devoted to M.-P., it is fitting to return to our general perspective, which articulates at its center the Saussurean, Husserlian, and Merleau-Pontian systems of questioning. In what concerns the correspondences between the Saussurean morphodynamic device and the Husserlian phenomenology of the strata of verbal consciousness, they have already been covered (cf. 6.4).

At this stage, it is the relations between existential phenomenology and Saussurean structuralism which need to be discussed. The relation between these two approaches does not have the form of a morphism (correspondence between objects, relations, and levels of structure), as was the case for the views of Saussure and Husserl. The reason is obvious. Whereas the epistemic perspectives of Saussure and of Husserl overlap, such is not the case with M.-P. Saussure and Husserl share the objective of an explicit and regulated determination of the forms of their object of study: the sign in its regimes of empirical functioning, for the one, and the sign in the structures of its appearing, for the other. We have seen how these approaches ultimately overlap, so as to also confirm the views of M.-P. who foresaw a convergence between phenomenological and scientific researches: "As soon as we distinguish, alongside of the objective science of language, a phenomenology of speech, we set in motion a dialectic through which the two disciplines open communications [...] the 'subjective' point of view envelops the 'objective' point of view [and reciprocally]."[142] Furthermore, "the phenomenon of expression belongs both to the scientific study of language and to that of literary experience,

---

[140]*PW*, p. 42.
[141]Cadiot and Visetti (2001).
[142]*Signs*, p. 86.

and […] these two studies overlap. How could there be a division between the science of expression […] and the lived experience of expression […]? Science is not devoted to another world but to our own; in the end it refers to the same things that we experience in living."[143] It is true, however, that the theory of the strata of verbal consciousness does not aim to elucidate the living forms of language, those which accomplish themselves in order to fill a gap in terms of thought and to install a world of meaning. The overlapping of Saussure and of Husserl, as instructive and rich as it may be in terms of consequences (notably epistemic), is therefore not exactly that contemplated by M.-P. between a "science of expression and an experience of expression".

In order to achieve this end, as we have seen, M.-P. referred to Saussure, but at the cost of certain conceptual distortions, which are not so much abuses as they are enhancements, and which enable to measure the gap between a phenomenology of speech and a science of language and, at the same time, to specifically situate the objectifying pretensions in terms of linguistics, within their limits as well as in their possibilities.

To pass from the negative sign (Saussure) to the diacritical sign (M.-P.) is to situate oneself upon both a pre-objectal and pre-qualitative plane, that is, within the originary envelope of solicitations, and to instruct a regime of upsurging of a figure as a response to these solicitations which nevertheless remains attached to them in their significations and their prolongations. We have seen that this diacritical structure captures an essential part of what M.-P. recognized as a source of ensoulment for the living world. We have also seen how diacriticity, as it commands any sort of perception and of expression, as it constitutes their shared vital forms, opens the possibility for an assimilatory superimposition of the various dimensions of experience—hence the miracle of primordial speech, in which the signs have the capacity to exceed the system which institutes them. Thus, the diacritical is in part the essence of a living language and a speaking world.

We therefore conceive that the reduction of the diacritical to the differential empties the sign of its originary and vital capacities. This is because it consists precisely in transforming the background of originary solicitations, in which are prepared the existences of a world, into substance, that is, into a homogeneous set of inert occurrences, and, correlatively, in restricting the dynamic mode of a "surge" of salience in the categorization of substance by means of a set of differentiating thresholds. Thence devitalized, the background becomes mute, and even disappears when invested by the differential form.

We have therefore passed from a structure of animation to a structure of organization, in other words, from a structure which institutes a world of expression into a structure which configures a universe of objects of knowledge, because in the latter scenario, the "response" to the animating force is reduced to the instantiation of a form within a substance.

---

[143]*PW*, p. 15.

And because this formalizing conversion abandons the greatest part of linguistic power, namely the power of opening onto its exteriorities, it will indeed be necessary for the device employed again at this level of determination to be "axiomatically" enhanced with "exchange" connections which it had lost. Also consequently, it is sedimented language, in which the form/meaning connections are already available, that becomes the legitimate object of linguistic objectivity. And this is exactly what is accomplished by the Saussurean morphodynamics of the sign, which we have seen to promote the relations of "exchange" into relations of control. In other words, it finds support in connections instituted between forms and significations, connections which are sedimentations, in order to instruct the indivisible unity of the sign through the differential constitution of signifieds. Likewise, as has already been emphasized, the systemic constraints constitutive of a sedimented language are rendered in the morphodynamic model on the mode of a stabilization path conditioning the emergence of boundaries and the existence of signifieds, which respond to the differential of admissibility.

To conclude, we can say that the passage from the diacritical to the differential constitutes a "good" reduction to what can be known of the animating forms at work in verbal gesturality. The reason for the epithet "good" is at least twofold.

Firstly, the passage from the diacritical to the differential maintains the order of difference. Morphodynamic modelization therefore respects the formal nature of the regimes of the activity of speech. Then, if the "lateral" relations (differential, between signs) do not overlap with the "vertical" relations (between signs and meanings), and that the elementary verbal gestures (the phonemes) do not intrinsically carry the power to signify, the morphodynamic approach enables however to "recuperate" the indivisible unity of the signifier and signified, albeit in a dissymmetrical mode. We may simply recall that the existence of signifieds is controlled by signifiers: The former are therefore inseparable from the latter. The Saussurean morphodynamics of the sign therefore appears to be an "appropriate" conceptual rectification of the signifying powers at work in expressive life, under all of its forms.

But there is still another angle, and not the lesser among them, though it does intersect with the preceding ones, under which the Saussurean morphodynamic model finds an existential rationality: that of semiogenesis. We have seen that among the main phenomenological characters of the activity of speech, which on the long run install a sedimented language, figures *consummation* (correlatively to *sedimentation*).

Thus, the expressive fact, that is, the sensible presence of meaning, is traversed with consummating propensities, which polarize expression into a signifier and signified in order to, *in fine*, crystallize into consciousness a pure content of thought, on the one hand, whereas on the other, the word, devitalized, returns to a state of inert vector. If the expressive fact escapes the forms of the morphodynamic model by direct consequence of the plane of intelligibility retained, conversely, the operation of consummation (and of sedimentation) which causes to pass from a sign which is polarized (as signifier and signified) to elements of substance of autonomous expression and content, indeed figures in this model, as shown by its phenomenological interpretation and its functional composition.

We have seen indeed that the various functional planes of the morphodynamic device successively receive phenomenological meanings in terms of verbal consciousness: of availability, of engagement…up to thematic consciousness (of the signified), and, still furthermore, in its logical prolongation, the consciousness of fulfillment. Yet, we have seen that the operation of fulfillment consists in selecting an element of the substance of content by detaching it from the control function (of the dynamics) it receives within the morphodynamic framework. In other words, fulfillment consists in neutralizing the differential device which institutes negative identities (the signifieds), and, correlatively, in retaining from the control function (complex) only the relation of "exchange" which is its first component. We thus retrogress here from the "meaningful" sign (merging a signifier and a signified) to an "indicative" sign (conventionally linking an identity of substance of expression to an identity of substance of content). This is indeed a case of consummation.

References

Benveniste, E. (1971). *Problems in general linguistics*. (M. E. Meek, Trans.). Coral Gables: University of Miami Press.

Cadiot, P., & Visetti, Y.-M. (2001). *Pour une théorie des formes sémantiques: motifs, profils, thèmes*. Paris: PUF, coll, Formes Sémiotiques.

Kearney, R. (2013). Écrire la chair: l'expression diacritique chez Merleau-Ponty. *Chiasmi International, 15,* 183–196.

Merleau-Ponty, M. (2011). *Le Monde sensible et le monde de l'expression*. Notes du cours au Collège de France, texte établi et annoté par E. de Saint Aubert & S. Kristensen, Genève: MétisPresses.

Rosenthal, V., & Visetti, Y.-M. (2008). Modèles et pensées de l'expression: Perspectives microgénétiques. *Intellectica, 3*(50), 177–252.

Chapter 8
Neurophysiological Homologation

8.1 Introduction

In this chapter, we propose to put the previously obtained results (theoretical elaboration and adjacent theses) to the test of the neurosciences, in particular, to the test of electroencephalographic (henceforth EEG) observation.

We will first show that the circumstances of generation of the "evoked potential" N400 favor its phenomenological interpretation (cf. 8.2.9 *sq.*), precisely in intentionalist terms: We will see that the amplitude of the N400 indeed lends itself to be approached as a measurement of the "unfolding" within consciousness of a meaning-intention. But if the N400 lends itself to phenomenological interpretation, and offers in return an empirical guarantee regarding the Husserlian theory of the strata of verbal consciousness, it is also the morphodynamic device which is validated, in that it delivers the functional architecture of a phenomenology of the sign. Now, we know that this morphodynamic device also claims to account for the forms of semiolinguistic objectivity as they are conceived by Saussurean structuralism.

We have already emphasized (cf. 1.2.2) the singularity of this epistemic conjuncture which, in all of its generality, is the following: The plane of neurobiological observation where the phenomenological component of semiolinguistic knowledge finds its empirical support is also likely to deliver data establishing the experimental validity of the forms of this kind of knowledge. In sum, the neurobiological data constitute a basis of empirical validation as much for semiolinguistic phenomenology as for a "theory" delivering the forms of semiolinguistic objectivity.

In such a scenario, the semiolinguistic theorization would then have two empirical foundations: the one being provided by its phenomenology, which has the value of an "auxiliary system", in other words, of an "observatory" (cf. 1.2.1.3), and the other being provided by neurobiology. We will nevertheless acknowledge,

© Springer International Publishing AG, part of Springer Nature 2018   197
D. Piotrowski, *Morphogenesis of the Sign*, Lecture Notes in Morphogenesis,
https://doi.org/10.1007/978-3-319-89848-3_8

at the risk of annihilating the unity of empirical semiolinguistic knowledge, that the forms of semiotic knowledge grounded in semiolinguistic phenomenology must be the same as those founded upon the neurosciences—a condition which will dually imposes a homologation of the forms of phenomenality by neurobiological data.

Now, as we have anticipated, this condition will be satisfied. But a new difficulty arises, one stemming from the possession by the morphodynamic device of an objective value to the same extent as a phenomenological signification. The whole question is then to know if this overlapping of the orders of objectivity and of phenomenality is strict or inclusive.

In the first scenario, which is eminently problematic, the orders of phenomenality and of objectivity perfectly overlap, and it is then of organicity and of teleology of which it is question, as addressed in the second part of Kant's Third Critique. We may indeed recall that the comprehension of morphologies as holistic conformations which are autoregulated and morphogenetic, and, more generally, the fact of an intrinsic significance of such morphologies, require the modality of a reflective judgment, therefore the recourse to a teleological principle—a principle which, in its greatest generality, constitutes the response to the problem of the contingency of forms and natural laws (with respect to their necessary unity, then transferred to a "superior" plane of understanding). It is that in the order of teleology, we should also recall, all happens "as if"[1] the *idea* of a totality was the *efficient cause* of a concrete form, as determining its constitutive parts and their specific interlinks. The idea of a totality is then as much the effective source as the manifest sense of the phenomenon which empirically accomplishes it and within which the orders of the sensible and of the intelligible thus overlap: A morphology is significant in that it delivers its concept in the very forms of its appearing. We may finally recall that, more radically because liberated from the "*als ob*" referring to a hypothetical superior understanding, the principle of a community of existence of the concept and of its achieved form finds itself to be at work in Goethean entelechy, as an "intuitive concept and effective idea."[2]

The second scenario is the one where the forms of objectivity participate in the forms of phenomenality. In such a scenario, then, there is no longer strict assimilation of the sensible and the intelligible in the body of a morphology which is significant as such, but the implication of forms regulating phenomena (that is, the forms of objectivity) in the very order of this phenomenality.

Only neurobiological experiments can decide which of these two cases holds. The first scenario supposes that the neurobiological correlates of the observable functionings on the plane of phenomena, and which a semiolinguistic science applies itself to account for, are quite exactly the same as those attached in their own right to the constitution of such phenomena.

This second scenario will be established under two conditions. On the one hand, it is required for the neurobiological correlates of the forms of linguistic phenomenality to be doubled with correlates which manifest the functionings of

---

[1]"Als ob."

[2]Petitot (1985b, p. 33).

semiolinguistic phenomena in compliance with the order of the system which objectivates them. On the other hand, the theoretical system needs to adequately account for these various neurobiological correlates.

We will now show that it is this second scenario which is attested by means of experimentation. In order to do this, we will proceed in three stages. First (Sect. 8.2), we will present a brief reminder concerning the N400 "evoked potential" and we will argument in favor of its phenomenological signification. Then (Sect. 8.3), we will introduce the "syntactic" potentials (LAN and P600) and we will show that the circumstances of generation of these two waves as well as of the N400 (then interpreted in a "classical" manner as a marker of semantic activity) invalidate the break between syntax and semantics. Finally (Sect. 1.4), after having proposed a "generic" interpretation of the LAN and P600 waves, we will show how the morphodynamics of the sign accounts for these evoked potentials and precisely in what manner, in a reciprocal assimilation of the semantic and the syntactic, it interweaves the orders of semiolinguistic phenomenality and objectivity.

Regarding this latter point, and anticipating somewhat, we have seen (cf. 1.3.2) that the normative constraints on semiolinguistic combinatorics are, in a morphodynamic approach, rendered under the form of conditions upon stabilization paths of singularities, paths which institute at their end differential boundaries in a substance of content. Now, as these are boundaries which establish signifieds (which are negative identities), and which, consequently, produce the sign-phenomena, we readily conceive that the forms of semiolinguistic objectivity (i.e. the forms of the semiolinguistic system), as they administer semiolinguistic morphogenesis, participate intrinsically in the forms of its phenomenality. It is precisely this which must be empirically and meticulously established.

8.2 The N400

8.2.1 Introduction to EEG Observation and to the N400

8.2.1.1 Introduction

Very schematically, EEG observation consists in recording upon the scalp potential differences relating to currents indirectly induced by neuronal activity: The electrochemical process which triggers the nervous influx indeed produces a disequilibrium of charges at the synaptic interface, a disequilibrium which elicits compensation currents in the extracellular space (cf. Piotrowski 2009a, Appendix 1 Sect. 2.3). The neuron can be modelized as an oriented dipole (following the dendritic-synaptic axis), and when the disposition of the neurons is favorable (axes having a same orientation), their potentials add themselves to one another and achieve observable values (of the order of the microvolt). By placing captors at the surface of the scalp (positioned following international norms), we can record at the position of each captor the "natural" electrical activity or that which is induced by the presentation of stimuli of various types. When referring to EEG responses to stimuli, we speak of "evoked potentials" (or ERP: event-related potentials).

8.2.1.2 Qualification of the Evoked Potentials

The EEG observation data are A = f(t) plots which relate variations over the course of time (t) of the amplitude (A) of the potential gathered from various cranial sites (Fig. 8.1).

The amplitude of the wave recorded is measured with respect to a reference electrode, and we elect to offset the plot following a baseline having for value the average of the potentials recorded a few milliseconds prior to the presentation of the stimulus. Furthermore, the origin on the axis of time is set at the moment of the presentation of the stimulus.

We choose to orient the axis of negative amplitudes "upwards", and "naturally" describe the EEG tracings as a succession of waves encoded under the form Nxxx or Pxxx, where N (resp. P) indicates negative polarity (resp. positive) of the crest (resp. bottom) localizing this component, and where xxx designates the latency of the wave, that is, its culminating moment (abscissa of the crest or of the bottom on the axis of time) with respect to the temporal beginning of the measurement. For example (Fig. 8.2), the "N400" is a negative wave of which the peak is located around 400 ms after presentation of the stimulus.

Moreover, each wave is specified by the location of the zone of the scalp where it has been recognized to be dominant. For instance, the N400 tends to have a centro-parietal distribution (as shown in Fig. 8.4).

Fig. 8.1 Electrode position from the international 10–20 system on the cap used for EEG recording and delimitations of regions for statistical analysis (*LF* Left frontal, *LT* Left Temporal and *LP* Left Parietal; RF, RT and RP, homologous regions on the *Right*; *FC* Fronto-Central; *CP* Centro-Parietal (Thierry et al. 2003, p. 538)

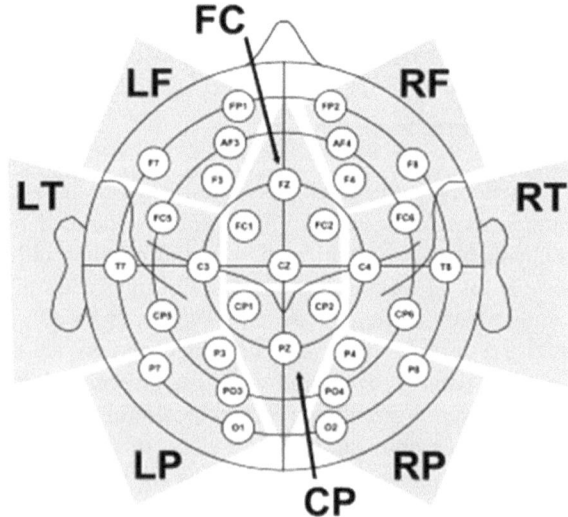

Fig. 8.2 The N400
component of an event-related
brain potential

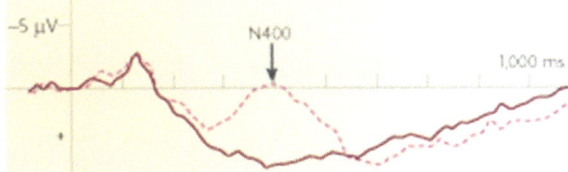

8.2.1.3 Difficulties

It is important to mention a few practical difficulties pertaining to EEG observation
and experimentation.

The high temporal resolution of EEG observation enables a detailed analysis of
the waning of cerebral process. However, its power to locate (spatially) generative
(primary) sources is very weak ("inverse problem"). It indeed appears that the
volumic currents recorded are very sensitive to the conductive configurations of
their environments. Thus, in the traversal of the layers having distinct properties of
conduction (and of which little is known)—successively: cephalorachidian fluid/
bone and bone/skin—the lines of current undergo major deformations which are
translated by a diffusion of the observable potentials (most particularly, at the level
of the bone of the cranium whose conductivity, which is weaker ($\sim 1/80$) than that
of other tissues, is anistropic). It may, furthermore, occur that the potentials
resulting from distinct generators are superimposed and confounded.

But the difficulty of spatial "reconstruction" of primary sources does not have for
sole cause the physiological complexity of the group formed by the brain and its
enveloping tissues.[3] To this, must be added that (i) on the basis of the equations of
electromagnetism, (ii) of the data of potentials recorded, and (iii) in the framework
of a model of generators as a distribution of dipoles, the "inverse problem" posed is
under-determined: It does not admit a unique solution. Hence the necessity of
introducing constraints upon the form of solutions: on the number and spatial
distribution of the generative dipoles. However, limitations persist inasmuch as, in
order to obtain a unique solution, the number of parameters to be determined[4] is
bounded by the quantity of data recorded. Also, for example, "EEG data gathered in
the international assemblage using 20 electrodes may not be explained with more
than three dipoles."[5]

In order to overcome these difficulties, we will combine EEG observation with
other methods, namely MEG, which observes the magnetic fields generated by
dipoles and which proves to be complementary to EEG, for three main reasons:
(i) the origin of the magnetic signal (primary currents) is the same as that of cellular
currents which interest EEG, (ii) as the magnetic field is not perturbed by the

---

[3]A complexity which also makes the "direct problem" (generators → surface potentials) very
difficult to address.

[4]6 per dipole: position (3), orientation (2), amplitude (1).

[5]Renault and Garnero (2004), Sect. 3.2.

Fig. 8.3 Functional regions
involved during
lexico-semantic processing
within the generation of the
N400

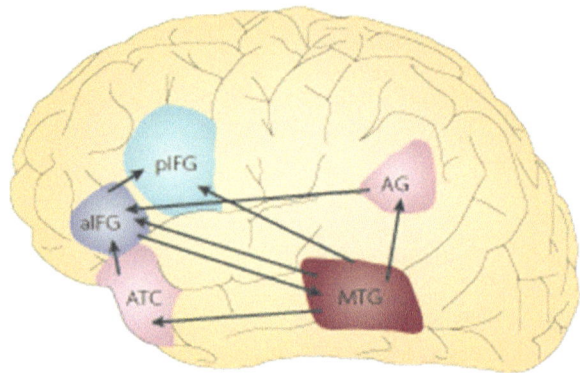

conductivity of tissues, MEG offers quite superior spatial resolution, and (iii) the
attenuation over distance of the amplitude of the dipolar potential being lesser than
that of the dipolar magnetic field (respectively $1/d^2$ and $1/d^3$), EEG enables a better
observation of deep electrophysiological sources. The fact remains that a precise
localization of the MEG and EEG sources "can only be obtained if a small region
contributes to the MEG or EEG data (which is rarely the case)."[6]

In what concerns the N400, and in the actual state of knowledge, the various
functional regions supposedly involved during lexico-semantic processing in the
generation of this potential are, essentially, the following[7]:

– Mental Lexicon: middle temporal gyrus (MTG), superior temporal sulcus, and
  inferior temporal cortex
– Integration of new information: anterior temporal cortex (ATC) and angular
  gyrus (AG)
– Operations of controlled memory retrieval: anterior inferior frontal gyrus (aIFG)
– Operations of selection (cf. 8.2.5): posterior inferior frontal gyrus (pIFG)
  (Fig. 8.3).

Being a matter of observing the EEG response to specific stimulations or to
specific cognitive tasks, other difficulties must be considered.

The first stems from the existence of spontaneous[8] neuronal activity of which the
intensity (of the order of the tens of μV and higher) is superior to the EEG
expressions of the cognitive processes (of the order of the μV) and which therefore
overlaps the latter.

The electrical manifestations of cognitive processes, called "evoked potentials",
and which EEG seeks to highlight, are therefore drowned in a mass of activations to
which are often added parasite contributions—for example, the signals generated by
various sensorial stimulations (odor, noise...) or by muscular activity (such as eye

---

[6]Ibid., Chap. 1.

[7]Lau et al. (2008).

[8]The *alpha, beta* rhythms, etc.

movements). Pathological manifestations must also be taken into account. The extraction of evoked potentials will therefore involve statistic processing.

The procedure consists in repeating numerous times the task defined. The signals thus produced, and offset with respect to a temporal origin which is the moment of the presentation of a stimulus, are then summed and averaged. The components of the signal which are synchronized with the stimulus overlap and reinforce one another whereas those which are independent from the stimulus are dispersed and their sum tends towards zero. We will generally proceed with two calculations: first, the average of the stimuli for each individual, then the average for the set of participants in the experimental trials ("grand average").

This methodology is not unquestionable. Firstly, when it is a matter of complex cognitive processes, we may legitimately question the validity of the hypothesis—upon which the method of extraction of evoked responses rests—of a quasi-mechanical systematicity[9] of an individual's responses to a same type of stimulus.

Indeed, *on the one hand*, we cannot elude possible interferences with learning or habituation mechanisms: As has indeed been observed, when we repeatedly submit an individual to trials where regularities between "primary" and "secondary" stimulations can be noticed, the subject may develop a state of habituation and anticipation which can manifest as the appearing, upon the presentation of a "primary" stimulus, of a specific EEG wave (the CNV: "contingent negative variation") which translates the cognitive state of wait for its associated "secondary" stimulus: "The CNV is a paired stimulus situation wherein an imperative stimulus follows a warning stimulus by a fixed interval."[10]

On the other hand, certain cognitive faculties are likely to involve processing attitudes which vary from one subject to another—putting into question the relevance of the interindividual "grand average". We will evoke in this respect an experiment[11] where the analysis of the interindividual average concluded that there is a biphasic linguistic process, i.e. manifested by a wave comprising two components recognized to respectively express (very schematically) semantic and syntactic operations. However, a more attentive examination demonstrated two types of individual responses: the ones comprising only the first component, and the others only the second one.

Finally, one of the most fundamental difficulties stems from the fact that the experiments are not neutral with respect to the choice and composition of the data

---

[9]As emphasized by J. Vion-Dury (personal communication), "we usually postulate a stability of the response outside of any logic of habituation, of plasticity, of distraction, of demotivation...".

[10]More specifically: "The prototypical CNV paradigm is a paired-stimulus situation wherein an imperative stimulus follows a warning stimulus by a fixed interval; a negative potential, the CNV, develops during the interstimulus interval. The CNV has been shown to increase in amplitude with increases in motivation and task complexity [...] The earliest formulations of the cognitive events underlying the CNV characterized these as anticipation or preparation for incoming information" (Van Petten and Kutas 1991, p. 105).

[11]Osterhout (1997).

submitted to the subjects. It is that the homogeneity of the stimuli presented, which is supposed to guarantee the similarity of responses, and therefore their additivity, does not proceed from their own fact, but indeed from their description from the perspective of a selected theoretical framework. Quite particularly, as the repetition of a same stimulus modifies its cognitive processing, it is necessary to present stimuli which are different but of the same type (paradigms). Yet, in all evidence, such paradigmatic series are not "raw" empirical data, but indeed theoretical constructs. EEG observation is thus revealed to be dependent on the choice of certain descriptive categories, which will determine the paradigms of experimental data and, consequently, the EEG "observables" themselves. This risk of theoretical bias, quite real, remains limited, however, because if these classes of stimuli are defined following parameterings which are specific to a certain theoretical apparatus, the ensuing EEG response is not in its neurophysiological characteristics qualified following the terms of the theoretical device at the source of the experimental set-up. Thus, neurophysiological observation can legitimately claim validity an as "auxiliary" observation device (cf. 1.2.1.3).

Being a matter of the N400, and as the functional signification which can be attributed to an EEG response is conditioned among other things by the nature of the stimuli (or of the variations of stimuli) which are at the source of it, it will be appropriate, in a first stage, to examine which stimuli, or which variations of stimuli, are at the origin of this EEG signature.

8.2.2   The N400

8.2.2.1   History and Generalities

It is in 1980, in an article[12] published in *Science*, co-signed by Kutas and Hillyard, and entitled *Reading senseless sentences: brain potentials reflect semantic anomaly*, that mention was first made of an "N400 effect", that is, a *variation* of the EEG activity within a window of 400 ms correlated with the presentation of words occupying the final position of a same sentence and alternatively appropriate or inappropriate to their context in terms of their meaning. This variation presents itself as a significant increase in negativity within the window of 300–500 ms, precisely under the form of a wave beginning around 250 ms, reaching its peak at approximately 400 ms and disappearing around 500–600 ms [cf. Kutas et al. (2006), and for a panorama Kutas and Federmeier (2011)]. This wave, the N400, has a centro-parietal (posterior) dominance and presents a light preeminence in the right hemisphere when the presentation is visual. As an illustration, the following Fig. 8.4 shows in different points of the scalp the EEG responses emitted with a lexical item ("bake", resp. "eat") which completes in a terminal position the

---

[12]*Science* (1980: 207), pp. 203–205.

sentence "The cats won't…" The dotted lines (*resp.* continuous) are those obtained in the case of semantic inconsistency (*resp.* consistency) with respect to the context.

As the title of this historical article clearly indicates, the N400 was readily presumed to reflect processing of a semantic nature. This interpretation was then largely admitted, as attested by the terms usually employed in order to qualify, when a N400 effect is recorded, the relation between the manipulated stimuli and their contexts. For example, while giving an account of the discovery of the N400, one researcher[13] reported the existence of words (stimuli) which are semantically appropriate or inappropriate to their context and another researcher[14] reported words which are semantically congruous (or not). A third researcher[15] reported semantically anomalous words, whereas yet another[16] reported words which adequately coincide or not with the remainder of the sentence, etc. Also, as summarized by the following citation, it seems to be understood that behind the N400 effect are cognitive processes which have to do with the sense of the units which they concern: "Kutas and Hillyard (1980) demonstrated that electrophysiological recordings of brain activity covary with meaning-related manipulations to language stimuli."[17]

We will note that the N400 effect has been attested for all languages in which experiments have been conducted, regardless of the modality (visual, auditory, sign language…) of the presentation of stimuli. However, the modalities of presentation have the effect of modulating the morphology of the N400. Thus, it is acknowledged that "the N400 is larger, begins earlier and lasts longer in the auditory than in the visual modality"[18] (cf. Fig. 8.5).

Finally, we will emphasize the "systematicity" of the N400: The N400 almost never fails to appear when data that are frankly abnormal in terms of semantics are presented: "[T]his component, the N400, has been the most robust finding with regard to ERPs and language."[19]

A certain interpretation of the N400, as a neurophysiological reflection of semantic operations in language, therefore seems to prevail. But the consensus no

---

[13]"Kutas and Hillyard (1980) reported that semantically inappropriate words (e.g., *He spread the warm bread with socks*) elicited a large-amplitude negative ERP component with a peak latency of 400 ms (the N400 component), relative to the ERPs elicited by semantically appropriate words (e.g., *It was his first day at work*)" (Osterhout and Holcomb 1995, p. 2).

[14]"N400 was first described as a difference waveform between ERPs to semantically congruent final words in sentences and their anomalous counterparts" (Salmon and Pratt 2002, p. 368).

[15]"[N400] was first observed by Kutas and Hillyard in response to a semantically anomalous word in a sentence context" (Kutas and Federmeier 2000, p. 464).

[16]"In a seminal paper 'reading senseless sentences: brain potentials reflect semantic anomaly' (*Science*, 207, (1980): 203–205), Kutas and Hillyard reported a negative wave with an onset of approximately 250 ms and a peak latency of about 400 ms in response to written words that semantically did not fit the preceding context" (Münte et al. 2001, p. 92).

[17]Osterhout and Holcomb (1995, p. 2).

[18]Salmon and Pratt (2002, p. 368).

[19]Münte et al. (2001, p. 92).

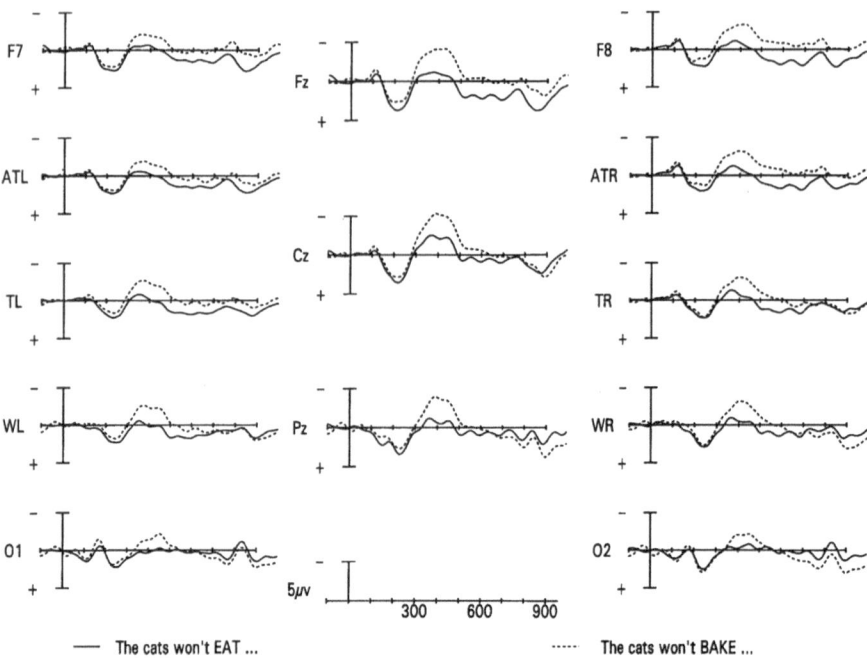

Fig. 8.4 ERPs to semantically anomalous verbs (*dashed line*) and nonanomalous verbs (*solid line*) (Osterhout 1997, p. 504)

Fig. 8.5 N400 effect for written and spoken sentence final word (Kutas and Federmeier 2000, p. 465)

longer holds from the moment we abandon the terrain of generalities and approximations. From the moment we explore more finely the conditions of generation of the N400, the consensual signification of this wave reveals its weaknesses: It proves in many respects to be insufficient and lacking, even inappropriate, and therefore questionable. And, without surprise, discordant voices may be heard —quite particularly in two respects: (i) that of the specifically linguistic character of the process reported by the N400, and (ii) more broadly, that of its functional signification.

8.2.2.2 Controversies

Concerning the first point, whereas Sakamoto et al. (2003), for example, give credit to the dominant conception according to which "it is widely accepted that the N400 is sensitive to semantic processing in language comprehension",[20] others, defending contradicting results, firmly oppose it: "[W]hereas some authors have advanced a language specific interpretation of the N400, some findings seem to be difficult to reconcile with this view."[21]

More concretely, in order to support the thesis of an N400 of a specifically linguistic nature, we will first establish that this wave is not the neurophysiological trace of the cognitive processing of "auxiliary" attributes of linguistic data, that is, the physical properties, psychological contents, etc., which participate in the factuality of a linguistic stimulus.

The most conclusive experiment in this respect consists in presenting the subjects with stimuli composed of graphical or phonematic elements of which the arrangement contravenes the linguistic shapes. When the sensorial modality of experience is vision, the subjects are usually presented with random concatenations of consonants. When the modality is auditory, they are presented with recordings of words played "backwards". The results of such experiments are very robust[22] and significant: We systematically observe the absence of the N400 for the "non-words". This would confirm the specifically linguistic nature of the processing manifested by the N400.

But, *a contrario*, numerous experiments highlight the presence of an N400 when the stimuli are not linguistic. For example, N400 effects are recorded in response to words combined with images which do not correspond semantically to them.[23] Likewise, the N400 is commonly observed in response to inconsistencies seen in faces or images.

Conversely, the semantic dimension of the processing which the N400 signals seems indisputable. We could indeed consider attributing a psychological signification to this EEG wave, as manifesting a cognitive process triggered by an unforeseen event. But experiments show that the presentation of stimuli

---

[20]Sakamoto et al. (2003, p. 387).

[21]Münte et al. (1998, p. 217).

[22]Münte et al. (2001, p. 92).

[23]"[...] several priming studies have shown N400-like effects to [...] a combination of words and semantically appropriate and inappropriate pictures (Nigam et al. 1992)" *in* Osterhout and Holcomb (1995, p. 17).

contravening a regular series (figures, sounds...) does not elicit an N400, but rather a P300 wave which is recognized to be triggered by unforeseen "inputs". The N400 is therefore manifestly the EEG correlate of a semantic processing.

In short, it seems well established that the "N400 is associated with almost any type of meaningful, or potentially meaningful stimulus [processing]: words, photographs, faces, environmental sounds, odors."[24] But, then agreeing that "the N400 seems to reflect a set of [...] neural processes that are common to the analysis of all sensory inputs [...] for the purposes of meaning construction",[25] the specifically linguistic nature of this wave is refuted.

Discussion. It seems difficult to not recognize the generic character of the semantic processing which the N400 manifests. The precedingly mentioned results are incontrovertible. But, this being acknowledged, it would be mistaken to conclude that the process which this wave signals, albeit very general, does not have linguistic relevance. This is because the functional signification of the N400 still remains very approximate, and nothing certifies that the operations to which this EEG response refers are not "logically" linked to the process of the elaboration of meaning in language. Indeed, at the risk of anticipating, care must be taken not to superimpose the processes ("meaning-intention", cf. 3.2.2) which institute an orientation of consciousness towards a certain signified, a fully linguistic object, and which *underlie* the actualization in consciousness of a particular meaning, with those, partial and terminal, which effectively accomplish this stage of actualization while leaving the realm of language (act of "fulfillment" of the "meaning-intention", cf. 3.2.7). But in order to defend this view, it will be necessary to cover numerous stages—the first being to provide the N400 with a functional signification compatible with the diversity of its attestations. It is this point which we will now document and discuss.

8.2.3   N400: Functional Significations

In the literature devoted to the EEG correlates of linguistic processing, five main functional significations of the N400 have been reported.

8.2.3.1   Compatibility Index

The first interpretation, by far the most widely shared, has already been exposed: The N400 would be an index of the semantic compatibility between a word and its context. For example, "the N400 [...] correlate in amplitude with the semantic fit between a target word and the [...] sentential or word context."[26] And in

---

[24]Kutas and Federmeier (2000, p. 467).

[25]Ibid.

[26]Frisch et al. (2004, p. 194).

quasi-identical terms: "In any case there is a consensus that N400 amplitude is a function of the 'semantic fit' between the target word and preceding context."[27]

This compatibility is often described in terms of "selectional restriction": The N400 effect being the neurophysiological signature of a transgression of constraints on semantic features, "selectional restriction violations [...] have been shown to reliably elicit N400 effects."[28]

It is important to emphasize that this interpretation confers to the N400 a status of EEG correlate of the forms of semantic legality in language. In this respect, also, one has been able to establish a significant correlation between the semanticity judgments and the variations in amplitude of the N400. For example, in an experiment presenting name-adjective combinations of which the contents refer to different sensorial modalities (for example: red-visual; rough- tactile...), and where the subjects are asked to grade (from 1 to 5) the semanticity of the combination (Sakamoto et al. 2003) obtain the following results (cf. table below), and conclude "that the N400 is sensitive to violation of linguistic (semantic) selectional restrictions."[29]

Adj/N	Visual	Tactile
Visual	Ex.: *Red color* Note: 4.4 *no* N400 effect	Ex.: *Red touch* Note: 1.7 N400 effect
Tactile	Ex.: *Smooth color* Note: 3.3 N400 effect	Ex.: *Smooth touch* Note: 4.1 *no* N400 effect

8.2.3.2   Index of Expectancy

Other interpretations of the N400 emphasize the variability of its amplitude in function of the anticipation of the word-stimulus. It may indeed be observed that more a word is "anticipated", the weaker is the amplitude of the N400 it induces.

This anticipation of the "target" word being commanded by the context which precedes it, it will be possible to describe the variations of the N400 as being inversely proportional "[to the] expectancy level produced by the preceding sentential context."[30] More specifically yet, since the N400 is recognized to manifest semantic processes, it is with respect to this dimension that the regimes of anticipation are to be qualified, thus: "[N400] component amplitude [...] reflects the accumulation of the

---

[27]Osterhout and Mobley (1995, p. 741).

[28]Kolk et al. (2003. p. 10).

[29]Sakamoto et al. (2003, p. 393).

[30]Salmon and Pratt (2002, p. 368).

semantic context or the degree to which a word is expected within the context."[31] Or, in more concise terms: "N400 is a good index of semantic expectancy."[32]

8.2.3.3 Integration Process

Among the functional significations which are most commonly attributed to the N400, we have that of correlate of an integrative process: "a [...] possibility is that N400 reflects a higher-level 'integrative' process (Rugg 1990; Rugg et al. 1988; Holcomb, in press)."[33]

The amplitude of the N400 would, then, measure the cost of the cognitive processing required for assimilating a unit in an encompassing context: "N400 reflects amount of effort involved in integrating semantic information into a higher order text or discourse representation."[34] From this angle, the greater the ease in integrating a unit, the weaker the amplitude of the N400 emitted: "[A]mplitude of N400 reflects the ease with which words are integrated with what went before."[35]

"*Context*" is to be understood here in a very broad sense: It can be only the "prime" word of an experimental process. In such case, when the "target" word is "primed", the decrease of its N400 can be explained as follows: "[There is a] smaller N400 because [the critical word] is easier to integrate with the context established by the prime word."[36] The *context* can also designate the syntagmatic sequence which precedes the word to come. In such case, "the N400 amplitude is determined by how easily the target word can be integrated into the semantic representation of the sentence or discourse in which it occurs."[37]

Be it as it may, the process manifested by the N400 always operates upon the semantic dimension: "[M]odulations of the N400 amplitude are generally viewed as [...] related to the processing costs of integrating the meaning of a word into the overall meaning representation that is built up on the basis of the preceding language input."[38]

8.2.3.4 Construction of Constraints

Following a fourth interpretation, the N400 would be the neurophysiological signature of a process of actualization of structural constraints: "[I]t was suggested that

---

[31]Ibid., p. 377.

[32]Ziegler et al. (1997, p. 768).

[33]Osterhout and Holcomb (1995, p. 12).

[34]Rourke and Holcomb (2002, p. 124).

[35]Weisbrod et al. (1999, p. 294).

[36]Holcomb et al. (2005, p. 165).

[37]Osterhout (1997, p. 496).

[38]Hagoort et al. (2003, p. 38).

the N400 reflects the amount of semantic or lexical constraints from the preceding context for a given target."[39] Supposing that the linguistic components indeed condition, to various degrees and following various modalities, the types of units likely to be adjoined to them, they may be associated with a field of virtual constraints, constraints which, in certain cases, will be actualized ("constructed") at the moment of the presentation of the completive unit, hence: "N400 amplitude reflects the buildup of semantic constraints imposed by preceding context."[40] More generally, the contextual constraints have a preparatory effect. They "favor" a set of units which are semantically congruous with the context: "[C]ontextual constraint directly influences the semantic features primed by a sentence context."[41]

8.2.3.5 Lexical Access

The last interpretation of the N400 is the following: "N400 is related to lexical access",[42] and the amplitude of the N400 measures the difficulty of this access.

This interpretation sensibly differs from the preceding ones inasmuch as it does not focus attention on the interactions between the unit whose processing elicits the N400 and its context. The N400 is considered here as the neurobiological trace of a process of "lexical access", that is, as the manifestation of a cognitive processing at the term of which a certain stimulus is acquired (or not) in its identity as a lexical unit, and, correlatively, actualized as such within consciousness: "N400 reflects processes involved in recognizing isolated words or pairs of words."[43] In a computational perspective, this process of lexical access, which the N400 thus manifests, consists in testing a correspondence between the stimulus and the entries in the lexical system, then conceived as a dictionary. Following this, the various pieces of information attached to an entry are subject to be transmitted to others modules in view of higher-level processing: "[N400] is sensitive to activation of word representations (lexical access or some other word-level recognition process)."[44] More explicitly still: "[N400 is function of] the ease of accessing information from long-term [semantic] memory."[45]

---

[39]Friederici and Frisch (2000, p. 479).

[40]Osterhout and Mobley (1995, p. 741).

[41]Titone (1998, p. 364).

[42]Bentin (1987, p. 308).

[43]Rourke and Holcomb (2002, p. 124).

[44]Osterhout and Holcomb (1995, p. 10).

[45]Kutas and Federmeier (2000, p. 465).

8.2.4   Synthesis of Functional Significations

These various interpretations obviously pose the question of their unity.

If it seems clear enough that in this picture, certain functional qualifications of the N400 are indisputably convergent, others are in part incompatible. Each, in the end, is manifestly insufficient for accounting for the totality of this wave's circumstances of generation.

Now, it is imperative to dispose of a univocal qualification of the N400, one which is synthetic and minimally consensual. Because if we do not acknowledge for this wave a specific functional identity, at least one which is well circumscribed, all the experimental results involving it may only ever serve in the debate regarding its signification, and not in the context of debates regarding the empirical validity of such or such theoretical hypothesis.

A minimal synthetic determination of the N400 is also the condition for this EEG response to entail *problems*, that is, so that it may serve as an observational angle upon empirical configurations which "are difficult to explain", which partially escape or which contravene the order of theoretical representations, thereby serving as their touchstones.

8.2.4.1   Convergences

Firstly discussing the homogeneous part of this set of functional qualifications, we will identify without difficulty numerous convergences.

The overlapping of significations #1 and #3 is most obvious: The integration of a unit within a context necessarily supposes a structural convergence between the unit and its context: "N400 reflects the ease with which a given lexical element can be *integrated* into the preceding semantic context. It appears that the amplitude of the N400 is a function of the *semantic fit* between the target word eliciting the N400 and a prior context."[46] And conversely, the fact of structural incompatibilities between a unit and a context leads directly to difficulties regarding integration: "[T] he N400 elicited in the processing of expressions seems to reflect this kind of integration process as a way of dealing with *incongruity*."[47]

Likewise, one may concede without difficulty the quasi-equivalence between interpretations #2 and #4. Indeed, the contextual constraints "prepare" a determined set of units and, in a certain manner, thereby anticipate on their actualization, and reciprocally: "[S]entence constraint, [is] a factor thought to affect the specificity of semantic expectations, [which is] operationalized as the cloze probability of the most popular response to each sentence."[48]

[46]Friederici and Frisch (2000, p. 479).
[47]Sakamoto et al. (2003, p. 392).
[48]Coulson and Kutas (2001, p. 72).

Fig. 8.6 The N400
amplitude as a function of the
word position (Kutas and
Federmeier 2000, p. 465)

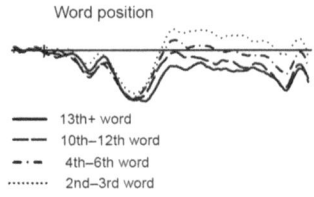

Word position

—— 13th+ word
— — 10th–12th word
— · — 4th–6th word
········· 2nd–3rd word

In order to empirically support the equivalence of interpretations #2 and #4, we may refer to an experiment which consisted in measuring the amplitude of the N400 emitted by the processing of words successively encountered during the unfolding of a sentence. The amplitude was observed to decrease in function of the position within the sentence. And this variation was interpreted in terms of "buildup of contextual constraints", specifically: "[W]e have observed a decrement in N400 amplitude with ordinal word position and interpreted this as a reflection of the buildup of contextual constraints across the course of a sentence"[49] (cf. Fig. 8.6).

We indeed understand that over the course of the progress of the sentence, the field of lexical possibilities will be progressively narrowed down, and, therefore and correlatively, the level of anticipation of the words to come will be increased, which explains the decrease in the amplitude of the N400.

In what concerns the fifth functional signification, that is, of the N400 as the cost of a process of lexical access, it is very close to the second one, but, for lack of other precisions, it is only partly assimilable to it.

Firstly, quite evidently, functional signification #2 involves signification #5: If a term is expected and anticipated in a given context, one may reasonably deduce that its access will be greatly facilitated. However, the passage from #5 to #2 does not make sense: The assertion "T is easily accessible, *therefore* T is anticipated" presents nothing analytical. In truth, what poses an obstacle to this implication is the vagueness of the concepts it coordinates. Because it is necessary to admit, as do numerous authors, that it is not very clear what is meant exactly by "access" to a lexical unit. For example, Perea and Pollatsek (1998) recognize that different definitions of lexical access are possible: "For some, lexical access means access of the visual or orthographic code [...]. To others, lexical access could mean identification of the phonological code, access of semantic codes, or access of all of the above."[50]

We therefore understand that in order to advance in the area of a synthetic and coherent interpretation of the N400, it is necessary to have theoretical frameworks where the notion of lexical access receives a specific determination. To this effect, we will have recourse to the computational theories of the lexicon, of which we will schematically recall the main features.

---

[49]Van Petten and Kutas (1991, p. 96).
[50]Perea and Pollatsek (1998, p. 768).

8.2.4.2 Computational Models of the Lexicon

There are two great types of computational models of the lexicon. Both admit the principle of modularity, that is, the existence of a layering of linguistic qualifications into homogeneous strata. In general, a distinction is made between the levels of phonological features: the level of word forms (lexemes), that of lemmas (abstract linguistic units), and that of semantic representations. Both types also relate the "actualization" of a lexical property to the mode of an "activation" of the cell(s) which represent(s) this property within the model.

On the one hand, there are "atomistic and sequential" models, on the other, models stemming from the works of PDP[51]: "interactive and distributed" models.

The atomistic models are characterized by the following features (cf. preceding Fig. 8.7): The semantic representations of the lexical items are atomic, in the sense that each lemma is associated to a sole concept, itself formally represented by a cell (or atom) of a network to which it participates at its modular level. These models are also sequential in that the connections between the models are oriented: The propagation of the activations of the units of the network is unidirectional, and there is no effect of retroaction. Conversely, within a same module, the units share links, these having a stimulating or inhibitive effect: Between units of the lemma-level, it is a matter of "associative relations" of which the force represents a frequency of co-occurrence, whereas between conceptual atoms, it is a matter of "semantic links" of which the weight measures the proximity of meaning.

In the atomistic model, a linguistic sequence is represented at each level of analysis by a pattern of activation. At the semantic level, the integrated character of the sequence is rendered in the form of an activation of a related set of conceptual atoms.

In what concerns the "interactive and distributed" models, as their name indicates, they allow for relations of interaction between various modules, and, in a given context, the sense of a lemma is represented by the activation of a complex of weighted (semantic) features—activation which stems from a process of stabilization of the dynamic of the semantic network initially activated by the lexemic constituents of the processed linguistic sequence (cf. Fig. 8.8).

What we shall retain here, is that within the framework of modular theories, the process of "access" is related to a process of "activation": To "access" a lexical unit amounts to increasing the activation state of its representation beyond a certain threshold which marks the "conscious recognition" of the unit by the cognitive system—the cost of the access is then measured by the quantity of activation necessary for attaining a level of conscious recognition.

Given this theoretical characterization of the notion of lexical access, we immediately derive the equivalence between significations #2 and #5. Indeed, we already know that #2 implies #5. But we now have at our disposal the concept of preactivation which gives new meaning to the function of anticipation, and which

---

[51]*Parallel Distributed Processing*, cf. (Rumelhart and McClelland 1986).

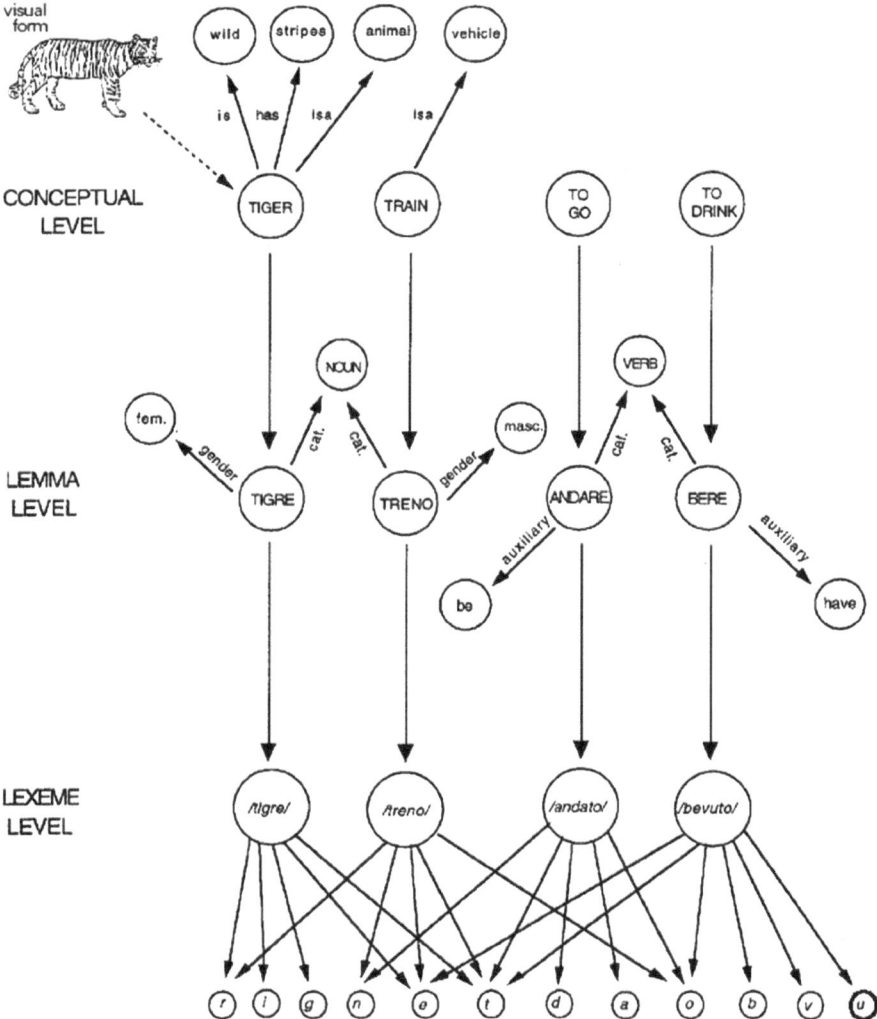

Fig. 8.7 Part of the lexical system showing the relation between lemma and other levels of lexical representation (Caramazza 1997, p. 182)

enables to relate this function to a process of lexical access. Indeed, a term T is so much easier to access if it is preactivated, and it is clear that this preactivation is a preparation for actualization of T, which thus is in a certain manner anticipated—hence: #5 implies #2.

Remark: In an interactionist (non-sequential) conception of lexical processing, where the identities of expression and content are determined following processes of reciprocal reinforcement of inhibition, we observe a balancing of the patterns of content activation (superior layer—cf. Fig. 8.8) and of expression (inferior layer),

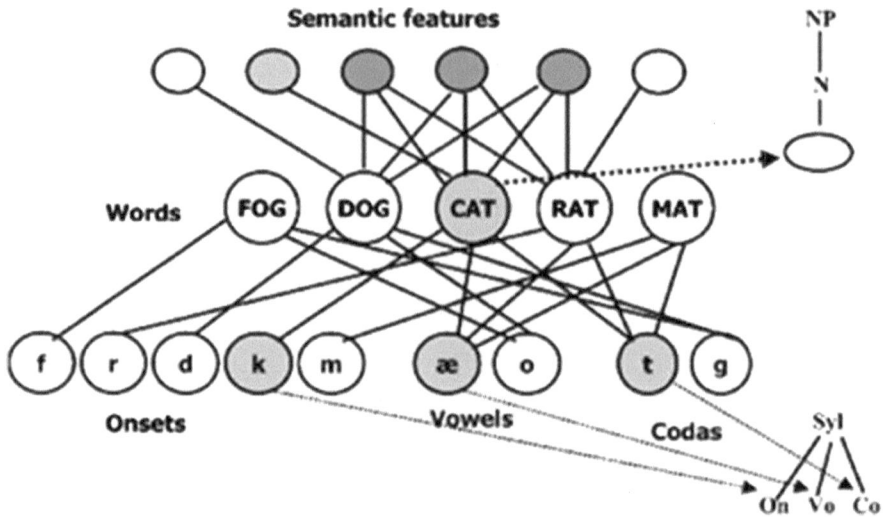

Fig. 8.8 Structure of the interactive and distributed model (Schwartz et al. 2006, p. 230)

via the intermediate layer of "lemmas" in order to, *in fine*, arrive to the actualization of a certain word. But this schema of functioning, in which the signifiers and signified are configured in a dynamic coupling, fails in the case of logatomes. Indeed, in such a system, the logatome does not have its place as an integrated phonematic composition: The activation of a phonematic pattern "outside of the lexicon" either produces "nothing", i.e. actualizes no cell at the level of lemmas, or activates a paronym (the system tends toward the closest attractor to the input configuration). In both cases, the result is unsatisfying. Because either the processing actualizes a paronym and furthermore its lexical content, which is undefendable, or the diffusion of the activations towards the layer of contents is blocked at the level of lemmas, and the semantic dimension, moreover attested, of the N400 which can be observed in these cases (cf. 8.2.11) is rejected. This, furthermore, reactivates the post-lexical versus pre-lexical problem and related difficulties (cf. 8.2.5 *sq.*).

8.2.4.3 Supplements

At this stage of our undertaking, the interpretations of the N400 are distributed over two disjoined classes: on the one hand, interpretations #1 and #3, and on the other hand, interpretations #2, #4, and #5. In order to complete the synthesis, there remains to merge these two classes. But significations #1 and #2, on the one hand, and #3 and #5 on the other, seem, at least in their apparent characters, incompatible. We will now show that this difficulty is only superficial and that the various functional significations of the N400 are naturally subsumed under the fifth interpretation.

Firstly, in what concerns interpretations #1 and #2, they relate the N400 to processes which may have some functional correspondences indeed, but which nevertheless pertain to distinct operational logics. Indeed, the degree of incongruity measures the structural (linguistic) incompatibility of the items in the course of assembly whereas anticipation, beyond the phraseological or syntactic determinations, has for background knowledge of an encyclopedic type or, more broadly, a set of socio-cultural representations. It is therefore abusive to relate incongruity to a lack of anticipation, and the synthesis of interpretations #1 and #2 appears to be in an impasse.

In order to escape this conjuncture, we will begin by signaling the fragility of the two interpretations discussed herein.

Regarding the first, although it is usually admitted, it is no less regularly contested. Also, it was quickly observed that "semantic incongruities are *neither sufficient nor necessary* to elicit the occurrence of an N400 component."[52] Moreover, it indeed seems "that any word elicits the occurrence of an N400 component", regardless of its degree of adequation with the context, and even, as we will see, when presented out of context.[53] This is at least what is asserted by Kolk and Osterhout: "[R]elevant is the observation that although *nonanomalous* open class words elicit a robust N400 component",[54] and "it is well known that an N400 component is elicited by each open class word."[55] Likewise, Kutas, to whom we therefore owe the first interpretation of the N400, will later concede that the N400 "is not simply an index of anomaly but rather a part of the brain's normal response to words (in all modalities)."[56]

Moreover, the experiments appear to support the second interpretation because, if the N400 is indeed observed for words which are in perfect congruence with their context, the amplitude of the N400 is influenced by their originality: "[S]everal studies have reported N400 effects to semantically correct but less expected words."[57] As shown by the two following plots,[58] the influences of the more or less "predictable" character of the target word upon the N400 are manifest in experiments which consist in presenting a text in which certain words are repeated. The amplitude of the N400 is lesser for the words having already been encountered. Moreover, we observe that the amplitude of the N400 covaries significantly with the frequency of the words present in the text (cf. Fig. 8.9).

---

[52]Besson et al. *in* Renault (2004, p. 192).

[53]Ibid., p. 193.

[54]Osterhout (1997, p. 498).

[55]Kolk et al. (2003, p. 15).

[56]Kutas and Federmeier (2000, p. 464).

[57]Kolk et al. (2003, p. 7).

[58]Kutas and Federmeier (2000, p. 465).

Fig. 8.9 The N400 amplitude as a function of word repetition and frequency (Kutas and Federmeier 2000, p. 465)

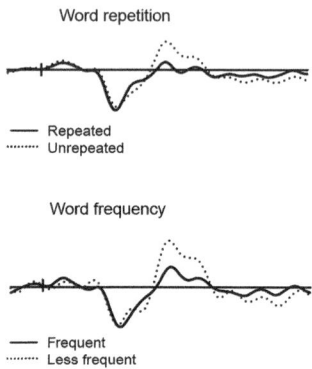

Word repetition

—— Repeated
········ Unrepeated

Word frequency

—— Frequent
········ Less frequent

In the end, we may therefore concede, on the one hand, that the N400 varies "even when the target word is not semantically deviant", which amounts to saying that, contrarily to the first interpretation, "violations are not necessary to elicit N400 effects",[59] and, on the other hand, in accordance with the second functional signification, that the N400 is modulated by the "levels of expectancy" regarding the presented word—this degree of expectancy being generally measured in terms of "subjective probability" (or "cloze probability"[60]).

However, we can not deny that the first interpretation has some measure of explanatory relevance: The degree of semantic compatibility, although it does not condition the emergence of an N400, remains a determinant factor. This is what is illustrated by the following experiment[61] which consists in presenting words in view of completing a sentence ("they wanted to make the hotel look more like a tropical resort. So along the driveway they planted rows of...") in a more or less congruous manner. The "target" words are characterized by two factors: their subjective probability of occurrence, and their semantic adequation to the context of the sentence. Three types of words are submitted to the subjects: T_1 words ("palms") which are congruous with the context, T_2 words ("pines") which are incongruous but which belong to the same semantic class as T_1, T_3 words ("tulips") which are incongruous and which belong to a different class than T_1. Moreover, the incongruous words are chosen (T_2 and T_3) so as to have a same subjective probability.

This being, if the second interpretation was fully valid, if the amplitude of the N400 was only a function of anticipation, we should have[62]:

---

[59]Kolk et al. (2003, p. 7).

[60] "Cloze probability is determined by requiring a large group of subjects to fill in the missing final word in a set of sentences" (Osterhout and Holcomb 1995, p. 19); "[it is] the proportion of individuals who provide that particular word as the most likely completion of [a] sentence fragment" (Kutas and Federmeier, p. 464).

[61]In Kutas and Federmeier (2000, p. 466).

[62]|N400| T is the amplitude of the N400 for target T.

'They wanted to make the hotel look more like a tropical resort.
So along the driveway they planted rows of ...'

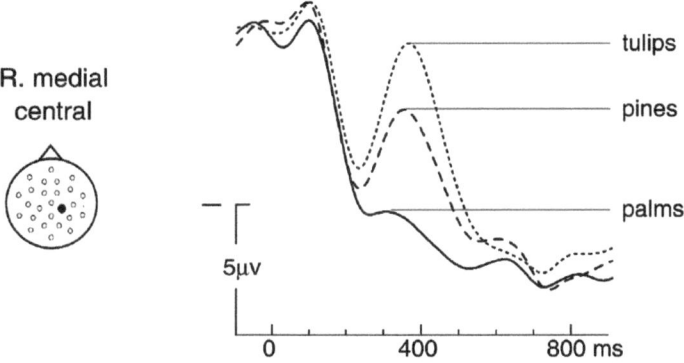

Fig. 8.10 The influence of plausibility [...] on the N400 response [...] (Kutas and Federmeier 2000, p. 466)

$$|N400|T_2 = |N400|T_3$$

Now, as shown by Fig. 8.10, we notice that:

$$|N400|T_1 < |N400|T_2 < |N400|T_3$$

Which amounts to saying that semantic congruity conditions as much the amplitude of the N400 as the subjective probability of occurrence.

The first functional signification of the N400 is therefore not without relevance. But the second remains indisputable. The N400 is as much a "good indicator of the semantic relations between a word and [its] context"[63] as it is a "function of the degree of resolution of the semantic anticipation created by the context."[64] Also, it will be necessary to admit each interpretation as some authors prudently do: "[W]ords are *easier to process* because they are expected in a context *or* are related, semantically, to recently presented words elicit smaller amplitude N400 relative to the same words out of context or in weak or incongruent context."[65]

But this consensual position is not without difficulty. As we have seen, interpretations #1 and #2 pertain to functional logics which are foreign to one another. In order to resolve this predicament, it will be a matter of placing oneself at a level which subsumes them and, preferably, without going beyond the framework of

---

[63]Besson et al. *in* Renault (2004, p. 192).

[64]Ibid., p. 193.

[65]Kutas and Federmeier (2000, p. 464), our emphasis.

attested interpretations. In fact, it is in the third functional signification that the solution resides.

Because if a word's processing, as emphasized by Kutas et al. (cf. *supra*), is facilitated by its anticipation *and* by its adequation with the context, it clearly appears that the processing of which it is question here is integrative. In support of this conception, we will recall that integrative signification #3 is homogeneous with signification #1, which allows to consider the process of integration as pertaining to a dynamic level subsuming the static fact of a structural compatibility. Now, the integration sought in its dynamic dimension also subsumes the second functional signification of N400. Indeed, the integration of units in a context will obviously be that much easier if the units to integrate are preactivated—in other words, anticipated.

Interpretation #3 of the N400, as an integrative process, therefore covers qualifications #1, #2, and #4 of this wave, and, thereby, enables to overcome the lack of homogeneity of interpretations #1 and #2. The third interpretation of the N400 thus appears to be a good candidate in what concerns a unified functional signification. It remains however to adjoin the N400 as a process of lexical access.

At this point, a new difficulty arises, because the interpretation of N400 as a process of access to a lexical unit apparently contravenes its integrative signification, at least in a compositional view. Simply, if the operation of integration presupposes an access to the constituents, such is not reciprocally the case. In order to integrate units, access to them is indeed required, but conversely, to access units, *a fortiori* one unit, by no means entails the construction of a totality by assembling successively acquired units.

Again, two apparently irreconcilable conceptions affront one another, without experimental data being capable of discriminating. Because it has been demonstrated[66] that *isolated* words generate an N400, meaning that this wave can therefore not be understood as a signature of an integrative process. But conversely, as we have seen, the integrative signification of the N400, as it subsumes interpretations #1, #2, and #4, adequately accounts for a great diversity of experimental configurations which involve contexts.

In order to escape the impasse, and to observe how artificial it is, it will suffice to recall that in terms of language, structures are not assemblages of predefined and mutually compatible elements, but integrated totalities, that is, unitary configurations where the parts and the whole interdetermine one another, in both their existence and in their identities.

Consequently, access to the constituents is not separable from the process of integration: It is in a sole operational moment, that of the actualization of an integrated totality, that the elements which are involved within it are doubly instituted and "accomplished" in their individual forms and qualities.

In this gestaltist and emergentist (dynamic) perspective regarding structures and relations between parts and whole, "integration of units" and "unit actualization"

---

[66]Cf. for example, Rourke and Holcomb (2002) or Ziegler et al. (1997).

are simultaneous and reciprocally determined processes: processes which overlap. It follows that the phases of actualization (access) and of integration (structural) are mutually assimilated: The dynamic of lexical access is no longer simply a stage of a process of integration but comprises an integrative sense. In other words, integration does not constitute an operation pertaining to previously acquired units, but resides in the principle of unit actualization—because to "access" an item is not to simply bring this item to a degree of activation sufficient for making it present to consciousness, but is essentially to configure this item as an actual part of a higher-level unit. And the apparent incompatibility of the third and fifth interpretations of the N400 is thereby lifted.

8.2.4.4   Synthesis and Conclusion

In the framework of a holistic conception of structures, and, conjointly, in the framework of modular architectures of the lexicon, the N400 is therefore subject to a unified functional signification, one which synthesizes and subsumes the various interpretations this wave may have received—that is, as a neurobiological signature of a process of actualization of linguistic components as parts of an integrative configuration.

We therefore elect to "put forth" the interpretation in terms of "access", because despite that same processes of integration and actualization overlap, a certain form of primacy can be conceded to this cognitive moment of lexical "accomplishment", such as, namely, seems to prevail in the processing of isolated words.

Certainly, the principle of decontextualized processes may be contested: Nothing effectively certifies that the lexical units provided out of explicit context are preserved from any contextual interaction, and that, being presumed to be autonomous, their access undergoes no "side effect". We will argue that the experimental protocol always induces implicit contexts (as for example the series of words successively presented) or, more fundamentally, that the actualization of a word always solicits the linguistic system's background: "In reality the idea evokes not a form but a whole latent system that makes possible the oppositions necessary for the formation of the sign",[67] or always involves contrastive regulation: "the terms *a* and *b* as such are radically incapable of reaching the level of consciousness—one is always conscious of only the *a/b* difference."[68]

By choosing to favor the interpretation in terms of "access", we will not reject the preceding objections; only, we will elect to emphasize a certain aspect of a global process. It will be a matter, in doing this, of providing ourselves with the means of grasping and of accounting for very numerous experiments employing presumably "empty contexts"—and which, in order to work around a whole series of difficult questions, we will limit ourselves to qualify as "indeterminate."

---

[67]*CLG/B*, p. 130.
[68]Ibid., p. 118.

8.2.5   Sub-lexical and Post-lexical Processes

Now, by situating ourselves within a framework centered upon the issue of lexical access, we directly inherit the distinction, established by experimental psychology and restituted in the computational models of the lexicon, between subconscious (or preconscious) and postconscious processes. That is, being a matter of lexical units, on the one hand we have the processes conducing to a level of lexical consciousness and which work beneath this cognitive plane, and on the other hand, we have those which pertain to lexical units acquired by consciousness.

It will now be important to submit this "sub/post-lexical" opposition to the trial of critical examination, and this, on the basis of the experimental results which track it. Our intention here is double and straightforward: It will be a matter of questioning the relevance of the "sub/post-lexical" distinction, but especially of preparing the grounds for a phenomenological qualification of the N400. The experimental results which will first be presented for refutational purposes will indeed reveal themselves to support an interpretation of the N400 as the deployment of a meaning-intention.

But first, it will be useful to bring a few precisions to the sub/post-lexical distinction, and, beforehand, to the dual notion of lexical access.

As has already been emphasized, the notion of lexical access is somewhat vague: It may be a matter of identifying a word-form, a corresponding lemma, the concept associated to this lemma, or both simultaneously… In currently prevailing theoretical perspectives, lexical access is factorized into three sub-operations: access, selection, integration. The first ("pattern matching"), by retaining from the stimulus a set of relevant physical or sensorial characteristics, consists in producing a mental representation and in testing the existence of a concordance between this representation and the entries of the lexical system, in order to activate those which correspond to it as well as its associated morphosyntactic and semantic properties. The second consists in selecting amongst a set of pre-activated units the one which is currently relevant. The third designates the integration phase into a higher order unit.

Concerning now the sub- and post-lexical processes, the first (sub-lexical) are passive mechanisms, being automatic and subconscious. As such, they are not affected by the strategies which the subjects put into play in order to fulfill various tasks, or by eventual attentional modulations. They are also rapid and spontaneous processes which take place prior to the actual recognition of stimuli.

It is in this respect that the propagational processes of activation appear and these, as we have already seen, have the effect of increasing the level of activation of the units linked to the prime word and of bringing these units closer to their threshold of actualization, thereby facilitating their "access": "When a prime word's is [presented] activity *passively* spreads to semantically related items"; "this type of automatic priming has generally been assumed to occur *prior* to the *actual recognition* of the target word—hence the term pre-lexical priming."[69]

---

[69]Osterhout and Holcomb (1995, p. 6).

The post-lexical processes are, for their part, conscious processes which thus suppose a prerequisite knowledge of the terms they involve. These processes follow strategies of action elaborated by the subjects and, as such, are under their attentional control.

As opposed to sub-lexical processes, which are swift and immediate, the post-lexical processes are triggered with greater delay and their execution requires more time. It is generally agreed[70] that post-lexical processing begin between 250 and 400 ms after the presentation of the target. In the preceding interval of time, the processing which unfolds is of a sub-lexical nature: It is automatic and unconscious.

Concerning the lexical decision task (LDT), two principal post-lexical strategies have been acknowledged: The "anticipation" strategy and the "relatedness" strategy.

The first ("expectancy induced priming", "attentional expectancy", or "expectancy driven mechanism") consists, after identification of the prime word, in imagining semantically associated words. The subject thus prepares him or herself to respond more efficiently upon presentation of a target word, one which is therefore voluntarily preactivated. This strategy is favored when the rate of primed target words is high.

The second strategy, called "semantic matching" (or the "relatedness strategy") consists in basing the lexical decision not on the recognition of the status of the stimulus, but on the evaluation of a relation of meaning between the target and the prime. The response then tends to be oriented towards the recognition of a true word (vs. a pseudoword).

In the context of sequential models, the matching between a prime and target is expressed in terms of connection. "Neely and Keefe (1989) have suggested that [...] to help make the lexical decision [...] subjects might ask themselves, 'Are the prime and target related?' If yes, respond 'word,' if not then consider responding 'nonword'."[71] In the context of distributed models, the matching takes place following the mode of a recovery of features and is thereby similar to an integration trial ("lexical coherence checking"). It is clear that this strategy presupposes an access to the meaning of the processed items.

It is usually admitted that the mechanisms of activation propagation and of semantic matching are closely linked with the respective lexical operations of access and integration. Of course, as for any operation of access, the operations of selection and of integration comprise an "automatic" dimension, in that they are intrinsically part of the natural waning of the linguistic activity and thereby present a mandatory character. But conversely to the operation of access or to the mechanism of activation propagation, the processes of selection and of integration are likely to interfere with conscious strategies which at a greater or lesser degree modify their natural functioning.

---

[70]Anderson and Holcomb (1995).

[71]Osterhout and Holcomb (1995, p. 7).

8.2.6   N400: A Sub-lexical or Post-lexical Process?

If we refer to the most commonly advanced explanation in accounting for the N400 priming effect, and if we also accept the evidence that in the trials without priming, the processing which gives access to lexical identity is of a sub-lexical nature, we can then reasonably conclude that the nature of the process related by the N400 is sub-lexical.

Indeed, to explain the decrease in amplitude of the N400 by the preactivation of the stimulus under examination amounts to saying that a non negligible part of the work of recognition of the item has already been accomplished by means of sub-lexical propagation. Therefore, conceding a principle of "homogeneous prolongation", the supplement required in order to access the identity of the target is likely to be of a same nature, as is the case for non-primed targets. The N400 would thus indicate a sub-lexical processing.

But such a conclusion is too fragile. Because nothing certifies that the identity of the prime is not solicited by the subject in order to perform the task, which would thus take the form of a post-lexical processing. This is the case when the response strategy bases itself upon an evaluation of the semantic compatibility between target and prime. This procedure, which quite of course presupposes the prior acquisition of the identities of the target and of the prime, is all the more facilitated when the prime is semantically close to the target, translating as a decrease in the N400's amplitude. In such case, the N400 signals post-lexical processing.

But various experiments contradict this view.

Indeed, when the "pondered" strategy is prevented by impeding the access to the lexical identity of the prime, the N400's priming effect subsists even when any conscious procedure is impossible. Specifically, a decrease in the amplitude of the N400 is observed when the presentation time of the prime is too short for its identification to be possible (masking),[72] or when the target is presented only 200 ms after the prime (henceforth: SOA[73] = 200 ms).[74] In these two experimental protocols, no post-lexical strategy is applicable, and the nature of the N400 is concluded to be sub-lexical.

In another experiment[75] which partly reproduces the preceding one, that is, at a low SOA (250 ms) and with presentation of a masked and of a non-masked prime, the existence of a priming effect is equally observable for the N400. But, moreover, the "masking" factor is revealed to be irrelevant: The amplitude of the N400 is the same regardless of the masking or not of the prime. Again, the interpretation of the N400 as the reflection of a post-lexical processing is empirically invalidated.

---

[72]Kiefer and Spitzer (2000).

[73]"Stimulus Onset Asynchrony", that is, the time between the beginning of the presentation of the prime and the moment of the presentation of the target.

[74]Kiefer and Spitzer (2000).

[75]Deacon et al. (2000).

But, *a contrario*, several experiments attest that the amplitude of the N400 increases when the conditions of experimentation favor a high level of attention. It can indeed be observed[76] that, in the case of a lexical decision task, for a high percentage of primed targets (80%)—which, as we have seen, induces sustained attention and post-lexical strategies—the amplitude of the N400 priming effect is superior to that obtained when this percentage is low (specifically 20%). Moreover, under identical experimental conditions, when the task required solicits no particular attention in order to achieve cognitively demanding results, that is, when the subjects are required to simply accomplish a silent and "passive" reading, the N400's priming effect is the same regardless of the percentage of primed targets. Moreover, an increase is recorded in the amplitude of the N400 when the SOA increases, therefore when we "open" the access to the lexical units. Such results manifestly plead in favor of a characterization of the N400 as a post-lexical process.

8.2.7   First Synthesis

The results described above therefore fuel a conflict which seems unresolvable—at least within the framework in which it arises. The experiments which attest to the conjointly pre- and post-lexical character of the processing manifested by the N400 are numerous, varied, and solid. Nothing at this point in time seems to indicate that the conclusions regarding the ones could be repudiated to the benefit of the others. The actual state of affairs may be qualified as a state of resignation before the heterogeneity observed in the functional signification of the N400. Thus, and taking into account the pre- or post-lexical character of the processing signaled by the N400, one is promptly led to pay heed to its other facet: "[T]he N400, while sensitive to the buildup of automatically established contexts such as those provided by a masked prime, nevertheless is a direct reflection of a later post-lexical process that requires attention to the semantic properties of the eliciting target stimulus"[77] or, furthermore, "while it is clear that the N400 effect [is] enhanced by the attentional manipulation [...], it does not appear that attentional processes are a necessary ingredient for generating an N400."[78]

In order to overcome this oscillation between two specifications, the art of synthesis remains—the N400 is then acknowledged to signal a process which can as just as well be spontaneous or pondered: "N400 component is not solely a direct reflection of the cognitive processes underlying lexical access. Rather, N400 amplitude appears to be quite sensitive to relatively late-occurring processes",[79] or,

---

[76]Weisbrod et al. (1999).

[77]Holcomb et al. (2005, p. 171).

[78]Holcomb (1988, p. 80).

[79]Osterhout and Holcomb (1995, p. 17).

"if N400 effects are found at both short and long SOAs, it would suggest that the N400 is sensitive to both types of processing."[80]

The question to which we are now confronted is the following: Under which conditions is it possible to admit that a same neurobiological response relates to two distinct cognitive processes? If we remain in strict compliance with the methodological principles which govern the investigations in the field of cognitive neurosciences, that is, principles of covariance, then it must be concluded that the theoretical distinction between pre- and post-lexical processes is not empirically founded. But, on the one hand, the gap between automatic, unconscious processes and those which are pondered and controlled appears difficult to overcome, while, on the other hand, it seems difficult to contest the existence of these two sorts of processing. The impasse is therefore patent.

We then understand that in order to be able to overcome this dead-end, the only solution would be to homogenize the processes qualified as pre- and as post-lexical: to regard them as forming a continuity within an integrated process of lexical recognition where these two processes may then be to be considered as successive phases of a same and single act of actualization of a word-consciousness, an act which the N400 would then signal.

In order to progress in this direction, we will resume the investigation while adopting the matching angle of the opposition between automatic versus controlled (processes), and this shall be done while emphasizing the attentional factor according to its various orientations and its intensity.

8.2.8   N400: An Automatic or Controlled Process?

Within the problematic framework retained here, and based on the evidence provided by the N400 phenomenon, the central question regarding the relevance of the sub/post-lexical opposition translates as a question regarding the automatic or controlled character of the process related by the N400.

In response to this question, a great number of experiments have been conducted, very often following the priming protocol (masked if required), and which consisted in focalizing the attention of the participants on the sole "surface" characters of the stimuli in order to block the access to meaning [essentially, LSTs (lexical search tasks) or PTs (physical tasks)]; the existence or absence of a priming effect then attesting to the respectively automatic or controlled character of the process signaled by the N400. The results are highly contradictory, and support divergent positions—as we will see right now.

We may begin with (Dombrowski et al. 2006) within the framework of the "standard" LST protocol. The task for identifying letters within the framework of a "standard" LST consists in presenting stimuli composed of a word or pseudoword

---

[80]Anderson and Holcomb (1995, p. 178).

S of n capital letters upon which n copies of a letter L in lowercase (over each letter of S) and which are to be recognized to be present or not within S. Evidently, by inviting to perform a column by column and letter by letter comparison, this protocol seems to very strongly confine attention to a "surface" level. The protocol by Dombrowski et al. (2006) consists in alternating LSTs and LDTs (lexical decision tasks), that is: (i) presentation of a S1/L1—LST(L1) task—presentation of a S2/L2 —LDT(S2) task—etc. Result: The existence of a N400 priming effect may be observed. When S2 is primed by S1, the N400 generated by S2 has a lower amplitude than that emitted when S2 is semantically "foreign" to S1. Immediate *conclusion*: "the theory of semantic activation being an automatic process should not be abandoned."

Deacon et al. (2000), for their part, examine the influence of masked primes on the N400. After having taken notice of the experiment conducted by Brown and Hagoort (1993) who observe that the masked primes induce no N400 effect, Deacon et al. (2000) renew the experiment while reducing the SOA (in order to avoid the loss of activations propagated by the prime), obtaining the effect anticipated—and they conclude that "the findings imply that the processing subserving the N400 is not post-lexical since the N400 was manipulated without the subjects being aware of the identity of the words."[81] In the same experimental paradigm of a masked prime, Kiefer (2002) also records (with SOA = 67 ms) an obvious N400 priming effect and concludes, albeit in a more nuanced manner, that the N400 is also involved in automatic processes: "[T]he N400 is modulated by automatic spreading activation and not exclusively by strategic semantic processes."[82]

A contrario to these various experiments which establish the automatic character of the process affiliated with the N400, Chwilla et al. (1995) observe that the execution of surface tasks entails no N400 effect. Taking account of the controversy surrounding the question, Chwilla et al. (1995) first refer to two experiments of the LST type where an N400 prime effect has been observed. Firstly, (Kutas and Hillyard 1989), which may be described as follows: Protocol: Presentation of triples (W1–W2–L); SOA (S1–S2) = 700 ms; SOA (S2–L) = 1200 ms; task: Decide if letter L is present in the words W1 or W2; *result*: Existence of an N400 priming effect. Then (Besson et al. 1992): Protocol: Presentation of pairs (W1–W2); SOA = 300 ms; task: Decide if the first letter of W1 is identical to the last of W2; result: Existence of an N400 priming effect. *Comments*: In order to account for these N400 priming effects, Chwilla et al. (1995) invoke in the first case a too long SOA which would "open the door" to semantic processing, in the second, the induction of a semantic process by effect of an implicit task of memorization. If the first argument deserves attention (we will return to it), the second, however, appears to be counterfactual: In fact, the memorization tasks reveal themselves to decrease the N400 effect (cf. *infra*). Be it as it may, in order to guard against these two experimental perturbations, Chwilla et al. (1995) propose the following experiment:

---

[81]Deacon et al. (2000, p. 137).

[82]Kiefer (2002, p. 27).

Protocol: (priming paradigm) presentation of pairs of stimuli (S1-S2) of the (W–W or PW[83]) type; SOA = 700 ms; tasks: T1 = LDT, T2 = PT: (identification of the case of S2); *results*: T1: Existence of an N400 priming effect, versus T2: Absence of an N400 priming effect; *conclusion*: When the processing is confined to the surface characters of the lexical representation (T2), the semantic levels are not invested and the N400 is not affected, and conversely, the lexical decision task (LDT) likely to requisition the plane of meaning and the operations which unfold within produce an N400 priming effect, hence: "N400 effect primarily reflects lexical integration process."[84] In other words: the N400 relates a post-lexical process. *Comment*: The result seems convincing, but in their discussion, the authors support their thesis by emphasizing "the absence of an N400 priming effect when the prime is masked", which is inexact (cf. *infra*). Moreover, voluntarily or not, in their expectancies they neglect the existence of an N400 effect of lexicality with respect to task T2, which their experiment indeed demonstrates, that is: The amplitude of the N400 for PWs is superior than with Ws—which considerably weakens their position. Indeed, the absence of an N400 priming effect for task T2 makes it acceptable to infer the non-automatic character of the underlying operation, and therefore the semantic and post-lexical character of the processing. One can even suppose that under the experimental circumstances of T2, the cognitive activity does not exceed the level of the individuated graphemic units, which it processes alternately. Such is not the case however since we observe an effect of lexicality which shows that the PWs are not processed in the same manner as are words—which throws into question the controlled nature of the process which the N400 would manifest. Indeed, the gap in amplitude between the N400 elicited by words and that elicited by pseudowords attests to a process which is sensible to the status of the stimulus. Yet, a priori, at least in the framework of the functional models of the lexicon, a true word distinguishes itself from a pseudoword in two respects: On the one hand, it figures as an entry in the mental lexicon and, on the other hand, a semantic representation is associated to it—and these two aspects are likely to be the cause of the intensification of the N400 emitted by pseudowords. As the absence of an N400 priming effect during the course of T2 ensures that the dimension of meaning is not solicited, the lexical effect therefore stems from the fact that the PW does not appear as a lexical entry. Indeed, the increase in the amplitude of the N400 of pseudowords is naturally explained by the excess of cognitive work associated with the search for an inexistent lexical entry. Consequently, the N400 would relate an operation of access, thus a sub-lexical operation, which is in contradiction with the preceding conclusion.

Remains that the controlled nature of the processes signaled by the N400 is incontestable: It may indisputably be observed that the attentional variations modulate the amplitude of the N400. This is namely what is demonstrated by the experiment conducted by Holcomb (1988) which, keeping in mind that the

---

[83]W = Word; PW = PseudoWord.

[84]Chwilla et al. (1995, p. 283).

participants are financially motivated to produce the best results, consists in inducing controlled strategies by increasing the rate of primed targets. When this rate is sufficient, the subjects benefit from using conscious "semantic expectancy" strategies (cf. *supra*), and Holcomb (1988) observes the N400 priming effect recorded in the experimental block with a priming rate of "50%" is superior to that of the block at "12%". However, and considering the number of experiments attesting to the automatic and "sub-lexical" nature of the processes supported by the N400, Holcomb (1988) prudently concludes that "[t]he occurrence of a larger N400 effect in the attentional than in the automatic block is not consistent with the proposal that this component reflects *purely* automatic semantic priming."[85] Likewise, amongst many others, Rossel et al. (2003), after having observed that the N400 priming effects are superior with a high SOA, thus when the subjects enjoy sufficient time for reflection, conclude that their results support the interpretation of the N400 as a correlate of controlled processes.

8.2.9   First Conclusions: Towards the Phenomenological Hypothesis

From this very vast and dense collection of experiments, illustrated here by means of a few representative results, there are mainly three things which stand out.

The *first* is that the distinction between sub- and post-lexical levels of processing sheds little light on the processes which the N400 neurobiologically express. The experiment indeed shows that the N400 signals just as much operations of an automatic nature as operations under attentional control, in other words: The N400 is linked to the entire series of lexical processes, from the access to the strict meaning of the term, i.e. the localization of an entry within the mental lexicon, up to the processes of semantic integration. All this, it should be acknowledged, informs us very little regarding the linguistic processing associated with this wave, and *in fine* casts doubt on the relevance of the sub- versus post-lexical distinction.

The *second* point, positive this time, is the relevance of the attentional factor, which should be addressed in greater detail, specifically through the examination of the neurobiological correlates of the various dimensions which develop within. We know indeed that the N400 is sensitive to attentional modulations, but care must be taken to distinguish two encapsulated components within the attentional regimes. On the one hand, there is attention as opposed to inattention, as it constitutes a ray of consciousness deliberately attaching itself to an object, thereby in opposition to an attitude of consciousness which neglects this object, and on the other hand, there is the specific aim accomplished by the attentive attitude, for example the interest focused on such or such feature of the thing grasped by consciousness.

---

[85]Holcomb (1988, p. 80), our emphasis.

The elucidation of the N400 phenomenon will therefore involve the examination of the modulations of this wave in function of the attentional modalities engaged.

Finally, *thirdly*, and lastly, what the preceding results reveal is that, save particular conditions we will explain later on, the semantic impulse of the priming is never without impact, and this is so even when the protocol forces the experiment to unfold on a cognitive plane which is in rupture with all content. In fact, the presence of a semantic prime seems to involuntarily produce an N400 effect. And whether this N400 is interpreted at the term of a sub-lexical process (strict access) or of a post-lexical process (semantic) has little import: Each time, it is on the plane of meaning that the triggering motive of the N400 effect lies, thereby exhibiting a certain facilitation of the processing on a plane or another of the lexical system. All occurs as if, from the moment a stimulus having the aspect of a word is presented to a locutor, a sort of irrepressible moment of meaning is induced, as if, albeit attention is turned towards strictly material characters, the stimulus, the moment it is approached, is woven with meaning, inasmuch as sufficient time is afforded. We have seen indeed that it is by decreasing the SOA that Chwilla et al. (1995) manage to suspend the N400's priming effect, hence to partially liberate from the semantic constituency of the sign.

The famous "stroop" effect similarly illustrates this state of affairs, as do the results of the recent experiment conducted by Orgs et al. (2008) where "all is done" to reduce the semantic "thickness" of the stimulus. Here is its protocol: In the framework of the paradigm of priming, the stimuli (successively words S1 and S2) are given in pairs. However, to prevent the priming effect as much as possible, S1 and S2 are successively submitted to "concrete" and distinct tasks (Ti), specifically: First, visual presentation of the S1 word. Then, T1: Evaluation of the color (blue or red) of S1, in order to decide of the presentation or not of S2 (go/nogo protocol), then, if affirmative, presentation of S2 by auditory means, T2: To decide whether S2 is given to the right or left ear. The results are without appeal: The N400s of the S2s semantically primed by S1s have lower amplitudes than those of unprimed S2s. Everything unfolds as if the simple encounter of the words, albeit attention is systematically oriented towards certain aspects of their conformation as sensorial data, comprised the dimension of a semantic orientation in a constitutive manner. In this experiment and in those precedingly reported, we therefore observe that the attentional orientation does not seem to be capable of suspending the access to meaning, if not under certain conditions we will expose further ahead.

In any event, this observation of a semantic component always ready to be expressed, to manifest itself and to be realized, despite efforts to muffle it, incites to conceive of semiolinguistic phenomena as phenomena *carried by* meaning, i.e. comprising in themselves a semantic content and directionality, rather than *carrying towards* meaning; in other words, to conceive of semiolinguistic processes not as following the principle of a series of processes, which are more or less interactive, grasping from the start certain sensorial (or physical) features of a series of stimuli in order to arrive, through successive stages, at an integrated semantic representation, but following a logic of apprehension *intrinsically* comprising the dimension of meaning, precisely as a regime of configuration of a sensorial diversity into semiolinguistic phenomenality.

From this angle, the tension observed towards an object of meaning is therefore no longer to be taken as the expression of a programmed succession of processing stages, but as the principle of constitution of sensorial material into a semiolinguistic phenomenon. The matter is therefore no longer of recognizing stimuli concretely present in their status as signs through the failure or success of the processes they undergo, but to constitute semiolinguistic phenomena (signs in the full sense of the term) through an act of apprehension (noesis) supported by a semantic directionality (intentionality), that is, through an act grasping a sensuous hyle and, by animating it with an intentional orientation towards an object of meaning, instituting its appearing as a sign.

8.2.10 Confirmation of the Phenomenological View and Implications

We have thus seen that increases in attention translate, on the neurobiological plane, into an increase in the amplitude of the N400. This result, established when the attentional modulation concerns the degree of attention obtained by a solicitation or not of a conscious strategy, is confirmed by various experiments where the attentional mechanism is addressed in its own right: in its opposition to inattention.

Thus, in the experiment conducted by Mc Carthy et al. (1993), the participants, after fixating the center of a screen for a moment, will see words appear in their right or left visual fields. They are directed to ignore the stimuli in the left visual field and to perform a task of semantic categorization upon those appearing in the right field. The experimental paradigm being that of priming, the usual N400 priming effect is observed for target-words appearing in the right field, but not for those in the left field. But most significantly, what is observed is that conversely to the stimuli of the right field, the ignored words of the left field do not generate an N400. It must be noted that Yagamata et al. (2000) obtain similar results. This experiment acquires its full meaning when gradated with respect to the two following ones: Firstly, there is the experiment conducted by Bentin et al. (1995) which replicated in part the one conducted by Mc Carthy et al. (1993) but using auditory stimuli. This experiment consisted in presenting words to the right or left ear with the directive to memorize those presented to the right and to ignore those at the left. Still within the paradigm of priming, Bentin et al. (1995) observe the absence of an N400 priming effect for the words which were ignored. However, this time, the ignored words generated an N400, and the trace of this semantic activity was revealed during tests of implicit memorization. The third experiment which must be noted was conducted by Debruille (2008). As with the preceding ones, a directive was given to ignore certain stimuli. The protocol is the following: The stimuli are triplets (D–P–T) successively presented at the center of a screen, and where D is a "distractor" word destined to be ignored, P a prime word, and T a target word. The participants must, during a first experiment, ignore D and emit a judgment regarding the existence of a relation of meaning between P and T, and, as a second experiment, they are required to perform a memorization task regarding D and to perform the same task as they previously did regarding the P/T relation. Referring to the results of Mc Carthy et al. (1993) and arguing that the attentional processes increase the amplitude of the N400s "as more integration should take place when subjects pay attention to distractors than when they try to ignore them", it is legitimate to foresee for the "distractors" a lower-amplitude N400 in the first experiment than in the second. *Now, what can be observed is exactly the contrary*: The N400 of the ignored distractors is of greater amplitude than that of the distractors which are object of memorization.

These three experiments, and several other related ones, clearly demonstrate the following: As soon as the attentional ray encounters the stimulus, even in the most tenuous manner, in other words, from the moment attention is no longer capable of ignoring such stimulus because, the structure of the perceptual field not allowing the attentional ray to devote itself to something else, this ray inevitably comes to palpate it, then the fact of meaning arises, thereby producing the N400 effects. Let's examine the first experiment: Well-intentioned participants are capable of respecting the directive to ignore the stimulus of the left visual field. To such end, it suffices for them to skew their visual attention towards the right part of the perceptual field and to lock their focus upon it, knowing that the visual cone has a very narrow angle. The left stimuli remain unheeded with respect to the acts of semiolinguistic apprehension, hence the absence of an N400 priming effect and of an N400 wave signaling any degree of semiotic involvement for the stimulus. However, in the second experiment, it is more difficult to orient auditory attention exclusively towards the right ear: The stimulus presented to the left inevitably fall, at least minimally, under the attentional ray, even when it is suspended. Thereon, the dimension of meaning, as structuring semiolinguistic attention, unfolds. This is what is attested by the generation of an N400 with ignored stimulus as well as by tests of implicit memorization. But this semantic involvement being minimal, no N400 effect may be observed. In the third experiment, the stimuli to be ignored are given in the full field, without any possible dismissal, and, although the subjects are directed to ignore them, their attention cannot keep itself from grasping them. The semiolinguistic processes then triggered despite their intent will be manifested by N400 waves of which the amplitude is even higher (this needing to be explained) than those of the waves produced during the memorization exercise during which attention is sustained.

These three experiments and the preceding strongly support the phenomenological conception: We see that regardless of the level of attentional intensity and regardless of the practical orientations of attention, from the moment signs are approached, they are approached in their indivisible unity, always with the thickness of a meaning which traverses them. Now, this radical impossibility of detaching meaning-consciousness from form-consciousness can only be explained if we recognize a regime of meaning-intention underlying signifying forms (signifiers), in their very characters of appearing as signs, therefore in the principles of their constitution as semiotic phenomena.

8.2.11   The N400: Phenomenological Signification and Empirical Corroboration

From the standpoint of the phenomenological framework, specifically, having advanced the hypothesis of an N400 as a neurobiological signature of an act of intentional apprehension and of a semiolinguistic aim, and, correlatively, of its amplitude as a measurement of the unfolding of this act, it will now be a matter of showing that the set of aforecited experimental results, as well as other complementary results, rigorously coincide with the hierarchy of the strata of verbal consciousness of which the morphodynamic device delivers an explicit and regulated functional architecture. In return, this functional device doubled with its phenomenological signification will find an empirical homologation (we will address here only the consciousness of engagement, of the signified, and of fulfillment—cf. 6.4).

We may recall that the strata of verbal consciousness which stem from a functional factorization of the morphodynamics of the sign successively relate consciousnesses (i) of availability (the word-form being apprehended as simply available for semiotic functioning in a context in the process of being formed), (ii) of engagement (consciousness of the existence or not of a semantic orientation (control function)), (iii) of motif (consciousness of the identity of a semantic orientation), (iv) of the signified (consciousness of the differential meaning), and (v) of fulfillment (construction of an actual representation of the significatory aim).

An EEG replication of this stratification of verbal consciousness should then be observable. In particular, implying the hypothesis of an N400 as a neurobiological signature of a noetic-noematic act (apprehension and intentional configuration of a sensuous manifold) which develops these five phases of consciousness, and correlatively the hypothesis of the amplitude of this wave as a marker of the unfolding of the significatory act, the amplitude of the N400 emitted during the processing soliciting a consciousness of fulfillment should be greater than that of processes based on a consciousness of the signified, as the latter should be greater than what can be observed when the sole consciousness solicited is motif-consciousness, and so on for consciousnesses of engagement and availability.

And this is exactly what is given by experience.

It is indeed very firmly established that the N400s recorded during tasks consisting in constructing a mental image of the word presented (fulfillment) (West et al. 2000) have amplitudes which are superior to those observed during tasks engaging signifieds, as for example during tasks of semantic categorization. To this, we should add that the N400s elicited by "abstract" words, that is, words of which the fulfillment is not obvious, have lower amplitudes than those observed for "concrete" words (ibid.). Likewise, it can be observed that the amplitudes of the N400s emitted when the consciousness of a signified is solicited are superior to those of N400s generated during lexical decision tasks for which a simple consciousness of engagement suffices (when the experimental protocol imposes limitations upon the intentive unfolding—cf. *supra*).

Moreover, it is now possible to explain: (i) the weak N400 produced during memorization exercises (Debruille 2008; Bentin et al. 1995), (ii) the neutralization of the N400 effect caused by short SOAs, and (iii) the various triggerings of N400s by pseudowords.

In what concerns the first point: The work of memorization forces attention towards the plane of verbal consciousness where the signifying form (the signifier) is located, that is, the plane of a consciousness of engagement. Also, memorization confines the intentive dynamics to the first levels of verbal consciousness, thus at the first stages of the unfolding of the semiolinguistic act—which translates as a low-amplitude N400.

In what concerns the variations in the N400 in function of the SOAs: It has been established that, in priming protocols, the amplitude of the N400 is an increasing function of the SOA between prime and target. Likewise, the N400 effect is smaller for short SOAs than for long ones (Rossel et al. 2003), and this holds until the neutralization of this effect (Chwilla et al. 1995). It is clear that if we recognize the amplitude of the N400 as a measurement of the unfolding of a "noetico-noematic" act, then the short SOAs have for consequence to interrupt early on the deployment of the linguistic act, which translates as a decrease, and, if required, as a cancellation of the N400 wave.

Finally, concerning the N400s associated to pseudo-words, we should first recall the terms of the problem.

The fact is the following: Pseudowords (or logatomes), like words, generate N400s. As reported by Kutas et al. in a previously cited excerpt which we reproduce this time in full: "N400 is not simply an index of anomaly but rather a part of the *brain's normal response to* words (in all modalities) *or word-like stimuli*, such as pronounceable pseudowords."[86] So be it.

It may be conceded, at least as a first approximation, that the specificity of a pseudo-word is to be devoid of meaning, it should not, logically speaking, entail neurobiological responses translating cognitive processes involving meaning.

Thus, the generation of an N400 by logatomes is evidently incompatible with the interpretation of this wave as an integrative process. For the simple reason that integrative processing supposes an access to the significations to be merged: "[O]n a purely plausibility based integration account, semantic processing cannot begin until a word is identified and its meaning accessed."[87] But these logatomes are reputed to be devoid of meaning: "Pseudowords are fundamentally different from words in that they do not specifically have meaning."[88] Thus, for words which do

---

[86]Kutas and Federmeier (2000, p. 464), our emphasis.

[87]Ibid., p. 466.

[88]Anderson, Holcomb (1995, p. 179).

not have meaning, the integrative processing is without avail, and no N400 should be observed.

More generally, it is the semantic nature of the process that the N400 would signal which is questioned here—at least with respect to a formalist conception of the linguistic sign. Because according to the formalist approach, the signifier is intrinsically in "rupture" with the signified: It is an autonomous symbolic configuration to which a signification, which is by nature exterior, is conventionally linked. In the computational approach to the cognitive sciences, the signifier is similarly a pattern of sensorial features likely to be recognized as an entry of the lexical system within which its associated meaning will be accessible: "[T]he brain makes sense of a complex inherently meaningless signal—that is a word."[89] From such a perspective, the N400 would reflect the processing that stimuli undergo for purposes of meaning construction: "N400 seems to reflect a set of neural processes that are common to the analysis of sensory inputs for the purposes of meaning construction."[90] It goes without saying that, being a matter of logatomes, and following such a conception, the generation of an N400 is singularly problematic. As put by Holcomb and Neville (1990), whereas for real words, "[...] the target item has been located in the lexicon, and [...] its lexically based information has been activated and passed on to stages further up-stream",[91] concerning pseudowords: "What type of information is being passed further on for non-existent entries?"[92]

These difficulties can however be overcome by attributing to the N400 the functional sense of lexical access. Indeed, if the amplitude of the N400 measures the quantity of cognitive work required for actualization within consciousness, for purposes of lexical decision for instance, of a linguistic unit, then the work required for fully scanning the directory (mental lexicon) in search of an unfindable term must be translated by an N400 with a non-null amplitude. Moreover, as the work accomplished in order to localize the inexistent entry of a pseudo-word must have a superior cost than the processing of a real word, the amplitudes of the N400s emitted with pseudo-words should be greater that those emitted with real words: "[W]ord-like characteristics also produce lexical activation, but because no complete match [is] achieved the amount of [N400] activation produced [should be] greater and more prolonged."[93] And effectively, it may be observed, for instance during tasks of semantic decision (Ziegler et al. 1997; Bentin 1987), that the N400 of logatomes is of superior intensity than those of real words (cf. *Max Planck Institute*, 1997 report: §3.1.14).

But *a contrario*, in several experimental circumstances, the processing of logatomes produces an N400 of lower or equal amplitude than for real words.

---

[89]Kutas and Federmeier (2000, p. 463).
[90]Ibid., p. 467.
[91]Holcomb and Neville (1990, p. 306).
[92]Ibid.
[93]Holcomb and Neville (1990, p. 306).

For example, Holcomb and Neville (1990) observes that during a lexical decision task regarding primed target-words, the amplitudes of N400s emitted with loga-tomes are equal to those produced with unprimed target-words (these being, as anticipated, higher than those emitted with primed target-words). The experiment reported in (*Max Planck Institute* Report, 1997, §3.1.14) produces similar results: The exercise consisted in recognizing whether the target-word is the same as the item previously presented as a prime. The amplitude of the N400 of the target word is then observed to be indifferent to their status (word or logatome).

Also, that the N400 is supposed to reflect the "amount of lexical search" or the "effort to activate stimulus meaning", the observation of this wave during the cognitive processing of logatomes remains inexplicable and unexplained: "[I]t is not known exactly what processes underlie N400 pseudoword effects."[94]

The experimental protocol elaborated by Hill et al. (2005) allows to elucidate the issue of the N400 with respect to pseudo-words. This protocol consists in pre-senting prime-target pairs and, following each pair, participants must decide whe-ther there are two real words in it or if a logatome is present. When the time interval (SOA) separating the presentation of the target and of the prime is "comfortable" (700 ms), the N400 induced by the primes is shown to be alike whether these are words or logatomes. Then, by varying the SOA, it can be observed that for an interval of 150 ms between a prime and target, the amplitudes of the N400s emitted with primes having this short SOA are significantly inferior to those of N400s produced with the long SOAs.

The lesson that can already be drawn here is that access to words seems to be "graduated": The cognitive process develops more "broadly" when it is afforded more time. Indeed, if access took place in one fell swoop, the amplitude of the N400 of the prime would not vary in function of the SOA. Also, the experimentators may conclude that two sorts of access exists, the one being "superficial" and the other being "deep", and, in their eyes, what may distinguish the one from the other is that superficial processing "does not require access to meaning." Also, since the lexical decision tasks do not seem to require a recognition of the sense of the words (a "task performed at lexical level, does not [necessarily] require understanding of the word"), the existence of two levels of word consciousness at the basis of lexical judgment may be conceded.

These results suggest in any case that the accomplishment of a lexical decision task can manage with characteristics at a level lower than an accomplished semi-olinguistic identification. Indeed, in the first part of the experiment, it is reasonable to think that the apprehension of the first item of the pair of stimuli limits itself to the characters necessary for lexical decision making, simply due to the immanent arrival of a second stimulus which pressures to abandon the first.

If we accept that short SOAs limit the engagement of consciousness to the minimum required by the lexical decision task, then this result would be in good accord with the general idea regarding the functioning of an N400 interpreted as a

---

[94]Anderson and Holcomb (1995, p. 179).

neurobiological marker of the deployment of meaning-intention: Deployments limited by the experimental constraints manifest N400s with a lower amplitude than those produced in circumstances where verbal consciousness is free. This also explains why the N400, if it relates the engagement of consciousness (the signifi-catory aim) that is required *a minima* by the task, can be equal for words and for logatomes.

From a phenomenological perspective, indeed, the judgment of lexicality may base itself on different levels of verbal consciousness. It is indeed possible to decide of the status of a stimulus on the basis of its position in a consciousness of engagement, therefore either at a very early moment of the intentive dynamic, either at later level of consciousness (motif consciousness or signified consciousness), therefore with respect to an eventual meaning-intention, or, finally, at the level of fulfillment, in view of an actualized meaning. These various engagements of consciousness, which lead the semiolinguistic act more or less afar, will give rise to N400s with varying levels of amplitude. If judgment has recourse to the sole consciousness of engagement, then the processing of the words and pseudowords do not differ in any manner whatsoever: It is executed on a same level of verbal consciousness and, as has often been observed (Hill et al. 2005; Holcomb and Neville 1990; Maess 1997), the N400s emitted have a same level of amplitude. However, if the subject decides to bring the judgment to the level of consciousness of the signified or of fulfillment, such a decision conflicts with the pseudoword's nature, which is traversed with no linguistic intentionality, and which then becomes an obstacle. The accomplishment of such an act will require a kind of intentional forcing which will then translate as an N400 of higher amplitude for pseudowords than for words, and this is effectively what, in various cases, is observed (Chwilla et al. 1995; Hutzler et al. 2004; Ziegler et al. 1997; Bentin 1987).

Supplement: In the experiment conducted by Hill et al. (2005) discussed above, it may be observed that for a 150 ms SOA, the amplitude of the N400 of *non-primed* targets is *inferior* to that of 700 ms SOA *primed* targets. Now, we know that the N400 of primed targets is generally inferior to that of non-primed targets. This result attests the possibility of a less "costly" lexical decision than that which would nevertheless benefit from a preparation (by means of lemmatic or semantic pre-activation). This also attests to the existence of a level of verbal consciousness where the recognition of the lexical status requires neither its identification as an entry of a mental lexicon, nor the examination of its hypothetical signification, but, at a lower level, of the simple consciousness of its generic opening towards meaning (consciousness of engagement). Moreover, it is reasonable to suppose that, for a short SOA, the modality of consciousness involved in testing the prime is maintained for the examination of the target. Thus, for a short SOA, the low amplitude of the N400s of both prime and target words would reflect a confinement to the first strata of verbal consciousness.

8.2.12 Complement: The Potential of Recognition

There remains to address the issue of the cognitive process preparing and conditioning the intentive act expressed by the N400. We may, a priori, require that such a process should satisfy the two following properties: On the one hand, for obvious reasons of coherence and economy in terms of theory, this process, albeit external to the linguistic system as such, since it prepares in view of it, must be inscribed within the layering of the planes of verbal consciousness. On the other hand, on the plane of neurobiological correlates, its EEG expression must situate itself within a temporal window preceding that of the N400, that is, between 0 and 300 ms.

The first condition is easily met, *modulo* a slight adjustment of Husserlian analysis. Indeed, the first level of verbal consciousness is the one where the perceptual morphologies are configured, morphologies of which the forms of appearing will be modalized when grasped through the view of semiolinguistic intentionality, which then phenomenalizes them into authentic signifiers. But whereas the objects of this plane of consciousness are for Husserl concrete objects as delivered through spatiotemporal perception, this plane should now be conceived as a plane of perception administered by the phonological or graphematic regimes. In this manner, the position of an object at the first level of verbal consciousness notifies of its adequation with the linguistic forms and its possible implication at the higher levels where, if required, it will be configured into an authentic sign. In what concerns the second condition, only experience enables to verify it. We will now investigate the existence of an EEG potential with a latency inferior to that of the N400 and which functions as a selector of potentially signifying forms, that is, as a discriminator between phonological or graphemical compositions that are well-formed versus ill-formed from the standpoint of a language's phonological or graphemical system. The "Recognition Potential" (RP) meets these expectancies precisely.

The RP is a positive potential with a latency of 200–250 ms which is essentially located in PO7 (for a synthesis, cf. Martin-Loeches 2007). Its observation is made difficult due to the fact that it occupies the window of the N1 and P2 waves of perceptual processing which thus overlap it. In order to be uncovered, the RP therefore requires particular experimental protocols, among which the RSS (Rapid Stream Stimulation) which has for principle to approach the processes examined following the mode of a figure/ground opposition. This protocol consists in presenting at high frequency (10–20 Hz) "background" stimuli (of control) followed by a "target" stimulus. The control stimuli are constructed by the disorganization of the target stimuli so as to carry the same basic features (for instance sensorial). When the control stimuli are presented in a rapid flux, the processing channels activated by their "low level" characteristics are neutralized by saturation and the neurocognitive processes entailed by the appearing of a "target" stimulus, comprising "in addition" a certain organization, can thus be revealed.

The RP manifests during recognition trials: It is emitted when visual forms (faces, drawings, letters) are recognized by subjects without their being asked to perform a specific task: "the RP is an indication of the moment at which the gestalt pattern of a visual stimulus is identified".[95] Thus, presenting anglophone subjects with (i) English words, (ii) images, (iii) words written in Arabic or Chinese script, and (iv) control stimuli (composed of the disorganization of letter segments), Rudell (1992) records RP responses for stimuli (i) and (ii), the amplitudes of the former being superior to those of the latter, whereas (surprisingly) the RP responses for stimuli (iii) do not differ from those elicited by the control [stimulus (iv)]. Moreover, it has been established (Martín-Loeches et al. 1999) that the RP amplitudes elicited by real words are superior to those of pseudowords, themselves superior to those relative to non-words.

We therefore see that the recognition potential is susceptible of functioning as a selector of expression morphologies, as required by the phenomenological conception: At a certain level of graphematic conformation, specifically that of pseudowords, the RP potential attains an activation level which has for triggering condition a true semiolinguistic process manifested by an N400. For inferior levels of activation, the semiolinguistic intentive dynamics are suspended: Such is the case in presence of aberrant graphematic configurations or, more generally, in presence of any visual stimuli.

8.3   The Syntactic Potentials: LAN and P600

8.3.1   Preamble

In the preceding section, we have therefore recognized the N400 as a neurobiological signature of the constitution and investment by consciousness of a sign-phenomenon, of which the main strata are rendered in the morphodynamic device of the Saussurean sign. In other words, the N400 appears as the neurobiological indicator of the elaboration within consciousness of a sign-phenomenon as is likely to be grasped at various levels of its signifying "thickness". The N400 thus has a phenomenological signification: It relates the double movement of elaboration and intentive presentation of a semiolinguistic phenomenon. This being, and as programmed (cf. 1.2.2) it would now be fitting to investigate the relation which the forms of semiolinguistic phenomena, here empirically validated by EEG observation, establishes with forms which administer the functioning of said semiolinguistic phenomena and objectivate them. We may recall that three scenarios are to be considered, and that only empirical examination can decide which is valid: (i) either the forms of phenomenality are one with those of their objectivity, which would place us within a problematics of signifying morphologies where phenomena actualize and deliver to intuition their meaning of object, (ii) or, conversely, the forms of semiolinguistic objectivity are fully external to those of their

---

[95]Martin-Loeches (2007, p. 91).

phenomenality, which would then place us within a "classical" framework of the epistemology of the empirical sciences which, similarly, marks a distinction between the model of data (description of phenomena) and the theoretical device (body of concepts with an objectivating scope), (iii) or, finally, there is a partial overlapping of the forms of objectivity and of phenomenality, and we are then led to a normative conception of the semiolinguistic phenomenality; in such case, indeed, the very constitution of the phenomena, in other words, the modalities of their effective presence, is determined by the constraints which regulate their "assembling". It is precisely this particular epistemic configuration, in which the constitution of phenomena is carried by the regulation schemas which these phenomena manifest (distributional, morphosyntactic and, more broadly, transformational constraints), which EEG observation will attest to.

In order to achieve this result, it will be necessary to use for foundation a set of experiments and data, respectively accomplished and acquired following theoretical perspectives from which our approach clearly distinguishes itself—notably these perspectives which base themselves on the representationalist and formalist currents of cognitive sciences. It will therefore be necessary for us to engage in some "recycling" work which will first consist (i) in *temporarily* establishing a basis by positing the distinction between syntax and semantics and by collecting the EEG correlates of the processes responding to them (respectively of a semantic and syntactic nature), (ii) in showing that these neurophysiological correlates are not frankly compatible with the formalist and representationalist hypotheses and, finally, (iii) in arguing in favor of a morphodynamic conception where the precedingly considered EEG responses are more accurately deemed, from the standpoint of the coherence of a global picture, interpretable as a signature of morphogenetic processes producing compositions of sign-phenomena. It thus appears that the EEG waves at first considered as neurobiological marks of morphosyntactic regulations will be, *in fine*, to be understood as neurobiological marks of the processes administering the emergence of signifying morphological complexes, more precisely, complexes which are composed under the guidance of a normative regulation of sign-phenomena. This is thus to say that the semiolinguistic schemas of functioning considered at first as a set of morphosyntactic regulations or constraints affecting sign-units will be interiorized to the order of the phenomena they administer—very specifically in that it is through normative requirements, that concern these phenomena and which in this sense "compel" their compositions into organic totalities, which the existence of such phenomena dually weaves itself.

8.3.2 Introduction

Curiously, over the ten years which have followed the "discovery" of the N400, which promptly became recognized as the signature of semantic processing, relatively few research works on the EEG correlates of syntactic processing have been conducted. But since then, works on the matter have increased and various "syntactic" EEG potentials and effects have been uncovered.

In this set of highly diversified and sometimes contradictory observations—due notably to the sensitivity of the EEG effects to a multitude of factors which are sometimes difficult to predict, such as, for example, the length of words, their frequency, the modes of presentation... —two "robust" components distinguish themselves, these being respectively "early" and "late": the LAN (left anterior negativity) and the P600.

8.3.3 LAN and P600: Description

The LAN[96] is a negative potential widely distributed over the scalp but with a clear frontal preeminence, and often with a dominance in the left hemisphere. The LAN shares a temporal window with the N400: Its latency is situated between 300 and 500 ms after presentation of the stimulus. However, an equally distributed negativity of which the onset is earlier, named ELAN for "Early LAN", is often observed—and it will be a matter here of establishing whether it is a variant of the LAN or a distinct EEG component (cf. *infra*).

The P600 is a positive late component. Its latency is approximately 600 ms and its bilateral distribution presents a clear posterior[97] (centro-parietal) dominance. Conversely to the LAN and to the N400 which have clearly marked peaks, its more distributed[98] morphology has a "plateau" form extending over a few hundreds of ms.

Remark: Likewise to the N400 of which the signification exceeds the sole semiolinguistic field, the P600, also called LPC (late positive component), can be observed under experimental circumstances which do not necessary involve semiolinguistic units. For example: "[The LPC] shows Old/New effects across a range of stimulus types."[99] Likewise, and as we will further investigate, the semiolinguistic operations which the P600 manifest go beyond simple morphosyntactic processing. Thus, "the LPC has been thought to reflect semantic integration and conscious understanding [34], confidence in the integration of a word within its context [35], semantic memorization and classification [36–39], post-decision closure [40], or 'repair' of an erroneous sentential structure [41, 42]."[100] In what follows, we will first focus on the morphosyntactic signification of the P600.

---

[96]For example: "[LAN is] a negative-going wave within the temporal window associated with the N400 component. Typically, this negativity has been widely distributed but largest over anterior (and sometimes anterior left H) sites" (Osterhout 1997, p. 497).

[97] "[P600] is a positive component with a mainly posterior scalp distribution, characteristically starting about 600 ms after the onset of the target word" (Kaan et al. 2000, p. 160).

[98]"[T]he P600 is much broader than the N400 with less of a defined peak" (Ainsworth-Darnell et al. 1998, p. 119).

[99]Danker et al. (2008, p. 784).

[100]Daltrozzo et al. (2012, p. 270).

Regarding their functional signification in terms of semiolinguistic operations, the LAN and the P600 are commonly recognized[101] as neurophysiological manifestations of morphosyntactic processing. These two waves are indeed very regularly recorded in the following cases of inconsistency:

- *Violation of the subcategory (of the verb)*: P600 (here with a N400: case of biphasic response, (cf. VIII.3.7a-3 •Biphasic responses), for instance (Fig. 8.11).
- *Violation of subject/verb agreement*: LAN and P600.
- *Violation of the syntagmatic structure*: LAN and P600.
- *Violation of the agreement of the reflexive pronoun*: LAN and P600.

For the three latter sorts of violations, we observe EEG plots which are similar to those recorded during the transgression of the subject/verb agreement, as in the following example (Fig. 8.12).

- *Garden-path sentences: P600.*

When the items composing a sentence are successively presented to the subject, some sequences of words are likely to orient the syntactic analysis process towards a privileged structure. If the word which immediately follows this sequence is not compatible with the anticipated structure, and if, not violating any rule of proper formation, it "forces" to review the previously assigned syntactic qualifications, the production of a P600 can be observed for this word. For example: "*The broker persuaded to sell the stock was sent to jail.* Initially the verb 'persuaded' is read as the main verb of the clause. At 'to' this interpretation appears not to be correct, because the obligatory direct object of 'persuaded' is missing. Instead, the correct analysis is the one in which 'persuade' is the verb in a relative clause modifying the subject noun, and the upcoming verb 'was' is the main verb of the clause. At 'to' a P600 is found, relatively to an unambiguous control sentence."[102]

To summarize: Concerning the LAN: "Left anterior negativities (LAN) present at around 200 ms or later have been observed in correlation with phrase structure violations (Friederici et al. 1993, 1996; Münte et al. 1993; Neville et al. 1991; Osterhout and Holcomb 1992, 1993) and with agreement violations in otherwise normal sentences (Coulson et al. 1998; Friederici et al. 1993; Gunter et al. 1997; Osterhout and Mobley 1995)."[103] And concerning the P600, on the one

---

[101] "ERP correlate of syntactic processing have produced a great variety of effects: *on the one hand*, ERP response to a disparate set of syntactic violations is dominated by a large amplitude centroparietal positive wave with an onset about 300 and 500 ms and a duration of several hundred ms. This positive wave [is] labeled P600 or SPS. *on the second hand* [...] are reports indicating that ERP response to certain types of anomalies [...] is dominated by a negative-going wave within the temporal window associated with the N400 component (between 300 and 500 ms). Typically, this negativity has been widely distributed but largest over anterior (and sometimes left H) sites [the LAN]" (Osterhout 1997, p. 497).

[102]Kaan et al. (2000, p. 160).

[103]Friederici and Frisch (2000, p. 479).

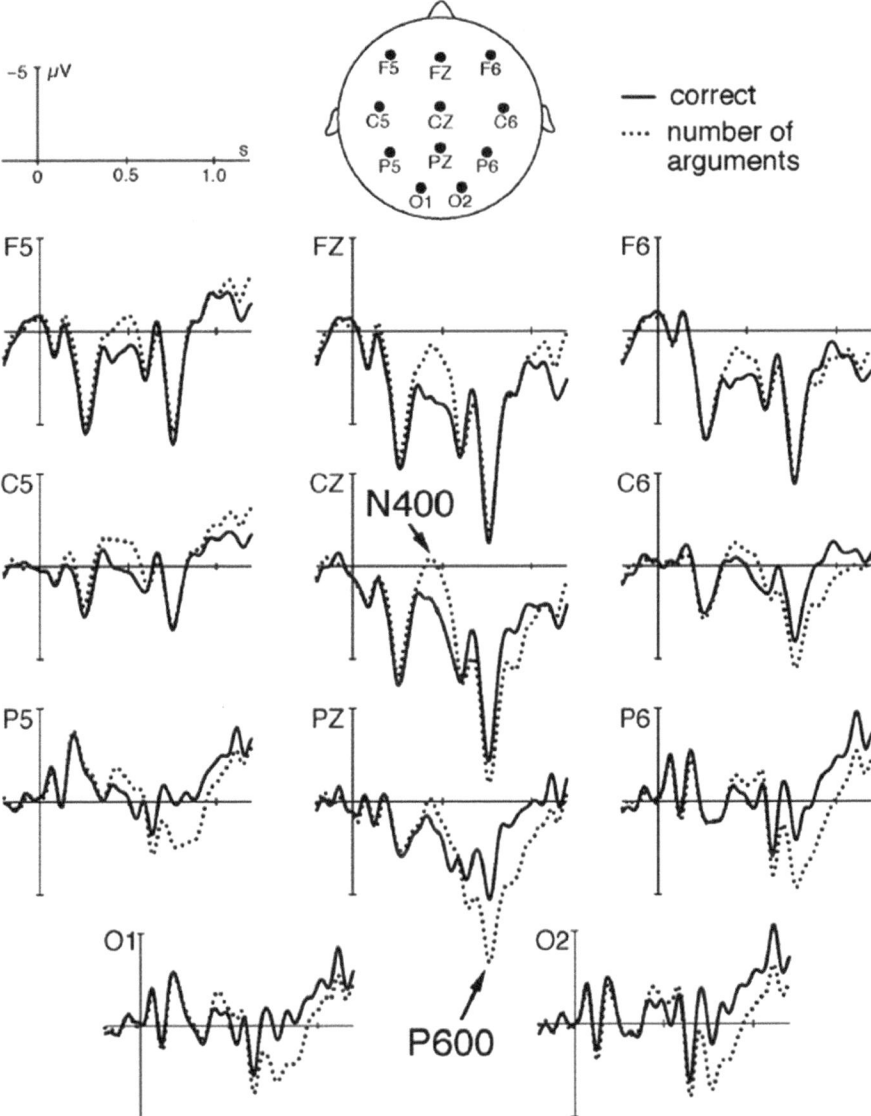

Fig. 8.11 "Averaged event-related potentials for the critical word (verb) for the correct sentences (*solid line*) and for the incorrect sentences (*dotted line*) in the number of arguments condition." (Friederici and Frisch 2000, p. 485)

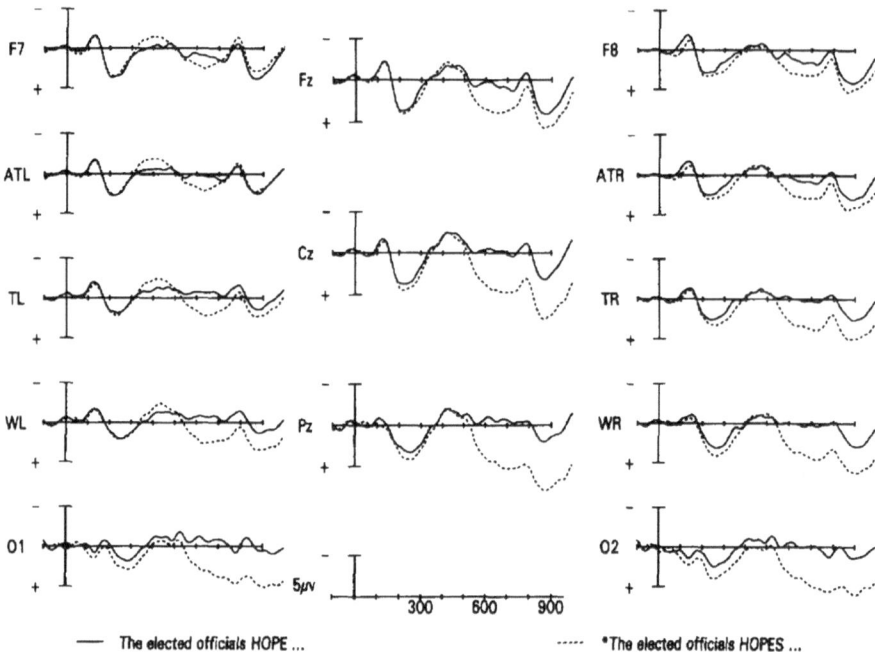

—— The elected officials HOPE ...   ····· *The elected officials HOPES ...

Fig. 8.12 "[ERP to] the critical word [*hope/hopes*] in non-violating (*solid line*) [*The elected officials* hope *to succeed*] and agreement-violating (*dashed line*) [*The elected officials* *hopes *to succeed*] conditions […]" (Osterhout and Mobley 1995, p. 745)

hand,[104] "P600 has been elicited by anomalies involving:—phrase structure ('the scientist criticized max's *of* proof the theorem'; ref, 1993, 1991)—verb sub-categorization ('the lawyer forced the man *was lying*'; ref, 1992, 1993, 1994)—subject-verb number agreement ('the doctors *believes* the patient will recover'; ref, 1993, 1995)—number and gender reflexive antecedent agreement ('the woman helped *himself* to the dessert')",[105] and also, "P600 was observed for the processing of sentences which either lead subjects to a garden-path or an incorrect structure."[106]

---

[104] And also: "The late positivity normally develops from 500 ms on (the so-called P600 component) and has been found to covary with a variety of syntactic anomalies such as garden-path sentences and other syntactically nonpreferred structures (Friederici et al. 1996; Hagoort et al. 1993; Mecklinger et al. 1995; Osterhout and Holcomb 1992, 1993), outright phrase structure violations (Friederici et al. 1996; Neville et al. 1991; Osterhout and Holcomb 1992, 1993), subjacency violations (Neville et al. 1991; McKinnon and Osterhout 1996), and agreement violations (Coulson et al. 1995; Friederici et al. 1993; Gunter et al. 1997; Hagoort et al. 1993; Osterhout and Mobley 1995)" (Friederici and Frisch 2000, p. 479).

[105] Osterhout (1997, p. 496).

[106] Friederici (1995a, p. 271).

And if we keep with this panorama of correlations between EEG components and types of linguistic inconsistencies and do not pay too much heed[107] to the cases of simultaneous[108] productions of N400 and P600 effects, we will readily conclude[109] the existence of neurocognitive processes specifically dedicated to syntactic processing, and correlatively, to a clear distinction in language of semantic and syntactic forms.

Discussion: Although the results presented here are not very debatable, and nonetheless robust, exposed in this manner, without further precision, they suffer from two insufficiencies which prevent their employment as empirical bases for any assertion regarding the order of language. This stems from two main reasons.

First, the above and swiftly drawn picture of correlations is incomplete: It omits a great number of other experimental results.

Also, this picture is somewhat ambiguous because it leads to believe that the causes of inconsistency under which the faulty configurations are regrouped here are "neutral" from a theoretical standpoint: in other words, that these "causes" adequately name and qualify "in themselves" the instances of transgression they subsume. But such is not the case, in truth (cf. example *infra*), and we will reproach this presentation with introducing a layer of theoretical presuppositions regarding the nature of the regimes of linguistic processing and the forms of failure which the LAN and P600 waves manifest. In order to illustrate this, we may consider the following incorrect utterance:

E: **der lehrer wurde* [has been] *gefallen* [fallen]

The inconsistency of E is clearly linked to the auxiliary "wurde" and the verb "fallen" which, being transitive, supposes the auxiliary "ist". But two readings of this are possible.

First, taking into account that the syntactic schema NP + Aux + V_{pp} is satisfied, it is possible to diagnose a transgression of verbal sub-categorization. This is the analysis proposed by Friederici and Frisch (2000): "Sentence [E] violates the number of arguments the verb can take, as the German verb *fallen (to fall)* cannot take an internal argument in the active voice and therefore cannot be made passive."[110] Or also: "[E is a] German passive sentence in which the context did not match a sentence-final verb with respect to its subcategorization frame."[111]

---

[107]"The available evidence [of simultaneous N400 and P600] does not support a clear model that relates semantic and syntactic levels to a pair of ERP effects that are unambiguously distinct in term of polarity, timing and topography" (Osterhout 1997, p. 497).

[108]Cf. §37a-3 •Biphasic responses.

[109]"Concerning the issue of the distinction between semantic and syntactic processes, ERP research has in fact shown distinct responses for the two linguistic processes. Semantically inappropriate words show a negative component, generally distributed all over but more intense in the posterior areas, peaking at about 400 ms after stimulus onset, called the N400 effect [...] with respect to syntactic processing two ERP components have been identified: An early left anterior negativity (LAN) and a late centro-parietal positivity. [...] To summarize, the ERP data show a distinct response for syntactic and semantic processes, which support models that include such a distinction" (De Vincenzi et al. 2003, p. 282).

[110]Friederici and Frisch (2000, p. 480).

[111]Ibid., p. 479.

But we can also attribute the error to the auxiliary "wurde", and the inconsistency would then be requalified as a syntactic violation. Indeed, admitting the distinction between grammatical and lexical words (closed- vs. open-class) we can inscribe "wurde" within a rule of syntactic derivation implying the category of transitive verbs V_t, that is: SV $\rightarrow$ "wurde" + V_t, and utterance E clearly transgresses this rule. And it is this analysis that Sakamoto et al. (2003, p. 391) report: "The syntactic violation in [E] [...] is caused by the incorrect choice of the auxiliary *wurde* 'is being.'"

In order to decide of the nature of the inconsistency examined here, it is then tempting to convoke[112] the EEG waves which this linguistic distortion elicits, that is, a LAN, and with less certitude, a P600. Indeed, if we accept the preceding table of correlations, we would be led to conclude the presence of a syntactic violation— but this, as we see well, at the cost of a denouncable circularity.

It would now be fitting to take into consideration eventual additional EEG waves, and to correlatively discuss the linguistic qualifications (i) of the LAN and P600 components and (ii) of the newly uncovered waves. In doing so, we will finally highlight the difficulties and contradictions facing the formalist approaches in their attempt to appropriate the results of neurophysiological experimentations.

Furthermore, neutralizing as much as possible the theoretical determination of the facts and of the motives of linguistic transgression, or, at least, attempting to reduce as far as possible the level of elaboration of the qualifications of the "admissibility" of a sentential configuration, we will endeavor to requalify the structural circumstances and operational economy of the production of EEG manifestations of semiolinguistic activity.

8.3.4 LAN/ELAN

First, concerning the LAN, an even earlier competitor has been found: The ELAN (Early LAN). Indeed, in cases of syntactic transgression concerning only grammatical categories, we have been able to uncover the presence of an anterior negativity presenting a significantly shorter latency than the LAN (between 120 and 180 ms). The question then arose as to whether it is a variant of the LAN or a distinct wave.

Taking into account the precedence of the ELAN over the LAN, it seemed[113] quite natural to correlate the ELAN with the processing of structures into

---

[112]"For this violation type Rösler et al. (1993) reported a left anterior negativity (LAN) between 300 and 500 ms [...] Rösler et al. also found a late positivity in sentences like [E], but did not discuss it any further, as it did not reach significance in the global statistical analyses" (Friederici and Frisch 2000, p. 480).

[113]"The different syntactic ERP effects have been taken to reflect different stages of syntactic processing during sentence parsing. The ELAN is viewed to reflect the stage of first-pass parsing during which an initial local syntactic structure is built on the basis of word category information, the LAN is taken to reflect syntactic processes in a second stage during which structural and thematic relations are assigned" (Friederici et al. 2004, p. 2).

components (construction of the derivation tree) and the LAN with ulterior morphosyntactic and sub-categorization processes.

The observation of an ELAN preceding the LAN therefore constitutes an argument in favor of modular architectures articulating, successively, syntactic, morphosyntactic, sub-categorical and semantic processes. Such is the case, for example, with Friederici's model: "[In] Friederici's model (2002), the two stages of syntactic analysis are reflected by distinct ERP responses: The first stage (Phase 1) is reflected by the ELAN, that is the early negativity found in response to phrase structure violations. The second stage (Phase 2) is tagged by both the LAN, which reflects morphosyntactic operations, and the N400, which reflects lexical semantic information and thematic role assignment operations. The two processing domains are thus viewed as functionally distinct but not distinguished on temporal parameters."[114]

But, as regards duration, the LAN/ELAN distinction does not prove to be relevant. ELAN effects have indeed been observed[115] when violations did not concern the categorical ordering of the sentence. Finally, it appeared[116] that the variations in latency observed for the LAN, which directly reflect the rapidity with which the information to be processed is acquired, are determined by two factors, that is: (i) the morphological composition of the word processed, and (ii) the modalities of presentation.[117]

It follows that the LAN and ELAN are to be seen as variants of a same and single EEG component manifested during various morphosyntactic transgressions.

---

[114]De Vincenzi et al. (2003, p. 390).

[115] "Early negativities [have been observed] with case, number, gender, and tense mismatch [and] in these violations the word categories is correct but the morphosyntactic features are wrong" (Hagoort et al. 2003, p. 39).

[116]"Input conditions appear to affect the latency of the early LAN as well. While the effect is observed early (when word category information is available early) in the auditory domain, the effect (for the same stimulus material) is present in the visual domain early only when words are presented fast and under optimal visual input conditions (high contrast), but not when presented under low visual contrast conditions [12] or when presented in a word-by-word fashion with longer pauses between each word [20]. Under the latter two conditions, the effect is reported to be present beyond 300 ms"; "When the structure of the critical word marks word category information is in the prefix, this information will become available early during auditory presentation. Under these circumstances, the effect is present early, i.e. 120–200 ms post word onset. When word category information is marked in the suffix, the crucial information becomes available much later with respect to the word onset, but still early with respect to the word category decision point (e.g. refined vs. refinement)" (Friederici et al. 2004, p. 1).

[117]Friederici et al. (2004, p. 2).

8.3.5 N280 and N400-700

Admitting the division between semantic and syntactic processing is valid and corroborated by distinct neurobiological processes, in all logic,[118] it should be expected[119] that the neurophysiological processing of linguistic units will clearly differ whether they concern units from closed-classes (grammatical or functional units) or from open classes (lexical units): "[O]pen- and closed-class words might be both functionally and neuroanatomically distinct."[120]

Orienting the investigation along this path, and in the hopes of more firmly supporting the first hypotheses of the formalist approach, it was possible to recognize[121,122] for some time the existence of two EEG components specific to closed- and open-classes—the N280 and the N400-700—with one dominating the anterior left regions and the other the parietal zone. It was also possible to advance that the anterior localization of the N280 is in accord with the regionalization of syntactic processing (anterior/LAN) and of semantic processing (posterior/N400).

These two components, the N280 and the N400-700, were the object of several experiments. They proved difficult to reproduce and were cause of debate. Thus, Osterhout et al. (1997a, p. 145) report that "not all researchers have observed an N280-like negativity in the ERP response to closed-class words and in several studies the two word classes have elicited negative-going components that were quite similar in timing and/or distribution."

---

[118]"In English, the distinction between syntax (sentence form) and semantics (sentence meaning) is roughly paralleled by the distinction between two lexical categories, the closed-class and open-class vocabularies" (Osterhout et al. 1997a, p. 144)—or also "Open class words carry the bulk of the semantic meaning of an utterance [...]. The closed class (function) words [...] serve the purpose of syntactically structuring a sentence" (Münte et al. 2001, p. 91).

[119]Although the distinction between open and closed class is far from forming a consensus: "[T]he list is open, we would say, for the lexemes; but in French, for example, the microsystem of the months of the year is closed, whereas that of the nominally-based relational terms (morphemes) is open: *à la demande de* ['at the request of'], *en référence à* ['in reference to'], etc." (Hagège 1990, p. 69).

[120]Osterhout et al. (1997a, p. 144).

[121] "Neville et al. (1992) found a variety of effects that appeared to differentiate closed and open class words: (1) negativity with a left anterior temporal maximum at 280 ms for the closed *but not* the open class words; [...] (3) a left late negativity (called N400–700) for closed class items. This was described to be the largest over left frontal regions, but in fact was seen with almost equally large amplitude over parietal areas" (Münte et al. 2001, p. 92).

[122] Or also: "Neville et al.'s (1992) observation that the two word classes elicited negative components that clearly differed in their temporal and spatial properties. Closed-class words elicited a negative component that was largest over anterior regions of the left hemisphere and that peaked at about 280 ms (N280). By contrast, open-class words elicited a large-amplitude, posteriorly distributed negativity that peaked between 350 and 400 ms (N400)" (Osterhout et al. 1997a) and "As in earlier reports (Kutas and Hillyard 1983; Kutas et al. 1988; Neville et al. 1992; St. George et al., 1994), closed-class words elicited an increased negativity between 400 and 700 ms, whereas open-class words elicited a late positive component (LPC) in this window" (Osterhout et al. 1997a, p. 164).

Before examining the case of the N280, and consecrating a few moments to a matter of prime importance, that is, of the informational content (differential or absolute) of the EEG components, we will note that the N280 has been shown by means of the waves evoked and has thereby been recognized as a component conveying *in itself*—and not differentially—information regarding neurolinguistic processing: "The N280 is an ERP component that is seen in an averaged waveform to words of one or more types. For instance, in the averaged waveform for closed-class words one can easily identify a component with a maximal amplitude at around 280 ms."[123] Furthermore, the author of this citation who nevertheless assumes a differential conception of the EEG information specifies that, from this standpoint, the LAN is to be distinguished from the N280: "The LAN effects are to be distinguished from the N280 that is reported in relation to the processing of closed- vs. open-class words. [...] The LAN, [...] refers to the amplitude *difference* between two conditions. It is identified by comparing the averaged waveforms of two conditions. That is, in one condition one sees an increased negativity in comparison with another condition."[124] In this respect, it would be interesting to note that ulterior experiments have established an equivalency between the N280 and the LAN, thus comforting the "positive" reading of the EEG information.

Indeed, let's return to the N280 of which the odyssey perfectly illustrates the difficulty in identifying an EEG component due to a difficultly predictable diversity of exogenous variational factors.

Taking note of the variability of the EEG effects generated by grammatical words, Osterhout et al. (1997a) implement a series of experiments aiming to examine in greater detail some exogenous parameters having already been recognized as variational factors for the waves associated with grammatical as well as with lexical words. Under "normal" conditions of presentation (within admissible utterances and for purposes [tasks] of comprehension), Osterhout et al. (1997a) observe that, essentially and in conformity with the habitual results, within the window of "late" latencies, the grammatical words generate an N400-700 whereas lexical words generate an LPC[125] (or P600). Within the window likely to contain the N280, "[...] i.e. between 250 and 350 ms, closed-class words did not elicit a clearly defined N280 component."[126] And within the window of 400 ms, we can observe a classical N400 for "full" words, whereas "empty" words produce a double N350-N450 peak—a double peak of which we will however only retain the first component due to the fact that "given the rate of stimulus presentation (each word displayed for 300 ms), the N450 might be caused (in part or entirely) by stimulus offset effects."[127] Moreover, the N350 and the N400 cannot be differentiated on the basis of their distributions: "[E]vidence of distinct scalp distributions

---

[123]Hagoort et al. (2003, p. 40).

[124]Ibid., p. 39.

[125]Late Positive Component.

[126]Osterhout et al. (1997a, p. 150).

[127]Ibid.

[…]: analyses on peak amplitude revealed highly unreliable interactions (N350 vs. N400): $F(4, 52) = 1.05$, $p > 0.4$)."[128]

But an examination of the results of this experiment through the prism of selected grammatical and lexical categories (resp.: articles, prepositions, pronouns, auxiliaries, names, and verbs) highlights a temporal distribution of the latencies of the first negativity after P2—a distribution which was "collapsed" by the computation of the mean consecutively to choice of the sole factors of "open-" versus "closed-class"—that is:

Peak = f(gram. cat.)	250–450 ms	Sites (250–400 ms)
Articles	Neg. Peak @280 ms	LH > RH Larger @Temp. and Wernicke Sites
Prepositions	Neg. Peak @320 ms	LH > RH Larger @Temp. and Wernicke Sites
Pronouns	Neg. Peak @350 ms	LH > RH Larger @Temp. and Wernicke Sites
Auxiliary Verbs	Neg. Peak @360 ms	LH > RH Larger @Temp. and Wernicke Sites
Nouns	Neg. Peak @400 ms	Slightly Larger @RH
Verbs	Neg. Peak @400 ms	Slightly Larger @RH

This somewhat surprising distribution of latencies which the preceding table illustrates, as a monotonous function of a series of grammatical categories, is quickly explained when considering the data no longer from a qualitative and formal standpoint, but from a quantitative and material one. Indeed, the computations performed by Osterhout et al. (1997a) have shown a strong correlation between the latency and the properties of duration and frequency of the stimuli. Moreover, concerning the cranial distributions of the first negativity after P2, no significant correlation has been shown[129] with the opposition between open and closed classes.

But these first results are perplexing, because they establish continuity between components which are distinct with respect to both their spatial distributions and their linguistic qualifications (from N280 to N400). As noted by the authors of a study following up on Osterhout et al. (1997a), "Osterhout et al. (1997a) reported a correlation of the latency of an early negativity and the mean normative frequency

---

[128]Ibid., p. 151.

[129] "The latencies of these negativities were highly correlated with the lexical properties of word length and frequency. Latency variation in these negativities therefore reflected quantitative differences in lexical properties rather than qualitative differences corresponding to the open- and closed-class distinction. There were also some reliable differences in the scalp distributions of these negativities. However, these differences appeared to be a function of grammatical category rather than word class and were restricted to categories that were farthest apart in terms of the latency of the negativity" (Osterhout et al. 1997a, p. 155).

Fig. 8.13 [M]ean amplitude in the 350–420 ms window (Cz derivation) plotted against mean logarithmic frequency of the different word categories (Münte et al. 2001, p. 98)—*Full/Dotted Line Open/Closed class*

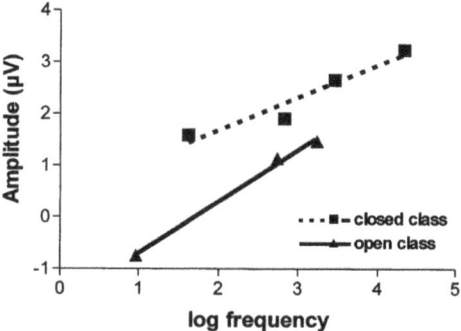

and mean length of the words regardless of word class. [They] did not distinguish between the N280/LPN and the N400 component."[130]

This continuity between the N400 and the N280 is of course problematic. But it has the benefit of attracting attention to the fact that grammatical words are not without generating a "semantic" component—as has often been observed. Therefore, "in accordance with previous works [...] an N400 effect was present for both classes of words. Thus the N400 cannot be viewed as being elicited by open class words only."[131] However, the N400 for empty words generally has a weaker amplitude than the N400 for full words. Furthermore, Münte et al. (2001) observe that the "law" of amplitude variation observed in function of word frequency holds just as well for the N400s associated with "empty" words as it does with full words —these all being shown by Fig. 8.13.

This study, of which the results "do not support the claim that the N280 component is a categorical marker of the closed class",[132] has later been refined, notably by Osterhout et al. (2002). The authors of Osterhout et al. (2002) recognize that the units of open and closed classes produce distinct EEG responses, but want to examine if "differences in brain response to open and closed class words are due to differences in the linguistic function of these words, or to differences in physical form."[133] In the experiment they conducted, the words of different classes and of varying lengths were presented in a context of correct utterances read for purposes of comprehension. The results are as follows:

First concerning the average responses to words belonging to open and closed classes, distinct EEG are indeed recorded (Fig. 8.14).

In these graphs, following an N1/P2 complex, *and for the two sorts of words*, it is possible to observe: (i) a negative component (approx. N300) with an anterior left dominance, but with an amplitude which is greater for grammatical words, (ii) a

---

[130]Münte et al. (2001, p. 92).

[131]Ibid., p. 97.

[132]Osterhout et al. (1997a, p. 154).

[133]Osterhout et al. (2002, p. 172).

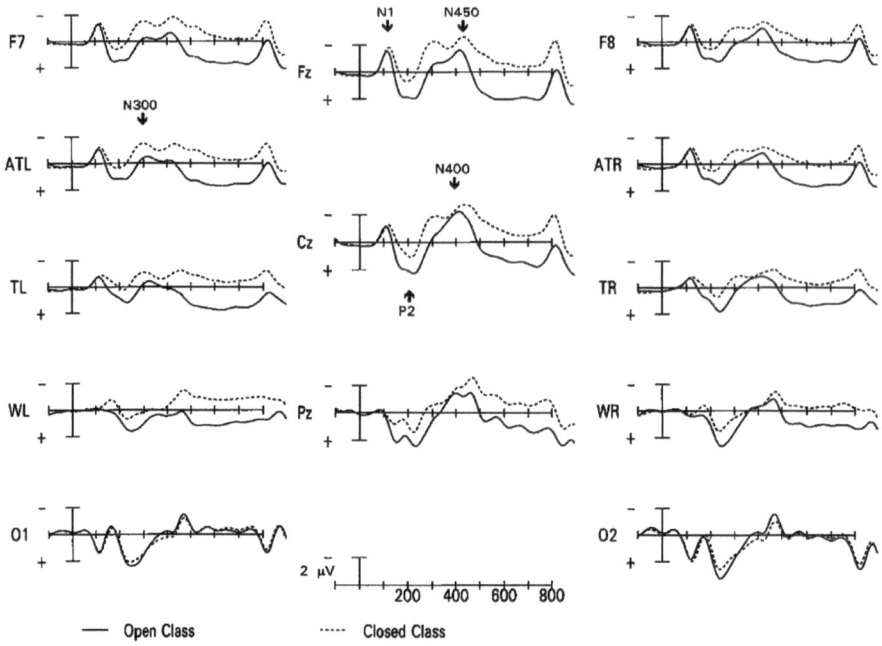

Fig. 8.14 "ERPs recorded […] to closed-class and open-class words." (Osterhout et al. 2002, p. 176)

non-relevant N450 because it is "probably reflecting word offset",[134] and a late component, resp.[135] negative [N400-700] or positive [P600] for closed or open-class words. Following Osterhout et al. (2002), the centro-parietal N400 component is only observed for full words, but ulterior results will lead to reconsider such a reading and to recognize a weak N400 for empty words.

The analysis of potentials evoked in function of grammatical categories (cf. Fig. 8.15) confirms the results obtained by Osterhout et al. (1997). The latency of the first negativity following P2 varies in function of the grammatical category.

A certain uncertainty must however be noted regarding the identity of the factor of variation. Indeed, Osterhout et al. (2002) observe that the parameters of frequency and of duration are highly correlated, and that it is indeed impossible to decide which is the cause of variation. Nevertheless, they deem that there are reasons "to believe that word length, rather than frequency, accounts best for the result reported here."[136] Without taking position, we present the argument here:

---

[134]Ibid., p. 178.

[135]Ibid.

[136]Ibid., p. 184.

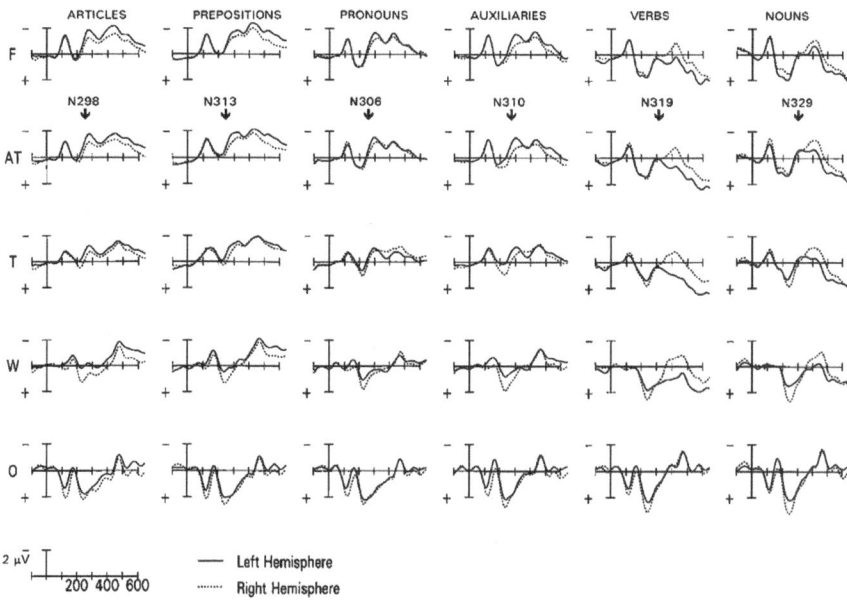

Fig. 8.15 ERPs averaged over grammatical categories [...] (Osterhout et al. 2002, p. 180)

- [first] we observed more *positive* going wave forms [...] as a function of length (=F(length));
- [yet] the correlation between length and frequency is of an inverse nature;
- [therefore] more *positive* going wave forms = $F_{inverse}$(frequency);
- [hence] more *negative* going wave forms = F(frequency);
- [yet] the expected effect for word frequency [...] is one in which lower frequency words elicit more *negative* going wave forms during the N400 epoch;
- [hence] there are reasons to believe that word length, rather than frequency, accounts best for the results reported here.

Finally, *and here is the conclusive result*, the evoked potentials are shown to continuously vary in function of the length (or frequency) of the words (Fig. 8.16).

"Inspection of Fig. 8.16 reveals that all words elicited a similar series of negative and positive going deflections (N1, P2, N300, N400, N450 [and Late Component])"[137]; Which amounts to saying that the waves evoked by the linguistic units, regardless of their grammatical category, all comprise components recognized to be "syntactic" (early negativity and late component) and "semantic" (N400). It follows that "the open and closed class distinction does not obviously and robustly manifest itself in the standard time domain analysis of scalp recorded event related potentials."[138] We will agree that these results cast a doubt on the

---

[137]Ibid., p. 181.
[138]Ibid., p. 185.

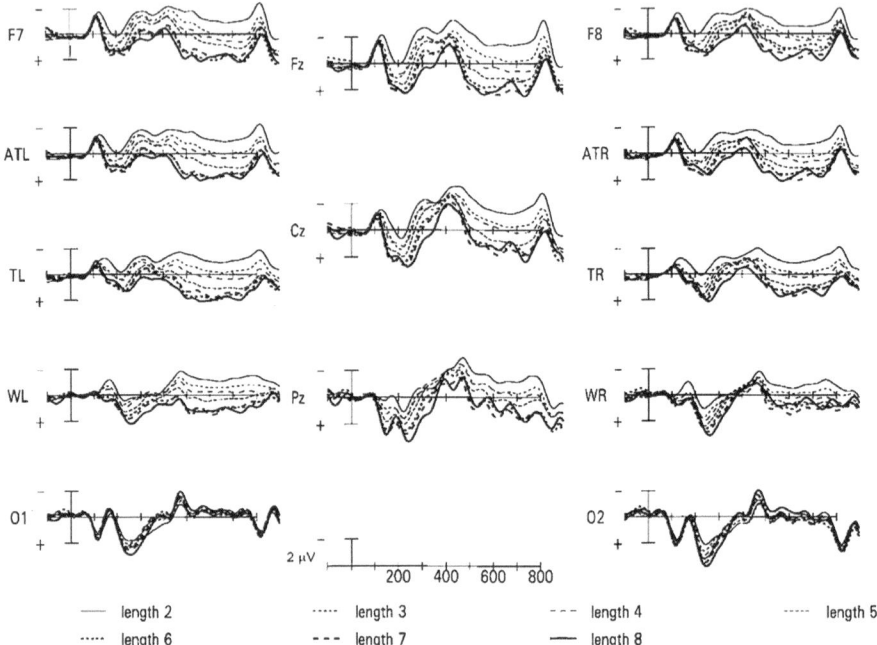

Fig. 8.16 ERPs averaged as a function of word length in letters [...] (Osterhout et al. 2002, p. 181)

validity of the opposition between syntax and semantics. It is, however, too early to draw conclusions, and it will therefore be appropriate to continue the examination of syntactic waves.

8.3.6   Conclusion and Transition

In order to uncover a few EEG invariants of linguistic activity, we were able to establish (i) that the LAN, ELAN, and N280 are three manifestations of a same early "syntactic" process, and (ii) that the P600 and the N400-700 reflect a same late "syntactic" process. On the basis of this, we will retain as neuronal correlates of syntactic activity in language only two components, that is, successively: a negative wave with anterior (left) dominance, the (E)LAN, and a terminal wave, charac-terized by a plateau morphology and with a centro-posterior dominance, the P600.

In order to have a complete picture of the history of linguistic processes—such as manifested in neuroelectric activity—there remains only to account for the "semantic" wave N400.

The N400 clearly precedes the late syntactic component. But it often appears in the same temporal window as the LAN. The preceding results nevertheless lead to suppose that it relates a processing which is ulterior to that of the LAN, and this hypothesis has been confirmed during a study concerning the EEG effects due to semantic and morphosyntactic violations with measurements taken at intervals of 10 ms. The precedence of the LAN over the N400 has thus been established[139] with precision.

Also, *we will retain three components* from the neurolinguistic activity manifested by the EEG: two components being correlated to syntactic processing, the LAN and the P600 (or their variants), and "flanking" a third, semantic component, the N400.

8.3.7 Syntax and Semantics

8.3.7.1 Interferences Between the N400 and Syntagmatic Structures

Introduction

We have, in the previous paragraph, roughly drawn the picture of the circumstances of the generation of the N400. It will now be a matter of completing it by noting the influence of the syntactic factors upon this wave.

The Case of the Final Position

It is known that in the grammatical tradition, the intonational description as well as numerous syntagmatic analyses agree upon a segmentation of texts into "sentences".[140]

The relevance of such segmentation is attested by numerous psychocognitive observations which have highlighted the specific role of the last word of a sentence. Indeed, the reading of the final word produces a "wrap-up effect": The terminal word is the object of a longer fixation period than are the words preceding it. This increase in the time of attention is often interpreted[141] as being associated with a

---

[139]"Finally, a further analysis was conducted on small (10 ms) intervals within the 300–500 ms epoch to measure the onset of the significant effects elicited by the target word (W1) for the two kinds of violation. The LAN was [...] starting at 340 ms till 400 ms [...]. The N400 [...] starting at 430 ms till 470 [...]. Thus, the detection of syntactic violation started 90 ms before the detection of semantic one" (De Vincenzi et al. 2003, p. 288).

[140]For example, for Hjelmslev, the "lexia" is the smallest amongst the "free" units.

[141] "The reading times show the so called wrap-up effect, that is an increase in reading time on the last word that affects all sentences, with or without violations [...]. The *causes* of the wrap-up effects have never been fully identified. Generally, they include all the processes of semantic interpretation of the sentence in a broad sense, such as establishing its true-value properties, establishing the referents of free pronouns, integrating the sentence into a discourse, establishing the speech act of the sentences" (De Vincenzi et al. 2003, p. 291).

Fig. 8.17 ERP (*full line*) elicited at Pz by a word (*hope*) in intermediate position (Osterhout and Mobley 1995, p. 745)

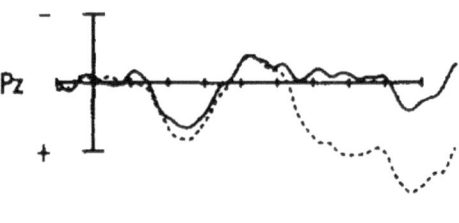

Fig. 8.18 ERP (*full line*) elicited at Pz by a word (*succeed*) in final position (Osterhout and Mobley 1995, p. 747)

process of recapitulation aiming to construct an integrated representation of the sentence at its various levels of semantic qualification. More generally, "it is well known that sentence final words are often strong attractors of global processing factors related to sentence wrap-up, decision and response requirement."[142] These interpretations are in accord with the point of view of linguists, who, likewise, see in the final word a "point of stoppage", that is, a marker in the place where "the semantic synthesis must be performed for all terms and relations having been exposed since the beginning of the unit of meaning which has thus been 'stopped'."[143]

Corroborating the particular status of words which wrap-up a sentence, EEG analysis records specific responses to them. It is thus well established[144] that terminal words produce great "positivity" (posterior).

In the following figures,[145] the drawings in *full lines* show the "great average" of the EEG recordings taken in Pz for words in an intermediate position (Fig. 8.17) and for words in a final position (Fig. 8.18), in utterances which are syntactically similar to *the elected officials hope to succeed*. With the authors, we observe that "sentence-final words [are] followed by a large amplitude positivity, which is often observed following sentence-final words."[146]

---

[142]Hagoort et al. (2003, p. 40); cf. also Van Petten and Luka (2012).

[143]Tamba-Mecz (1991, p. 48). Furthermore: "[I]n these places of 'caesura' of the linear sequence, occurs a kind of 'precipitation' which results in a global semantic result, by integrating all semantic data constitutive of the unit of meaning under consideration.".

[144] "Psychophysiological research has show that ERP elicited by final word are not the same than by sentence embedded word; for example, ending words are followed by a large positivity not typically observed following sentence embedded words" (Osterhout 1997, p. 498).

[145]Graduation in hundreds of ms.

[146]Osterhout and Mobley (1995, p. 744).

Fig. 8.19 "Grand average
ERPs (recorded over site Pz)
to sentence-final words in
sentences that contained a
semantically anomalous word
[e.g. *the boat* barked *during
the storm*] (*dashed line*) and
well-formed control sentences
[e.g. *the boat* sank *during the
storm*] (*solid line*) [...]"
(Osterhout and Mobley 1995,
p. 755)

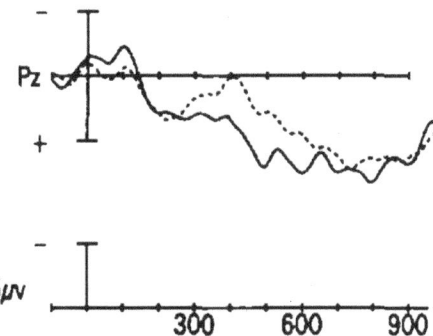

This late positivity is interpreted as "reflective of a 'syntactic closure' operation
and the realization that the sentence is over"[147]—an interpretation corroborated[148]
by the fact that "late positivity" is decreased in utterances composed by randomly
placed words, which are therefore in sequences without semantic or syntactic
coherence.

With regard to the particular function of the final word in the process of "re-
ception" of a sentence, it can be anticipated that the N400 effect manifests especially when the integrity of the unity of the sentence is disrupted.

It may then indeed be observed that a semantic inconsistency situated in some
place of the sentence has a repercussion upon the word in the final position, *when
this very word does not present any character of inadmissibility*. For example, in
the following Fig. 8.19, we can observe that "the final word [here the word 'storm']
in sentences containing [semantically] anomalous words elicited an enhanced
N400-like effect between 300 and 500 ms."[149] We will contrast this result with the
fact that in "normal" utterances, the amplitude of the N400 produced decreases with
the position of the word until it disappears: "[F]inal words of congruent sentences
differ from both intermediate words and the final words of the other sentence types
in eliciting no N400 activity".[150]

---

[147]Van Petten and Kutas (1991, p. 105).

[148] "The reduced amplitude of the late positivity for final words of the random strings [...] is
consistent with this broad description. These words did not complete a structural or semantic unit"
(Van Petten and Kutas 1991, p. 105).

[149]Osterhout and Mobley (1995, p. 753).

[150]Van Petten and Kutas (1991, p. 103).

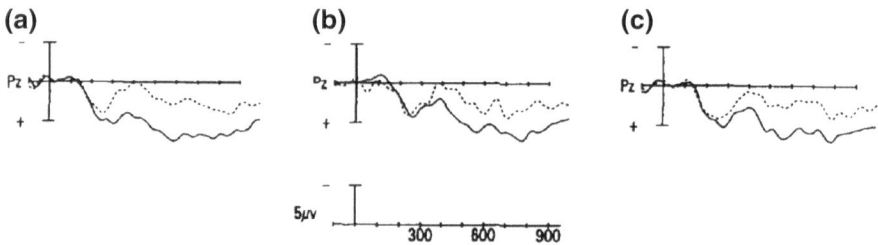

Fig. 8.20 "Grand average ERPs to sentence-final words (recorded over site Pz) in sentences containing agreement violations (*dashed line*) and non-violating sentences (*solid line*). (**a**) subject-verb number condition. (**b**) reflexive-antecedent number condition. (**c**) reflexive-antecedent gender condition (Osterhout and Mobley 1995, p. 747)

N400/Syntax Interactions

- *Final Position*

What is less expected, conversely, is that we observe the same thing when the transgression is purely *syntactic*: The final word also elicits an N400 effect.[151]

As an illustration, the following diagrams (Fig. 8.20) show the "grand average" of the EEGs recorded in a final position in cases of morphosyntactic transgressions (dotted lines) and in normal cases (full lines) concerning:

(A) Subject/verb number agreement:
  the elected officials hope/*hopes to succeed
(B) Number agreement of the reflexive pronoun:
  the hungry guests helped themselves/*himself to the food
(C) Gender agreement of the reflexive pronoun:
  the successes woman congratulated herself/*himself on the promotion

We indeed see that the potentials evoked by "the sentence-final words in the ill-formed sentences were more negative-going than those to well-formed sentences, most notably between 300 and 500 ms."[152] We also observe that "these effects extended into the 500–800 ms window."[153]

---

[151] "ERPs to a sentence containing an embedded syntactic anomaly have been show to deviate in two ways from those to well formed controls: the anomalous word elicits a P600-like positivity, and the non-anomalous sentence-final word elicits an enhanced N400-like response relative to ERPs elicited by the same words in sentences that do not contain an embedded anomaly" (Osterhout 1997, p. 498). Likewise, "the present ERP study, and in general all those ERP studies in which the syntactic violation is not the last word of the sentence, show an end-of-sentence negativity, that is they show that the final act of sentence interpretation is somehow disturbed by the presence of a syntactic mistake" (De Vincenzi et al. 2003, p. 291).

[152] Osterhout, Mobley (1995, p. 746).

[153] Ibid., p. 748.

• *Intermediate Position*

With respect to what precedes, it is necessary to add that N400 effects caused by syntactic inconsistencies have been observed[154] in other positions than the final position. However, the reproducibility of such results is not excellent. For example, the experiments conducted by De Vincenzi et al. (2003) do not highlight an N400 effect for the words which follow a violation of number agreement between S and V. Conversely, when the violation is semantic, we observe[155] that the N400 effect propagates over the various words which follow the incompatible word (cf. Fig. 8.21).

What is important to note here, is that the N400 effect entailed following a syntactic or semantic distortion occurs even when the words composing the ill-formed utterance remain coherent from the standpoint of their participation in an integrated semantic representation (cf. preceding examples and *infra*).

• *Biphasic Responses*

In order to support the preceding examples regarding the generation of an N400 even when the processed word presents no semantic incongruity, we will mention an experiment (Osterhout 1997) in which it was observed that the potentials evoked by a distortion of sentential structure "localized" on a lexical unit (open class), if it elicited a "classical" P600 for some subjects, it could elicit an N400 for others.

Specifically, it is a matter of utterances comprising a "reduced relative clause" which is syntactically deviant, for example: *The boat sailed down the river* sank *during the storm* (vs. the correct version: *The boat sailed down the river and sank during the storm*) where the critical word *sank* "[is] expected to be perceived to be syntactically anomalous."[156]

In the experiment reported, *sank* (in an intermediate position) therefore elicits, for some subjects, an N400 effect, and for others, it elicits a P600 wave. Observing this, the author[157] concluded in favor of a functional interconnection between the

---

[154] "Several researchers have noted that [...] words immediately subsequent to the [embedded syntactically] anomaly [...] or at the end of the sentence containing the anomaly [...] elicit an enhanced N400 response" (Osterhout 1997, p. 519).

[155] "[W]hile syntactic violations produce an immediate perturbation that quickly returns to base line, the effects of the semantic violations last until the end of the sentence" (De Vincenzi et al. 2003, p. 289).

[156] Osterhout (1997, p. 505).

[157] "The critical words of [this experimentation] are fully compatible with the semantic content of preceding context and were preceded by the same semantically related words in both the sentences. Even so, the critical word [in the anomalous clause] elicited a greatly enhanced N400 component in some subjects—this N400 effect was nearly identical [...] to that elicited by semantically anomalous words [...] for such an account to hold, one must assume that N400 amplitude can reflect the semantic anomaly engendered by an unparsable or misparsed sentence." And correlatively, therefore, "some have suggested that the [semantic] priming occurs via an automatic spread of activation through a conceptual or lexical network (ref)—the findings reported in [this] experiment, together with other recent works (1993, 1992) are clearly inconsistent with this notion" (Osterhout 1997, p. 519).

Fig. 8.21 Grand average
ERPs recorded […] to
subject-verb Semantic
agreement violations
sentences and controls. Onset
of the critical words in
non-violating (*solid lines*) and
number agreement violating
(*dashed lines*) conditions is
indicated by the *vertical bar*
(De Vincenzi et al. 2003,
p. 289)

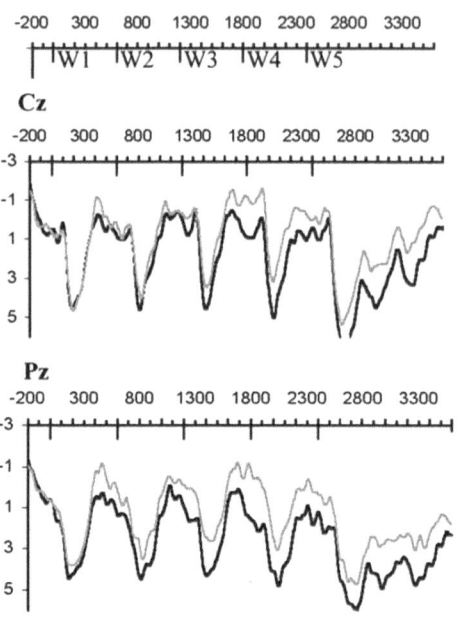

semantic and syntactic dimensions, and correlatively, they concluded the interpretation of the N400 as a semantic process developed within an associative network structure was invalid.

It is remarkable that in this experiment, the first results obtained by an average
calculated over all subjects revealed a biphasic N400/P600 wave, and it is only
following a more minute analysis that it was revealed[158] that this biphasic wave
superimposed two very distinct types of processing.

It remains that, in numerous experiments where the material presented comprised a violation of sub-categorization rules (namely concerning the verbal argument structure), the N400 and the P600 were generated together. For example,
Friederici and Frisch (2000) tested NP-NP-V and V-NP-NP configurations "using
verbs which obligatorily required only one argument in sentences which contained
two arguments."[159] Thus:

---

[158] "[S]yntactically disambiguating words appeared to elicit a *biphasic response*, i.e., an increase
in N400 amplitude followed by a large P600-like wave. However, subsequent inspection of
individual waveforms revealed that *no individual subject showed a clear biphasic response to the
'syntactic' anomalies*. Rather, for most subjects the response to these words was dominated either
by a monophasic increase in negativity between 300 and 500 ms (N400) *or* by a later-occurring
positive wave (P600). When averaged together, these monophasic responses took on the
appearance of a biphasic response" (Osterhout 1997, p. 509).

[159]Friederici and Frisch (2000, p. 498).

(a)   *Anna weiß, dass der Kommissar (NOM) den Banker (ACC) abreiste (V) und wegging.
      *Anna knows that the inspector (NOM) the banker (ACC) departed (V) and left.
(b)   *Heute trödelte (V) der Cousin (NOM) den *Geiger* (ACC) am Aufzug.
      *Today dawdled (V) the cousin (NOM) the *violinist* (ACC) at the lift.

In both cases, a biphasic N400/P600 wave was observed, and such was also the case during similar experiments[160] reported in Frisch et al. (2004) or Friederici et al. (2004).

This N400/P600 biphasic constitution generated by a term which is supernumerary from the standpoint of the actantial structure of the verb is interpreted as reflecting "difficulties in integrating this surplus noun phrase (N400) followed by the attempt to syntactically reanalyze the perceived input (P600)."[161]

- *Neutralization*

But the syntactic influences on the N400 do not limit themselves to the production of an N400 "effect": *neutralization* effects may also be observed. Indeed, it can be observed with a very good rate of reproducibility that a transgression of the syntagmatic structure of an utterance has the consequence of "suspending" the N400 effect entailed by a semantically inappropriate word, and even of annulling the "normal" semantic processing of a term which is congruous with its context.

This mechanism for the neutralization of the semantic process manifested by the N400 has been reported, for example, by Frisch et al. (2004): "The N400 observed in the condition with only a semantic violation [SV] was absent in the combined violation condition, such as [*Die Suppe wurde vom bemalt*; (*the soup was by-the painted*)], in which the sentences also contained a phrase structure violation [PSV]."[162]

Likewise, Hahne and Friederici (1999)[163] observe that the (E)LAN component generated by syntactic inconsistency blocks the N400—*except* when the task is that of a *comprehension exercise*.

In Hahne and Jescheniak (2001), the experiment consisted in comparing the potentials evoked by certain target words (participles), successively in (i) well-formed utterances composed of logatomes enhanced with morphosyntactic marks ("correct jabberwocky sentences"), (ii) standard well-formed utterances, (iii) utterances of type (ii) of which the syntagmatic order was perturbed ("Phrase

---

[160] For example: "Osterhout et al. (1994) found a biphasic N400-P600 pattern for subcategorization violations in sentential complements" (Frisch et al. 2004); "A biphasic N400-P600 pattern was also found by Friederici and Frisch (2000) for argument structure violations in active sentences with an intransitive verb and an internal object NP as in '(4)[(4) *Paul weiß, dass der Chemiker den Physiker emigrierte*; *Paul knows that the chemist the physicist emigrated*]" (Frisch et al. 2004); "Whereas a word category violation elicited (a LAN followed by) a P600 in the ERP, argument structure mismatches evoked N400-P600 responses" (Frisch et al. 2004).

[161] Friederici et al. (2004, p. 2).

[162] Frisch et al. (2004, p. 196).

[163] In Hahne and Jescheniak (2001).

Structure Violation"), and (iv) type (i) utterances which were perturbed in the same manner as (iii). We now observe that the semantically "appropriate" words of the "perturbed" standard utterances generate the same N400 component as the utterances composed of logatomes. In other words, the violation of the syntactic structure inhibits "normal" semantic processing: "In summary, an N400-like activity was only obtained for participles occurring in syntactically correct, regular sentences. It was neither obtained for participles occurring in syntactically incorrect sentences nor for pseudo-participles in jabberwocky sentences."[164]

In the same manner,[165] the N400 elicited in the event of a violation of argument structure will be neutralized when the transgression of the argument structure is preceded by a syntagmatic inconsistency.

- *Recapitulation and Conclusion*

    - *Primo*: Syntactic distortions can elicit, as a repercussion of in situ generation, an N400 wave;
    - *Secundo*: Syntactic distortions are likely to neutralize an N400 effect.

It is then possible to conclude that there exists a *functional interdependence between the N400 and the syntactic coordination* of the various lexical units.

8.3.7.2 The Modular Hypothesis

Confirmations

Although it is somewhat perturbed by the conclusions regarding the N280 (Sect. 1.3.5), the distinction between syntactic and semantic processes seems nevertheless conveniently corroborated by the results of EEG observation. We have seen that syntactic and semantic violations generate EEG responses with rather distinct latencies and topologies, and, moreover, other experimental trials confirm the existence and the autonomy of specifically syntactic and semantic processes.

Thus, in experiments concerning utterances of which the semantic dimension was neutralized by the use of logatomes, these being nevertheless endowed with morphosyntactic marks attributing them a "pure" grammatical identity, it was observed that a grammatically incorrect distribution of these marks elicits LAN and P600 waves. This result would therefore establish the existence of syntactic processes which are independent from any process involving a lexical content.

---

[164]Hahne and Jescheniak (2001, p. 206).

[165] "The main results of Experiment 1 are, first, that argument structure mismatches and phrase structure violations elicited qualitatively different patterns of ERP activity and, secondly, that an argument structure violation produced an N400 only when there was no additional phrase structure violation. A verb whose argument structure information did not meet the syntactic and thematic restrictions of the current sentence fragment elicited an N400-P600 pattern, whereas no N400 could be observed when the verb could not be structurally integrated due to an additional word category mismatch" (Frisch et al. 2004, p. 206).

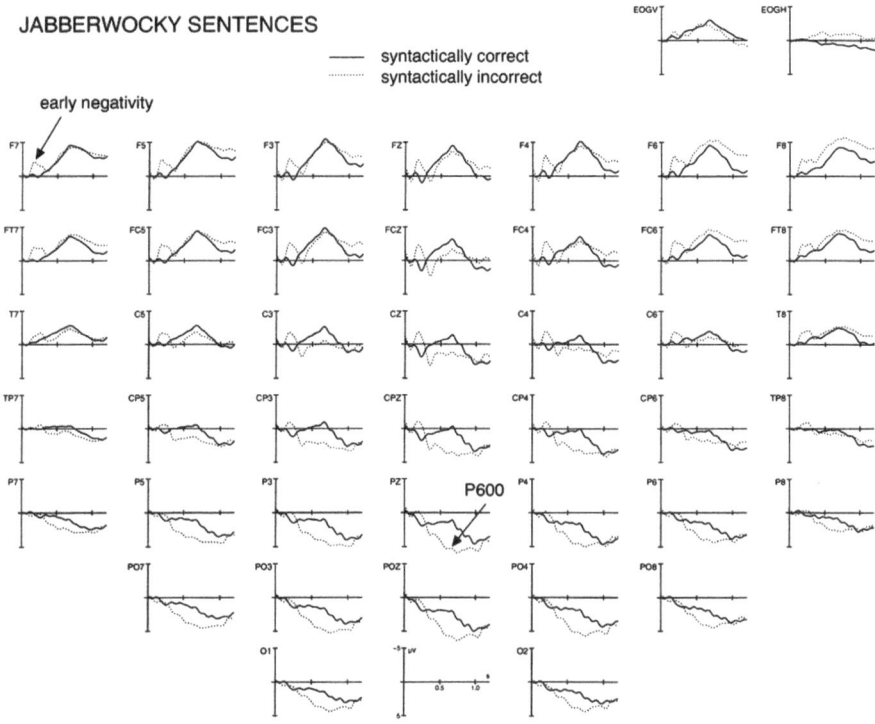

Fig. 8.22 Grand average ERPs from 20 participants for the critical participle for syntactically correct sentences (*solid line*) and syntactically incorrect sentences (*dotted line*) in [...] jabberwocky sentences, i.e. sentences consisting of pseudowords. [...] Syntactically incorrect sentences elicited an early anterior negativity followed by a P600 in regular as well as in jabberwocky sentences (Hahne and Jescheniak 2001, p. 205)

For example, Hahne and Jescheniak (2001) obtain the following recordings (Fig. 8.22).

In parallel, in an experiment[166] involving linguistic material specifically designed so as to avoid any telescoping between semantic and syntactic factors, the results attest to the existence of waves correlated with respectively syntactic and semantic processes and also establish their autonomy. This experiment consisted in testing sentential configurations comprising either (i) a syntactic transgression, either (ii) a semantic inconsistency, or (iii) the superimposition of (i) and (ii): "In the present experiments, we attempted to create sentence materials that contained relatively pure syntactic and semantic violations as well as those that presented compound violations."[167] In the end, one may observe that the "composed" transgression entails a biphasic response which is the sum of the responses to each

---

[166]Ainsworth-Darnell et al. (1998).

[167]Ibid., p. 113.

type of transgression. This result allowed the authors[168] to conclude that semantic and syntactic processes are independent.

Similar results have been obtained in the works of Hagoort et al. (2003), where the experiment consisted in varying the grammaticality of a semantically "deviant" utterance, for example:

(2a) *the boiled watering-can* smokes *the telephone in the cat*
(2b) **the boiled watering-can* smoke *the telephone in the cat*

We can observe that the ill-conjugated verb *smoke* of (2b) produces (in comparison to 2a) a P600/SPS effect. This enables to conclude that the P600 effect is independent from semantic factors, and, once again, that the syntactic and semantic processes are autonomous: "[D]espite the fact that these sentences do not convey any conventional meaning, the ERP effect of the violation demonstrates that the language system is nevertheless able to parse the sentence in its constituent parts."[169]

A contrario, indeed, we have seen that the generation of a LAN has the effect of neutralizing the semantic process manifested by the N400. But this interaction (cf. *supra*) is explained in the framework of modular theories by the precedence of syntactic analysis over the computation of a semantic representation.

Concerning the P600 specifically, in quite numerous experiments where linguistic units are presented outside of their syntactic configuration (isolated words or random succession of words), the P600 is shown to be significantly attenuated or to completely disappear—which confirms once more its specifically syntactic status. For example, in Van Petten and Kutas (1991), words in a final position within "random"[170] utterances produce a "smaller broad P600 maximal at parietal sites": "[T]he reduced amplitude of the Late Positivity for Final Words of the random string is consistent with [the syntactic] interpretation: these words did not complete a structural [...] unit."[171]

Likewise, we were able to observe[172] that grammatical words presented in an isolated manner (out of context) in lexical decision tests did not produce results within the P600 window—which is consistent with other observations.[173]

---

[168] "The semantically incongruous conditions evoked an N400, while the syntactically incongruous condition evoked a P600, and the doubly incongruous condition evoked a complex resembling an N400–P600 compound. These findings suggest that there are distinct cognitive processes associated with processing syntactic and semantic anomalies in a sentence were encountered simultaneously, the N400 and P600 did not cancel each other out, but remained distinct and were similar in amplitude and latency to the comparable components in the pure anomaly conditions. These results indicate that the generators of the brain responses that are measured by the N400 and P600 may be independent" (Ainsworth-Darnell et al. 1998, p. 126).

[169] Hagoort et al. (2003, p. 40).

[170] "Random word string with semantic or syntactic structure" (Van Petten and Kutas 1991, p. 105).

[171] Van Petten and Kutas (1991, p. 105).

[172] Münte et al. (2001).

[173] "Similar results were reported by Van Petten and Kutas (1991) who observed that N400-700 amplitude was largest for closed-class words in well-formed, nonanomalous sentences, and reduced in amplitude when the same words were embedded in random strings of words. Also

Conclusion. Should we disregard the numerous cases where the P600 and N400 display a troubling coincidence—such as for subcategorization violations which produce a biphasic N400/P600 response, or during experiments[174] where, instead of the N400 expected upon presentation of a pragmatically incongruous word in a grammatically correct utterance, we observe a clear P600 (for example *mouse* in *the cat, which the mouse chases, quickly escapes*)—we would indeed concede that the EEG manifestations of the linguistic activity support the hypothesis of a clear demarcation between syntactic and semantic processes. Temporarily accepting such a conception, it will now be a matter of examining which are, in this specific context, the "qualifications"—taken in a broad sense: the "interpretations"—attributed to the EEG waves deemed to be "syntactic".

Intermezzo: Informational content of the EEG waves

It is at this moment that the question arises concerning the "positive" or "differential" status of the neurocognitive information conveyed by the EEG components. But if the strictly differential conception of the N400 is unanimously rejected,[175] the case of the LAN (and of the P600) is less clear, and it will now be important to firmly establish in a definitive manner the "positive" character of its informational scope—which is not without consequences, since from this positive status, it will be necessary to infer that a LAN (of variable amplitude) is observable in all cases of syntactic processing.

Since it is a matter of LANs, we know that they are commonly associated with a process of syntactic analysis—and more often in terms of "effect", that is, as should be recalled, that it is the significant increase in the amplitude of the LANs which is acknowledged to signal a difficulty in linguistic processing due to the presence of a morphosyntactic transgression: "[LAN] refers to the amplitude *difference* between two conditions, [it is] identified by comparing the averaged waveforms of two conditions [:] in one condition one sees an increased negativity in comparison with another condition",[176] or "the Left Anterior Negativity (LAN) [is] detection of the (apparent) ungrammaticality."[177]

Firstly, we will note that in the framework of an approach which remains at the "surface" of the linguistic data so as to only retain their "correct" or "incorrect" character, the "differential" conception of the information conveyed by an EEG "effect" is quite coherent, valid, and sufficient. Indeed, if the horizon of analysis limits itself to the recording of correlations between, on the one hand and on the plane of linguistic manifestations, the variations in the admissibility of the

---

consistent with this finding is the fact that the N400-700 seems to be absent when words are presented in a list format for lexical decision (Garnsey 1985)" (Osterhout et al. 1997a, p. 164).

[174]Kolk et al. (2003).

[175]"[N400] is not simply an index of anomaly but rather a part of the brain's normal response to words (in all modalities)" (Kutas and Federmeier 2000, p. 464).

[176]Hagoort et al. (2003, p. 39).

[177]Kaan et al. (2000, p. 161).

utterances, and on the other hand, on the plane of the production of EEGs, the corresponding plots, then the EEG information relates to the fact of an alternation of admissibility, and, thenceforth, it is inscribed in the variation of the EEG plot—hence its differential character. It is this conception which, implicitly, is often followed—as shown in the few following citations (our emphasis): "For the syntactic domain two ERP components have been identified [...]—The first component is an anteriorly distributed negative potential, which typically *occurs to words that render the sentence as incorrect*",[178] or: "[T]here are a number of studies showing that LAN effects [...] are induced by syntactic violations."[179] And for the early variant of the LAN: "Negativities with a maximum latency below 200 ms have been found *to occur with phrase structure violations* in English [...] and German [...] and have been termed early left-anterior negativity (ELAN)."[180]

But when, in conformity with the established conclusions, the EEG components are acknowledged to have a determined functional signification—such as, for example, when we consider the "event-related brain potentials (ERPs) [as] a sensitive and powerful tool for the description of on-line processes in human parsing"[181]—the differential conception of the EEG information must be abandoned to the benefit of a "positive" conception, in favor of which at least three supplementary arguments can be given.

First, we will present observation reports which clearly support the "positive" perspective. For example, Osterhout et al. (2002) observe that the EEG curves for the "full" words and for the "empty" words present in themselves, i.e. independently from any variational trial, LAN, N400, and P600 components: "[I]nspection of [Fig. 8.16] reveals that all words elicited a similar series of negative and positive going deflections (N1, P2, N300, N400, N450 [and Late Component]."[182]

Then, we will turn towards the EEG recordings where we see that the LAN effect does not emerge from naught but accentuates a previous plot: The LAN observed in cases of syntactic violation does not distinguish itself from a "flat" plot which would correspond to the EEG of a successful process: It is transformed by a significant increase in the amplitude of a constituted component—which (cf. Fig. 8.23) manifests the processing of a correct configuration, as should appear:

Finally, we will argue the following: Firstly, we will reasonably concede (i) that the "motor center" engaged during the "early" phase of the processing of an utterance is the same whether this utterance is correct or incorrect. Also, from (i) it is legitimate to infer that (ii) the EEG manifestation of an exceeding cost in terms of syntactic processing (manifested by an increase in amplitude) will have the same topographical and temporal localization as that of a successful processing. Moreover, (iii) we will readily concede that the processing of a correct utterance is

---

[178]Kolk et al. (2003, p. 4).

[179]Schlesewsky et al. (2003, p. 118).

[180]Frisch et al. (2004, p. 194).

[181]Schlesewsky et al. (2003, p. 116).

[182]Osterhout et al. (2002a, p. 181).

Fig. 8.23 ERP effect as produced by a significant increase in the amplitude of an ERP component

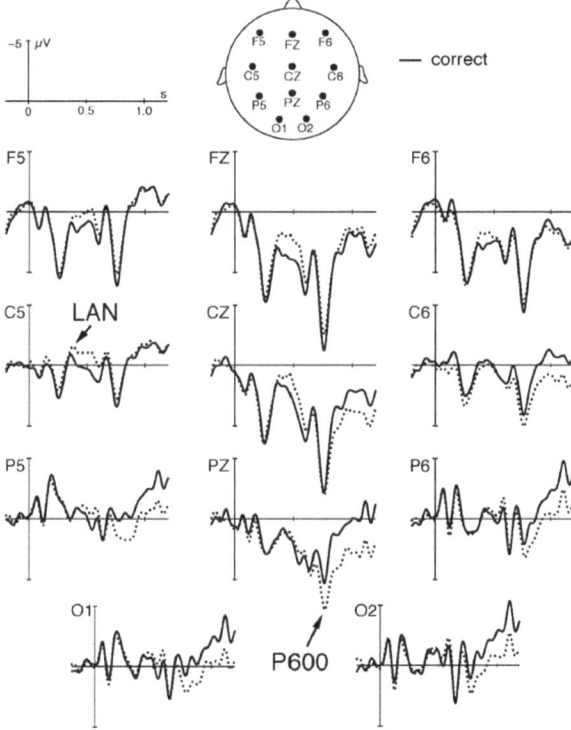

not a zero-cost cognitive procedure. Consequently, we conclude that the EEG component in contrast to which the EEG manifestation of the syntactic violation is identified is to be recognized as *a manifestation of a "normal" process*.

From this perspective, the LAN is generally interpreted as an EEG correlate of morphosyntactic processes: "These [experiments] and related data suggest that the early anterior negativity reflects an initial parsing stage during which phrase structure information is processed",[183] or, "there are a number of studies showing that LAN effects reflect first pass parsing processes."[184] This neurolinguistic qualification of the LAN does not entail particular problems. Quite to the contrary, we have seen that the LAN/N400 interactions or the production of LANs in the case of agrammatical utterances composed of logatomes lend themselves to a modular analysis. In fact, the problems arise when the P600 is taken into consideration— problems which we shall now address.

---

[183]Hahne and Jescheniak (2001, p. 209).

[184]Schlesewsky et al. (2003, p. 118).

The P600 and the Difficulties of the Modular Approach

- *Qualification*

Given, on the one hand, that the processes of syntactic analysis are manifested by the LAN, and on the other hand, that semantic processing is associated with the N400, it is legitimate to wonder what kind of processing the P600 signals.

For the P600, the experimental configuration which most distinctively exhibits the functional identity is that of "garden-path sentences". Indeed, the production of a P600 effect can be observed with this type of utterance of which the first elements orient the syntagmatic analysis towards a "preferential" structure upon the presentation of the word where this structure proves inappropriate: "The P600 has been found, […] for words that are unexpected given the preferred reading of the preceding context (garden-path sentences)"[185]—and this effect is unique: "[V]erb-based preferences for syntactic structure are correlated with a P600 *only* (Osterhout et al. 1994)."[186]

Regarding this empirical evidence, the P600 may readily be interpreted as the neuronal manifestation of a process of "reanalysis" which consists in returning to previously presented data so as to structurally qualify it in a manner which is adapted to the newly acquired information. Thus, in an utterance like *the woman persuaded to answer the door*, "language users choose for an active interpretation of the verb *persuade*, expect an object NP after the verb but read to instead. This forces them to *reanalyze* the sentence, in order to arrive at an interpretation in which the verb is taken as a passive participle. It is this *process of reanalysis* which is assumed to be responsible for the occurrence of the P600 effect. Amplitude (Osterhout et al. 1994) and latency (Friederici and Mecklinger 1996) of the P600 effect are thought to vary as a function of the difficulty of recovery from the garden-path."[187] Thus, the question "But what cognitive processes does the P600 reflect?" has for answer: "A common view is that the P600 [is] associated with reanalysis processes"[188]: "[T]he broadly distributed late positivity appears to be related to processes of structural reanalysis."[189]

But a functional characterization of the P600 must also account for the other circumstances of occurrence of this wave—thus, the case of morphosyntactic distortions where the P600 appears in association with a LAN or an N400. To do this, we will broaden the notion of "reanalysis" to that of "repair" (or of "revision"), and the P600 will be interpreted as the trace of a process which consists in returning to anteriorly presented data so as to produce a new assemblage which better satisfies the formal, semantic, or syntactic constraints of language: "The late centroparietal

---

[185]Kaan et al. (2000, p. 160).

[186]Friederici and Frisch (2000, p. 481).

[187]Kolk et al. (2003, p. 29), our emphasis.

[188]Kaan et al. (2000, p. 161).

[189]Friederici (1995, p. 277).

positivity (P600) has been described in association with a wide range of different syntactic anomalies including those requiring a reanalysis of the preceding structure and those requiring a repair of a syntactic violation",[190] or: "[T]he current characterization of the P600 component as an indicator of revision processes (reanalysis and repair) in sentence comprehension."[191]

Also, since it is a matter of the production of biphasic LAN/P600 waves, "if detection of the (apparent) ungrammaticality occurs before repair processes are started, it is reasonable to associate the LAN with detection and the P600 with repair processes"; "the P600 reflects repair processes *following* the detection of an (apparent) ungrammaticality."[192]

The conception of the P600 as a process of "revision" of which the cost is measured in terms of amplitude accommodates quite well the observation of a covariation of the P600 effect with the difficulty of the "revision" of the data presented: "Osterhout et al. tested sentences in which the target word signaled a syntactic structure which was either *ungrammatical, less preferred, highly preferred* or *obligatory* given the preceding verb. The P600 amplitude at this target word was largest for the ungrammatical continuations, smaller for the grammatical *but* less preferred continuations, and smallest for the preferred or compulsory continuations."[193] Hence the interpretation that: "These data suggest that the more difficult it is to construct a grammatical representation, the larger the P600; the interpretation of the P600 is that it reflects the cost of reprocessing."[194]

This interpretation is furthermore meant to explain a certain number of EEG phenomena. *Firstly*, there is the generation of an N400 and of a P600 in the event of a violation of subcategorization rules. Specifically: "[L]ate Positivity [is] related to processes of structural reanalysis which may become necessary when the initially build syntactic structure cannot be successfully mapped onto the semantic information and verb argument information provided by the lexical elements."[195] Thus, the N400 would reflect the difficulty of semantic integration whereas the P600 would reflect the process of "revision": "The N400–P600 pattern was interpreted to reflect difficulties in integrating this surplus noun phrase (N400) followed by the attempt to syntactically reanalyze the perceived input (P600)."[196] *Then*, the fact that, in the experiment reported in Osterhout (1997), a group of subjects produced a P600 and the other an N400: "[P]erhaps, subjects in the N400 group were less likely than subjects in the P600 group to attempt syntactic reanalysis."[197]

---

[190]Friederici et al. (2004, p. 2).
[191]Frisch et al. (2002, p. B83).
[192]Kaan et al. (2000, p. 161).
[193]Ibid.
[194]Ibid.
[195]Friederici (1995, p. 277).
[196]Friederici et al. (2004, p. 2).
[197]Osterhout (1997, p. 515).

In short, "This so-called *P600 component* is often seen as an indicator for greater syntactic processing cost due to a necessary revision of a (temporary or persistent) structural mismatch [...], which may either consist of an outright syntactic violation [...] or of a dispreferred disambiguation of an ambiguous string [...]."[198]

But this interpretation is not without raising very serious and numerous objections.

- *Objections*

First Objection: If the P600 manifests a process of (re)analysis, we cannot see why this process would not be the responsibility of the neuronal centers which prioritarily conduct it, that is, the centers which generate the LAN: "[I]n the view of the finding that structure building processes are correlated with activities in the left anterior cortex, one might expect these cortical areas [of LAN] to be involved in processes of syntactic reanalysis as well."[199]

Second objection: If the P600 reflects a process of "revision", the transgression to be "reviewed" must have been detected *before* the mechanism correlated to the P600 is engaged. However, in the case of subcategorization violations presenting a biphasic N400/P600 response, no process of "detection" is manifested. And, moreover, we cannot attribute to the P600 the function of anomaly detection since the N400 precedes the P600. Thus, commenting an experience concerning EEG production for the utterance (7d): *The doctor charged the patient was lying* versus (7c): **The doctor forced the patient was lying*, Friederici and Frisch (2000) emphasize that "the authors interpret this biphasic pattern [N400 P600] to reflect the sentence's ungrammaticality and the fact that the inability to form a coherent syntactic representation of the sentence might have 'Rapidly engendered semantic anomaly' [*but*] they admit that this interpretation appears somewhat paradoxical given the relative onsets of the N400 and the P600."[200]

Third objection: The interpretation of the P600 as a process of "revision" does not accommodate the "positive" (non differential) conception of EEG waves. Now, as is the case for both the LAN and for the N400, it is necessary to recognize a positive character to the information conveyed by the P600.

We may recall indeed that, for the N400 as for the LAN, the amplitude of the P600 varies with the cost of processing, and we have seen that for utterances satisfying the linguistic constraints, the P600 is small, but not absent. Also, the P600 does not reflect a process triggered in response to linguistic transgressions, but rather, as with the LAN, *a process which is accomplished under "normal" linguistic conditions* and which reveals itself to be "reactive" upon the encounter of certain structural malformations or difficulties.

It follows that *the interpretation of P600 as a process of reparation or of reanalysis cannot be upheld*, because a "revision" of the syntactic structure

---

[198]Frisch et al. (2002, p. B85).
[199]Friederici (1995, p. 277).
[200]Friederici and Frisch (2000, p. 480).

evidently supposes the prerequisite of a malformation (case of transgression) or of a stage of inappropriate construction (case of POS[201]).

- *Empirical Confirmation of the Objections*

This conclusion is reinforced by various experiments which highlight the existence of the P600 in "correct" sentence configurations for which no "revision" is required, and which also convey additional information regarding the conditions of generation of this wave.

We will mention the works exposed in Kaan et al. (2000) where, after having observed that "the P600 is elicited in grammatical, non-garden path sentences in which integration is more difficult (i.e., 'who' questions) relative to a control sentence ('whether' questions)", the authors defend the idea that "the P600 is not restricted to reanalysis processes, but reflects difficulty with syntactic integration processes in general."[202]

The experiment performed by Frisch et al. (2002) has, for its part, consisted in submitting "correct" utterances comprising grammatical ambiguities (located at the 1st critical word) which are resolved at the moment of the presentation of an ulterior item (2nd critical word). The data have the following form:

(2a) *Die Frau hatte den Mann gesehen*

[the woman]$_{amb}$ had [the man]$_{obj}$ seen

(2b) *Die Frau hatte der Mann gesehen*

[the woman]$_{amb}$ had [the man]$_{sub}$ seen

In (2a) and (2b), the determinant of the first SN comprises information of which the function is "ambiguous" ("die" in "die Frau" indistinctly marks the nominative or the accusative) and it is upon presentation of the following SN that the indecisiveness regarding the case value of "die Frau" is lifted.

The EEG recordings (Fig. 8.24) then show that (i) with respect to the plots generated by a first non-ambiguous item—for example "der" *or* "den Detektiv" in:

(4a) *Der Detektiv hatte die Kommissarin gesehen und...*

[the detective]$_{masc.sub}$ had [the policewoman]$_{fem.obj}$ seen and...

(4b) *Den Detektiv hatte die Kommissarin gesehen und...*

[the detective]$_{masc.obj}$ had [the policewoman]$_{fem.sub}$ seen and...

The potential evoked by the ambiguous item shows a clear P600 effect:

And (ii) for utterances comprising an ambiguous item, the resolution of the ambiguity at the level of the second critical word shows a P600 effect when the resolution contravenes to a supposed preferential orientation which privileges a subject-object grammatical distribution, respectively, at the first and second NPs (Fig. 8.25).

---

[201]Preferential orientation sentence.

[202]Kaan et al. (2000, p. 159).

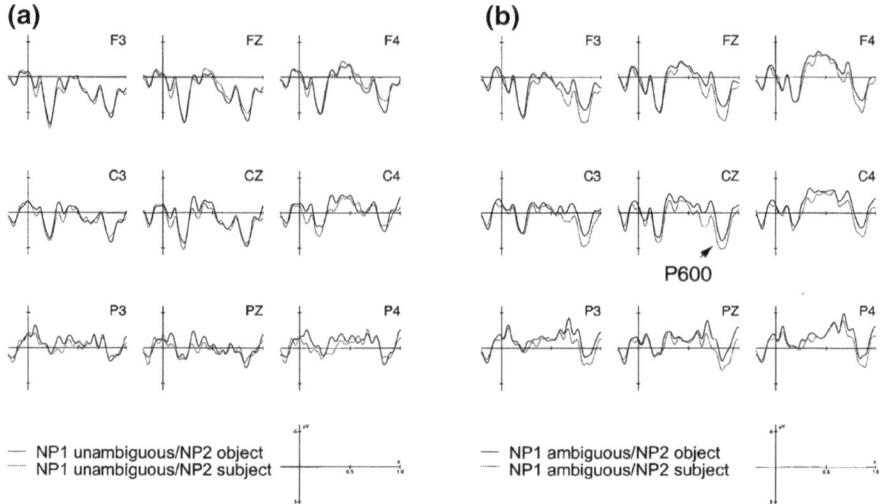

Fig. 8.24 ERP effects on the second argument [...]. **a** There is no difference on the second argument between the two conditions in which the first argument is already unambiguous. **b** The difference between the two conditions in which the second argument disambiguates. When the disambiguating second argument is the subject, then we see a more positive going waveform between about 600 and 1000 ms compared to a disambiguation towards object (Frisch et al. 2002, p. B89)

The conclusions drawn by Frisch et al. (2002) are the following.

The first experiment shows that the P600 does not reflect a process of "revision." Indeed, "this P600 effect for a sentence-initial, ambiguous argument compared to an unambiguous one cannot be taken as a marker of syntactic revision seeing that there is no preceding interpretation to be revised."[203]

Concerning the results of the second experiment, they are mobilized as part of the debate regarding the architectures of syntactic processing. The fact that the processing of the preferential structure is of a lower cost is an argument against "serial" architectures. Indeed, "Serial models [...] predict that an ambiguous structure should be equally easy to process as its simplest unambiguous counterpart."[204] However, we observe that the cost for processing "expected" and "unexpected" structures is unequal. And we can interpret this result in favor of a process of a "parallel" type: "The finding of a P600 on the second argument when this disambiguates the previous string towards a dispreferred structure shows that, (no later than) at this item, our parser has built up a measurable preference for one continuation (subject first) over the other (object first). This implies that not all alternatives are considered up to the point of disambiguation in an unweighted

---

[203]Frisch et al. (2002, p. B90).

[204]Ibid., p. B84.

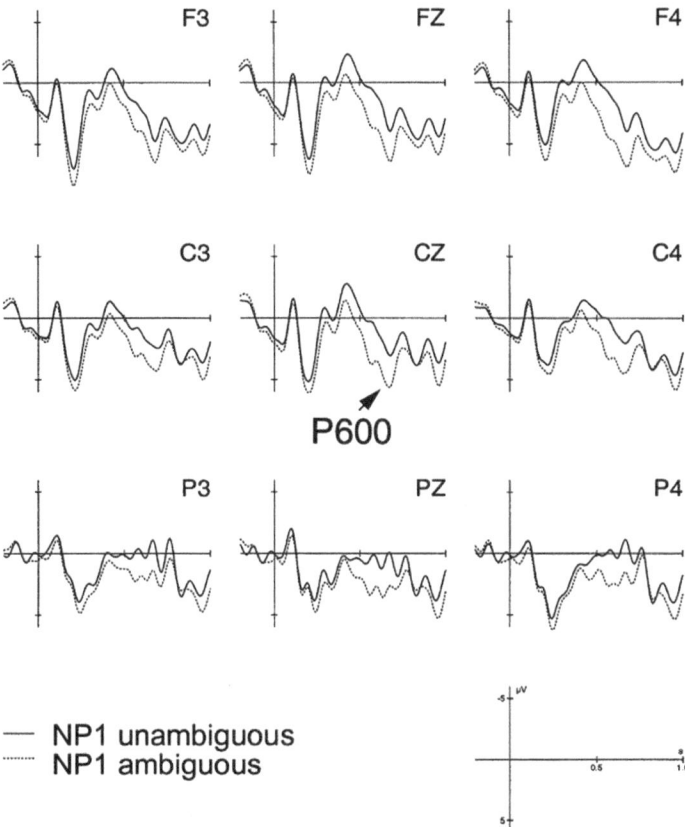

Fig. 8.25 Approximately between 400 and 800 ms, the ambiguous conditions are more positive going compared to the unambiguous ones […] (Frisch et al. 2002, p. B88)

manner, but that after the ambiguity is acknowledged, we consider some continuations to be more likely than others (in the sense of Gorrell 1987)."[205]

What stems from these two experiments, independently of any problematic framework, is indeed the insufficiency of the functional interpretation of the P600 as a "revision" process—and the authors, after having deemed[206] necessary to considerably broaden the functional signification of the P600, conclude by suggesting that this component measures a *cost of syntactic processing*: "As our data show, the P600 component must be taken as an *indicator of syntactic processing cost in general*. […] such processing cost was enhanced for an ambiguous initial argument

---

[205]Ibid., p. B90.

[206] "In sum, the finding of a P600 component for a syntactically ambiguous element shows that the view that this component reflects processes of syntactic revision or integration has to be extended" (Frisch et al. 2002, p. B91).

as well as for its later disambiguation towards the dispreferred argument order (object-first)."[207]

Considering that the "cost" of syntactic processing is, clearly, a direct function of the difficulty of the processing, we will note that the previous interpretation of the P600 is closely related to that defended by Kaan et al. (2000), that is, "the P600 as an index of syntactic integration difficulty."[208]

Assessment and Transition

From the preceding, it clearly appears that the paradigm of formal syntaxes is invalidated by the results of EEG experimentation. Specifically, the hypothesis of a linguistic system articulating two sorts of processing, *the one* being of a syntactic nature, aiming to construct the formal scaffolding of an utterance—as a structured assemblage of grammatical categories—the *other* being of a semantic nature, which would produce a semantic representation, is contradicted by the existence of two waves (LAN and P600) of which the observed neurocognitive determinations and functionings contravene the functional qualifications attributed to them in the theoretical frameworks discussed herein, these postulating a syntax/semantic separation.

Taking this conjuncture into account, it would be legitimate to renounce the principle of a disjunction between form (syntax) and content (semantics) and to consider a theoretical perspective which better accommodates the experimental data. But this supposes that the "observed" regimes of empirical functioning which the theoretical model must account for have previously been "uttered."

In what concerns the P600 and the LAN, and in order to provide their formal description in maximally "neutral" terms, *on the one hand*, we will retain that the LAN manifests a "standard" process and that its sensitivity to the grammatical admissibility of the items presented tends to show that this processing aims to produce a "form". *On the other hand*, for the P600, we will retain two things: (i) that it participates (for the same reason as the LAN) in the elaboration of a "form", and (ii) as is clearly shown by the attempt to synthesize various interpretations of this EEG component, that the P600 operates *retroactively* upon previously presented units.

It follows that a theory of linguistic forms must comprise two operational phases of "syntactic" construction—to be understood then in the broad sense of "morphological construction"—the one being *direct*, the other being *retroactive*—respectively manifested by the LAN and P600 components.

Now, as we will now see, the morphodynamics of the Saussurean sign evidently satisfies such functional and empirical requirements. The device of the Saussurean sign indeed composes two morphodynamic operations in a relation of reciprocity,

---

[207]Frisch et al. (2002, p. B91).
[208]Title of Kaan et al. (2000).

and they are correlatable to the LAN and P600 waves. These two operations are precisely those through which the normative semiolinguistic constraints are expressed and, moreover, through which are instituted the sign phenomena whose investment by consciousness is signaled on the neurobiological plane by the N400. This is to say that the empirical foundation delivered by the neurosciences support the principle of a semiolinguistic rationality as instituting signifying phenomenalities through normative requirements of which these phenomenalities render the expression possible. We shall examine this.

8.4   Phenomenality and Objectivity: The Functional Unit

We may return to the question of the linguistic processes to be associated with the LAN and P600 waves. In order to do this, we propose to consider an extremely simple example: that of number agreement between an article and a noun. Let's take for example the "word-forms" *le* ['the'], *les* ['the' plural], *cheval* ['horse'], *chevaux* ['horses']. In conformity with the construction *Det-N*, the combinations of these items with respect to combinatorial norms are: *Le cheval* ['the (singular) horse']; **les cheval* ['the (plural) horse']; **le chevaux* ['the (singular) horses']; *les chevaux* ['the (plural) horses'] (the asterisk indicates the incorrectness of the syntagm).

According to the morphodynamics of the sign, we know (cf. 6.3), on the one hand, that the oppositions between (semantic) values are achieved by means of a boundary instituting into the substance of content two subdomains which compete with respect to their extension and actualization, and, on the other hand, that the actualization of one of the opposing values proceeds from a stabilization path (from an originary instable germ) which is determined by an item (or by a complex of items) included within the semiolinguistic totality being formed. This stabilization path "expresses" the normative constraints which weigh upon the configuration, for instance sentential, of which it is question.

Therefore, in our example, the article *le* ['the' (singular)], as it participates in a totality involving the signifier *cheval* ['horse'] following the construction *Det-N*, determines the actualization of *cheval* ['horse'] in its opposition to *chevaux* ['horses']. Which amounts to saying that, from a dual point of view, the paradigmatic opposition between *cheval* ['horse'] and *chevaux* ['horses'] proceeds from a construction, that is *le* ['the' (singular)] *N*, where *cheval* ['horse'] is admissible from the standpoint of the linguistic norm, whereas *chevaux* ['horses'] is not. In other words still, during the unfolding of an utterance, the presentation of the article *le* ['the' (singular)] instructs an opposition between *cheval* ['horse'] and *chevaux* ['horses'] and commands a stabilization path which is normative in that it requires the actualization of *cheval* ['horse'] versus *chevaux* ['horses'], the one being admissible, the other not.

This first operational moment which commands the emergence of an oppositive boundary and which requires the actualization of one of the two terms of the

Fig. 8.26 LAN as reflecting
the emergence of a boundary
(oppositive relation)

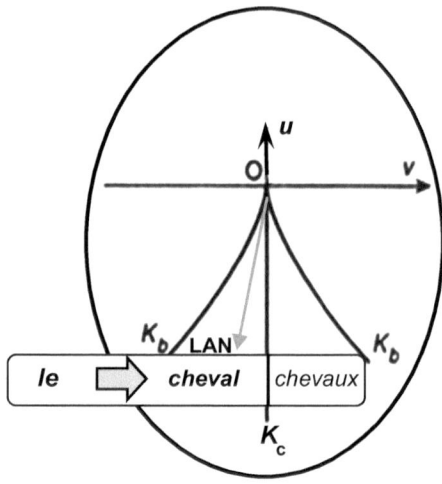

opposition is indeed to be put into correspondence with the "first" neuronal manifestation of the syntactic processes, that is, the LAN (Fig. 8.26).

But, in such a state, the syntagm *le cheval* ['the' (singular) 'horse'] does not
constitute an organic totality: For such to be the case, it would still be necessary, by
retroactive means, for the item *cheval* ['horse'] to determine the item *le* ['the'
(singular)]. Reciprocally, therefore, and under the government of the item *cheval*
['horse'], it is necessary for the article *le* ['the' (singular)] to be constituted as a
differential identity, therefore, to establish a relation of opposition with another sign
of its category. This is why, correlatively, the actualization of *cheval* ['horse'] will
have a "retroactive" effect on the paradigm of articles, and will trigger (on the plane
of content) a process of differentiation between the values attached to *le* ['the'
(singular)] and *les* ['the' (plural)]—a process which will thus take the form of a
stabilization path governing (normatively) the actualization of *le* ['the' (singular)] in
its opposition to *les* ['the' (plural)]. This complementary operational moment, due
to its "retroactive" character and its "wrap-up effect" is, for its part, to be associated
with the P600 (Fig. 8.27).

The requirement regarding the correlation of the LAN and P600 effects with
morphosyntactic transgressions is even satisfied: In the framework of the analysis
proposed here, the morphosyntactic inconsistencies in language elicit a conflict in
stabilization processes: The "forced" paths oppose the "required" paths (cf. 6.3.2)—
and this conflict is of course to be correlated with the increase in the amplitude of
the "syntactic" waves. We will note that the neutralization of the N400 observed in
cases of "syntactic" violations (cf. Sects. 8.3.7.1–8.3.7.1.3) is directly explained
here: Inasmuch as it is differential forms, of which the production is neurobiologically manifested by the LAN and P600 waves, which institute the
sign-phenomena (as indivisible connections between a signifier and a signified), it
is conceivable that the amplitude of the N400, which relates the acquisition in
consciousness of the sign-phenomenon at more or less high levels of its

Fig. 8.27 LAN and P600 as
reflecting a direct and a
retroactive process of
boundaries generations

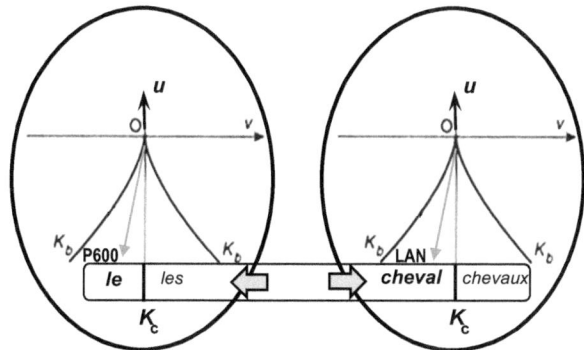

stratification, is diminished and even cancelled when the sign-phenomenon is not
accomplished in its totality, but instead presents itself in a partial and unaccom-
plished form.

We will also note that, in this descriptive framework, it is this same "retroactive"
mechanism of boundary generation which we see at work when, upon presentation
of an ambiguous item, the term which follows resolves the ambiguity. Figure 8.28
illustrates the process of determination of the case value of *die* when *den Mann* is
presented.

Indeed, and returning to the example examined by Frisch et al. (2002), the
determinant *die*, undefined from the standpoint of its case value (nominative or
accusative), receives a univocal characterization upon presentation of the Nominal
Syntagm (NS) *den Mann* which completes the sentence. Thus, the generic semantic
space which is primarily conveyed by *die*, and when *die N* enters into a syntagmatic
connection with *hatte den Mann*, it is the locus of a process of retroactive categorization which determines the two opposing case values and actualizes one (vs.
the other). In such a scenario, we *effectively* record the production of a P600
component—which empirically corroborates the structural analysis advanced here.

On a more general level, we will observe that the "graduated" character of the
process of formation of a global signifying morphology is supported by various
strata of verbal consciousness to which the constituents of this morphology are
alternately subject. Indeed, and returning to the preceding example, we see that in
the unfolding of the flux of speech, the utterance of *die Frau* is situated as such
beneath any differential determination (stratum of the signifieds), thus at a level of
"motif" consciousness or of "engagement" consciousness for example, and awaiting promotion to a higher degree of significatory consciousness. It is the retroactive
process related by P600 which will contribute to this, specifically in that it produces
and associates to *die Frau* an oppositional case value. We therefore see, in a very
general manner, that in a morphodynamic perspective, the production of semiolinguistic formations consists in successive elevations of the items involved in
terms of levels of verbal consciousness: The units originally solicited pertain at first
to the strata of verbal consciousness where their significations are, in various
respects, undetermined, and these signifiers, in their "native state" and of an

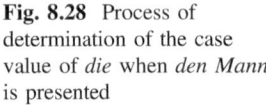

Fig. 8.28 Process of determination of the case value of *die* when *den Mann* is presented

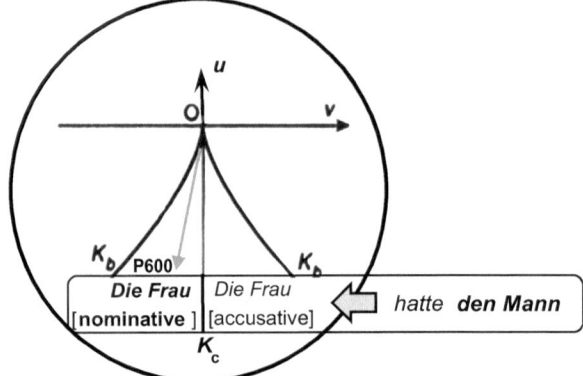

"uncertain" meaning, are the vectors of normative constraints which will govern the semantic (differential) promotion of the items which are adjacent in the totality they compose. Likewise, if we consider speech in its character as a syllabic flux of which each "articulatory segment" is in wait of its status: for example, as a morpheme or as the simple syllable of a word. Again, as in the preceding case, for this to be possible, it is necessary for the parts which are still undetermined in their role and status to be nevertheless endowed with a semiolinguistic existence which is, so to speak, preparatory, and as pending either to signify for themselves or to be taken as mute components of a complex of meaning encompassing them, for instance as syllables of a lexical unit. The theory of the strata of verbal consciousness, which is the phenomenological expression of the morphodynamics of the sign, satisfies this requirement.

In accordance with our expectations, the morphodynamics of the Saussurean sign thus verifies the condition of empirical adequation enounced above, that is: to comprise two operational phases of "morphological" construction, the one being direct, the other being retroactive. We have seen, indeed, that the components of the syntagmatic sequence see their paradigmatic identities (differential identities) determined by the normative constraints induced by their context, and these constraints operate in the double direction of a construction which is direct (reflected by the LAN) and retroactive (given by the P600).

We will finally note that, on the one hand and *in the end*, this double movement of normative construction realizes an organic totality of signification: a totality where the whole and the parts interdetermine one another. Indeed, as we have seen, the differential value of each term involved in a global configuration is configured by the normative constraints of the surrounding terms. And, on the other hand, we will note that the morphological processes which govern the semiolinguistic compositions are the very ones which institute sign-phenomena in the fullness of their semiotic existence—this fullness being degraded and semiolinguistic existence being reduced to inferior forms when the verbal gestures contravene the normative requirements too frontally. In such circumstances, then, the signifier, albeit turned towards the constitution of a signified (as a differential value of content), does not accomplish its purpose, and devoid of the value towards which it is functionally oriented, confines its semiolinguistic existence to the preparatory strata of a full consciousness of signification.

Correlatively, we now see precisely in what and how the forms of semiolinguistic objectivity and phenomenality are intertwined: The normative constraints, which weigh upon syntagmatic constructions and which relate an order of semiolinguistic objectivity, carry the preparatory phases of a semiolinguistic consciousness (preparatory phases in which such normative constraints are grounded) at the level of their accomplishment as authentic sign-phenomena. In this sense, the regimes of a semiolinguistic systematicity, these regimes which are manifested by the phenomena in their observable functioning and configurations and which pertain to a semiolinguistic objectivity, find themselves to participate in the final constitution of the phenomena they involve.

References

Ainsworth-Darnell, K., et al. (1998). Dissociating brain responses to syntactic and semantic anomalies: Evidence from event-related potentials. *Journal of Memory and Language, 38,* 112–130.

Anderson, J., & Holcomb, P. (1995). Auditory and visual semantic priming using different stimulus. *Psychophysiology, 32,* 177–190.

Bentin, S. (1987). Event-related potentials, semantic processes, and expectancy factors in word recognition. *Brain and Language, 31*(2), 308–327.

Bentin, S., et al. (1995). Semantic processing and memory for attended and unattended words in dichotic listening: Behavioral and physiological evidences. *Journal of Experimental Psychology: Human Perception and Performance, 21*(1), 54–67.

Besson, M., Fischler, I., Boaz, T, & Raney, G. (1992). Effects of automatic associative activation on explicit and implicit memory tests. *Journal of Experimental Psychology: Learning, Memory, and Cognition, 18,* 89–105.

Brown, C., & Hagoort, P. (1993). The processing nature of the N400: evidence from masked priming. *Journal of Cognitive Neuroscience, 5*(1), 34–44.

Caramazza, A. (1997). How many levels of processing are there in lexical access. *Cognitive Neuropsychology, 14*(1), 177–208.

Chwilla, D., et al. (1995). The N400 as a function of the level of processing. *Psychophysiology, 32* (3), 274–285.

Coulson, S., & Kutas, M. (2001). Getting it: Event-related brain response to jokes in good and poor comprehenders. *Neuroscience Letters, 316,* 71–74.

Daltrozzo, J., et al. (2012). The N400 and late positive complex (LPC) effects reflect controlled rather than automatic mechanisms of sentence processing. *Brain Sciences, 2,* 267–297.

Danker, J. F., et al. (2008). Characterizing the ERP Old-New effect in a short term memory task. *Psychophysiology, 45,* 784–793.

de Saussure, F. (1959). *Course in General Linguistics.* (W. Baskin by Trans.) (*CLG/B*).

De Vincenzi, M., et al. (2003). Differences in the perception and time course of syntactic and semantic violations. *Brain and Language, 85,* 280–296.

Deacon, D., et al. (2000). Event-related potential indices of semantic priming using masked and unmasked words: Evidence that the N400 does not reflect a post-lexical process. *Cognitive Brain Research, 9*(2), 137–146.

Debruille, J. (2008). Knowledge inhibition and N400: A within and between subjects study with distractor words. *Brain Research, 1187,* 167–183.

Dombrowski, J.-H., et al. (2006). Semantic activation, letter search and N400: A reply to Mari-Beffa, Valdes, Cullen, Catena and Houghton (2005). *Brain Research, 1073–1074,* 440–443.

Friederici, A. (1995). The time course of syntactic activation during language processing: A model based on neuropsychological and neurophysiological data. *Brain and Language, 50,* 259–281.

Friederici, A. (2002). Towards a neural basis of auditory sentence processing. *Trends in Cognitive Sciences, 6*(2), 78–84.

Friederici, A., & Frisch, S. (2000). Verb argument structure processing: The role of verb specific and argument specific information. *Journal of Memory and Language, 43,* 476–507.

Friederici, A., et al. (2004). The brain knows the difference: Two types of grammatical violations. *Brain Research, 1000,* 72–77.

Frisch, S., et al. (2002). The P600 as an indicator of syntactic ambiguity. *Cognition, 85,* B83–B92.

Frisch, S., et al. (2004). Word category and verb-argument structure information in the dynamics of parsing. *Cognition, 91,* 191–219.

Hagège, C. (1990). *La structure des langues.* Paris: PUF, coll. *Que sais-je ?,* 2006.

Hagoort, P., et al. (2003). Syntax-related ERP-effects in Dutch. *Cognitive Brain Research, 16,* 38–50.

Hahne, A., & Jescheniak, J. (2001). What's left if the Jabberwock gets the semantics? An ERP investigation into semantic and syntactic processes during auditory sentence comprehension. *Cognitive Brain Research, 11,* 199–212.

Hill, H., et al. (2005). SOA-dependent N400 and P300 semantic priming effects using pseudoword primes and a delayed lexical decision. *International Journal of Psychophysiology, 56,* 209–221.

Holcomb, P. (1988). Automatic and Attentional Processing: An Event-Related Brain Potential Analysis of Semantic Priming. *Brain and Language, 35,* 66–85.

Holcomb, P., & Neville, H. (1990). Auditory and visual semantic priming in lexical decision: A comparison using event-related brain potentials. *Language and Cognitive Processes, 5*(4), 281–312.

Holcomb, P., et al. (2005). The effects of prime visibility on ERP measures of masked priming. *Cognitive Brain Research, 24,* 155–172.

Hutzler, F., et al. (2004). Inhibitory effects of first syllable-frequency in lexical decision: An event-related potential study. *Neuroscience Letters, 372,* 179–184.

Kaan, E., et al. (2000). The P600 as an index of syntactic integration difficulty. *Language and Cognitive Processes, 15*(2), 159–201.

Kiefer, M. (2002). The N400 is modulated by unconsciously perceived masked words: further evidence for an automatic spreading activation account of N400 priming effects. *Cognitive Brain Research, 13,* 27–39.

Kiefer, M., & Spitzer, M. (2000). Time course of conscious and unconscious semantic brain Activation. *NeuroReport, 11,* 2401–2407.

Kolk, H., Chwilla, D., et al. (2003). Structure and limited capacity in verbal working memory: A study with event-related potentials. *Brain and Language, 85,* 1–36.

Kutas, M., & Federmeier, K. (2000). Electrophysiology reveals semantic memory use in language comprehension. *Trends in Cognitive Sciences, 4*(12), 463–470.

Kutas, M., & Federmeier, K. (2011). Thirty years and counting: Finding meaning in the N400 component of the Event-Related brain potential (ERP). *Annual Review of Psychology, 62,* 621–647.

Kutas, M., & Hillyard, S. (1980). Reading senseless sentences: Brain potentials reflect semantic incongruity. *Science, 207,* 203–205.

Kutas, M., & Hillyard, S. (1989). An electrophysiological probe of incidental semantic association. *Journal of Cognitive Neuroscience, 7,* 38–49.

Kutas, M., Van Petten., C, & Kluender., R. (2006). Psycholinguistics electrified II: 1994–2005. In M. Traxler & M. A. Gernsbacher (Eds.), *Handbook of psycholinguistics* (2nd ed.). New York: Elsevier.

Lau, E., Phillips, C., & Poeppel, D. (2008). A cortical network for semantics: (de)constructing the N400. *Nature Reviews Neuroscience, 9*(12), 920–933.

Maess, B., et al. (1997). *A high density auditory ERP study: The processing of words, pseudowords and non-words*. Max Planck Institute Annual Research Report, pp. 42–45.

Martín-Loeches, M., et al. (1999). The recognition potential: an ERP index of lexical access. *Brain and Language, 70,* 364–384.

Martin-Loeches, M. (2007). The gate for reading: Reflections on the recognition potential. *Brain Research Reviews, 53,* 89–97.

Mc Carthy, G., et al. (1993). Modulation of semantic processing by spatial selective attention. *Electroencephalography and Clinical Neurophysiology, 88,* 210–219.

Münte, T., et al. (1998). Brain potentials and syntactic violations revisited: no evidence for specificity of the syntactic positive shift. *Neuropsychologia, 36*(3), 217–226.

Münte, T., et al. (2001). Differences in brain potentials to open and closed class words: Class and frequency effects. *Neuropsychologia, 39,* 91–102.

Orgs, G., et al. (2008). N400 effects to task irrelevant environmental sounds: Further evidence for obligatory conceptual processing. *Neuroscience Letter, 436,* 133–137.

Osterhout, L. (1997). On the brain response to syntactic anomalies: Manipulations of word position and word class reveal individual differences. *Brain and Language, 59,* 494–522.

Osterhout, L., & Holcomb, P. (1995). Event-related potentials and language comprehension. In M. D. Rugg & M. G. H. Coles (Eds.), Chapter 6 *Electrophysiology of mind: Event-related brain potentials and cognition. (Available on line)*. Oxford: Oxford University Press.

Osterhout, L., & Mobley, L. (1995). Event-related brain potentials elicited by failure to agree. *Journal of Memory and Language, 34,* 739–773.

Osterhout, L., et al. (1997). Brain potentials elicited by words: Word length and frequency predict the latency of an early negativity. *Biological Psychology, 46,* 143–168.

Osterhout, L., et al. (2002). Words in the brain: Lexical determinants of word-induced brain activity. *Journal of Neurolinguistics, 15,* 171–187.

Perea, M., & Pollatsek, A. (1998). The effects of neighborhood frequency in reading and lexical decision. *Journal of Experimental Psychology: Human Perception and Performance, 24*(3), 767–779.

Petitot, J. (1985). *Morphogenèse du sens: 1, Pour un schématisme de la structure.* Paris: PUF, coll. Formes Sémiotiques.

Piotrowski, D. (2009). *Phénoménalité et Objectivité Linguistiques.* Paris: Champion, coll. Bibliothèque de Grammaire et de Linguistique.

Renault, B. (Ed.). (2004). *Imagerie cérébrale fonctionnelle électrique et magnétique.* Paris: Hermès, coll. Sciences Cognitives.

Renault, B., & Garnero, L. (2004). L'imagerie fonctionnelle EEG-MEG: principes et applications. In B. Renault (Ed.), *Imagerie cérébrale fonctionnelle électrique et magnétique.* Paris: Hermès, coll. Sciences Cognitives.

Rossel, S., et al. (2003). The anatomy and time course of semantic priming investigated by fMRI and ERPs. *Neuropsychologia, 41,* 550–564.

Rourke, T., & Holcomb, P. (2002). Electrophysiological evidence for the efficiency of spoken word processing. *Biological Psychology, 60,* 121–150.

Rudell, A. (1992). Rapid stream stimulation and the recognition potential. *Electroencephalogr. Clin. Neurophysiol, 83,* 77–82.

Rumelhart, D., & McClelland, J. (Eds.). (1986). *Parallel distributed processing: Explorations in the microstructure of cognition.* Cambridge: MIT Press.

Sakamoto, T., et al. (2003). An ERP study of sensory mismatch expressions in Japanese. *Brain and Language, 86,* 384–394.

Salmon, L., & Pratt, H. (2002). A comparison of sentence- and discourse-level processing: An ERP study. *Brain and Language, 83,* 367–383.

Schlesewsky, M., et al. (2003). The neurophysiological basis of word order variations in Germanc. *Brain and Language, 86,* 116–128.

Schwartz, M.-F., et al. (2006). A case-series test of the interactive two-step model of lexical access: Evidence from picture naming. *Journal of Memory and Language, 54,* 228–264.

Tamba-Mecz, I. (1991). *La sémantique.* Paris: PUF, coll. *Que sais-je ?,* 655.

Thierry, G., et al. (2003). Electrophysiological comparison of grammatical processing and semantic processing of single spoken nouns. *Cognitive Brain Research, 17,* 535–547.

Titone, D. (1998). Hemispheric differences in context sensitivity during lexical ambiguity resolution. *Brain and Language, 65,* 361–394.

Van Petten, C., & Kutas, M. (1991). Influence of semantic and syntactic context on open and closed-class words. *Memory and Cognition, 19,* 95–112.

Van Petten, C., & Luka, B. (2012). Prediction during language comprehension: benefits, costs, and ERP components. *International Journal of Psychophysiology, 83*(2), 176–190.

Weisbrod, M., et al. (1999). Electrophysiological correlates of direct *versus* indirect semantic priming in normal volunteers. *Cognitive Brain Research, 8,* 289–298.

West, W., et al. (2000). Imaginal, semantic and surface level processing of concrete and abstract words: An electrophysiological investigation. *Journal of Cognitive Neuroscience, 12*(6), 1024–1037.

Yagamata, S., et al. (2000). Event-related evoked potential study of repetition priming to attended and unattended words. *Cognitive Brain Research, 10,* 167–171.

Ziegler, J., Besson, M., et al. (1997). Word, pseudoword and nonword processings: A multitask comparison using event-related brain potentials. *Journal of Cognitive Neuroscience, 9*(6), 758–775.

Chapter 9
Conclusion

To attain an objective truth regarding languages, without breaking with their lived reality, is indeed the greatest ambition of semiolinguistic knowledge—or at least what it should be, in that it would respect the essential interiority of its object without sacrificing anything of empirical rationality. It would thus be a matter of recognizing languages as pertains to both their manifested systematicities and to their functional constraints which legitimize a nomothetic approach, as well as regarding their undisputable existential and praxical scope, because (cf. Chap. 7) it must indeed be acknowledged that language constitutes a world and that speech is gesture.

But this ambition, as reasonable and even as imperious as it may appear, is not without encountering some very serious obstacles. Very simply, firstly, because it demands to assemble gnoseological points of view and forms of intelligibility which are radically antithetical.

Thus, in short, knowledge through concepts concerns experience in what is generic within it, and it does so under the gaze of a transcendental subject *versus* the singular and subjective experience which equally inhabits and reflexively accompanies the realization of an act of speech. Furthermore, maybe we should acknowledge with Dilthey the specificity of a class of phenomena said to pertain to the mind—phenomena which, on the one hand, appear unlinked from the regulating principles of a material categoriality and are therefore irreducible to them, and which, on the other hand, pertain to an eminently subjective experience in which the relation to the object is grasped as a signifying one. In short, it is the Kantian distinction between determinant judgment and reflective judgment which is revived here: On the one hand, we have the concept as governing the synthesis of an empirical diversity and instituting under its sovereign legislation an order of objectivity, and on the other hand, we have the "sensible" concept that delivers the intuition of an organic totality as it thereby manifests its meaning as a principle of unity in what concerns the parts which contribute to its accomplishment in view of its own end (teleology).

© Springer International Publishing AG, part of Springer Nature 2018   283
D. Piotrowski, *Morphogenesis of the Sign*, Lecture Notes in Morphogenesis,
https://doi.org/10.1007/978-3-319-89848-3_9

The magnitude of the gap between objective truth, at least in its mechanistic and causal form, and living reality—organic and signifying—has been amply investigated and discussed. It may have been quite possible indeed to deny this gap, such being the case with the reductionist postures which seek to relate the phenomena of life and expression to the order of the materialities in which they are realized. But this posture has an ideological component, and regardless of the degree of elaboration of the systems of questioning mobilized to such an end (for instance, complexity theories), the reductionist solution operates *in fine* by means of denial and impoverishment—precisely in that it prohibits this specific form of intelligibility, which is nonetheless blatant, which emanates from signifying phenomena and which, dually, open onto them.

That is to say that the junction between the two sides of the epistemic gap must be elaborated following a mode which does not overlay the one onto the other, for example by conceiving a third term serving as articulation, or by elaborating a specific order of categoriality (no longer simply physicalistic). It is thus that the establishment of various bridges will have been attempted between the two sides, all the while preserving their respective regimes of intelligibility—for instance, Dilthey himself: combination between the regimes of explanation (causality) and of comprehension (semantics)[1]; transcendental structuralism, drawing from Goethe, d'Arcy Thompson and, today, J. Petitot; Turing in his attempts to mathematize the emergence of forms…

Such a passage is also one which, in fact, the present study will have laboriously opened, but, to put it as such, by way of a side effect. Because the greater question of the principles and modalities of a connection between the living forms and the categorical forms of language, if it is put forth in this work from the very beginning, it is addressed interspersedly while remaining in the background, so to speak.

More specifically, the path we have chosen is, firstly, that of a reconstruction of semiolinguistics as an empirical science. And while addressing the issues at hand and while uncovering solutions, new problems have emerged so as to direct the investigation along paths which were at first unsuspected. We have thus progressively introduced elements of which the assembling establishes *in fine* a bridge between the living forms of language and its objective reality.

It will therefore not be useless to devote these last pages, which are both conclusive and recapitulative, to putting into perspective the various problematic issues successively addressed as well as the treatments they were given, so as to most clearly reveal their essential unity and gnoseological scope.

Among the main points of entry into this study we have the question of the epistemological status of semiolinguistic knowledge, most particularly in what concerns its claim to a status of empirical science—in short, the question of the possibility of a rationalization of semiolinguistic facts. This question, which may be

---

[1]As emphasized by Mesure (1990, p. 203): "[Dilthey acknowledges] the necessity, in the field of the mind, of articulating the explanational or causalistic approach with a comprehensive or hermeneutical approach."

considered approachable by means of a general philosophical and epistemological examination, as is the case with the approaches evoked above, we have elected to address it on the basis of the epistemic referential which, more or less implicitly, is the object of consensus amongst the natural sciences, that is, the Popperian epistemology of refutation. In this light, the issue of the empirical rationality of semiolinguistic knowledge presents itself as a question regarding the adequation of the theoretical forms of this discipline with the architectural principles of the empirical sciences, as defined by Popper—that is, as the combination of a principle component, coordinating the specific concepts in the empirical field under study, with an "auxiliary" component which serves as an observatory inasmuch as (i) its descriptive forms are in part independent from the concepts defined in the main component, and (ii) it elaborates the space of empirical possibilities (phase space). Thus, the auxiliary component functions as an interface between truly theoretical concepts and empirical reality: It constitutes the device in the terms of which are expressed both the factual observations or experimental results as well as the material configurations of which theory anticipates the realization by means of a deduction by virtue of its own conceptual schemas, so as to establish the possibility for a contradictory confrontation. The auxiliary component thus safeguards from the circle of self-consistency, which closes itself the moment the theoretical concepts introduced to explain a certain empirical set are the same ones describing it.

The issue of the empirical character of semiolinguistic knowledge, in other words, of the positivity of this knowledge, is therefore equivalent to that of the existence of an auxiliary component conjugated with the conceptual devices of the discipline. However, it is indeed necessary to admit that semiolinguistic knowledge has for a long time been developed in the undeclared absence of an auxiliary component (observational) and that it is appropriate, in order to establish it in the capacity to which it aspires, to overcome this shortcoming. Now, two paths may be followed to this end, paths which will entail an epistemological configuration which is as original as it is problematic.

The first path is that of a recourse to a reduced and ancillary form of phenomenological analysis—a form which, in order to resolve the problems that will moreover be induced, will need to be surpassed in view of achieving a phenomenology of broader scope. For the time being, we will simply expect phenomenological analysis to deliver the forms which characterize the appearing of semiolinguistic phenomena, so as to define the observational component which lacks the devices of semiolinguistic knowledge.

Husserlian transcendental phenomenology has been mobilized to this end and following discussion of its descriptive options and of the difficulties which Husserl was required to overcome in his analysis of verbal phenomena, the word-sign found itself *in fine* to be described following the principle of a stratification of the attentional field of consciousness.

More specifically, and essentially, the phenomenological conformation of the word-sign is that of an organic coordination within the attentional field of consciousness, with, on the one hand ("primary level"), the object of a simply perceptual aim, that is, the sign considered as a concrete mark, and, on the other hand ("thematic" level), the object of a significatory aim, that is, a certain content of meaning. The interdependent statuses of the primary object and of the thematic object attribute an unequal importance to the material and semantic faces of the sign: Whereas consciousness "inhabits" the thematic object of signification, it only attributes to the primary object an accessory value, and thus steers away from it in order to fully accomplish itself through the object of meaning which it then thematically targets.

We are thus led to distinguish, from the standpoint of phenomenological characters, (i) the sign as the object of a perceptual aim, that is, as a simple sensible object closed upon itself, (ii) the modalization of this sensible object as it occupies a primary position in the attentional field of consciousness, a position which assigns it a specific phenomenological character, that is, the marking of "indifference", and (iii) this being concomitant to the installation of the thematic object, i.e. the object of a significatory aim, "fully invested" as such, within this same field of consciousness.

This phenomenological conformation of the verbal phenomenon delivers the forms of an "auxiliary" semiolinguistic device, so as to respond to the architectural requirements of the theories of experience and to establish semiolinguistics as an authentic empirical science. But in truth, things are not so simple, given that many internal and external problems arise.

Firstly, concerning the internal problems, the semiotic being of a signifier is only partially restituted in such an analysis—the concrete marking in a position of primary object may indeed, in the manner of a signifier, be turned towards an object of signification, but this object of signification is external to it. The significatory aim which installs this object is, in the Husserlian device, by no means carried and even less determined by a signifier; it does not form a whole with it. And in truth, in the Husserlian construction, the signifier finds itself to be turned not towards a signifier but towards a plane of consciousness (of meaning). The solution, as it was developed (Chap. 4), consists in transcribing within a purely semiotic device the system of attentional strata, specifically by introducing the principle of a generic aim towards meaning which, in that it orients a simply perceptual object towards something other than itself, allocates it its phenomenological character as an "accessory being" (without other information), and which, introducing here a structural line of questioning, finds its particular semantic orientation against the background of relations established between the signifiers.

Thus, Husserlian phenomenology, approached here as an auxiliary resource for knowledge concerning signs and meaning, determines in part the type of theoretical (structural) forms of which it may constitute the vicinal observatory. Following this, being a matter of developing a theoretical device as a main component correlated to Husserlian description, we will naturally turn towards Saussurean structuralism. But before pursuing in this direction, it is necessary to relate the "external difficulties".

Let's suppose a theoretical apparatus (principal component) of which the phenomenological analysis would constitute the "auxiliary" component. Very well. But another auxiliary device should be considered: that provided by the neurosciences. We then find ourselves in a bizarre epistemic conjuncture, where a theoretical apparatus relating a well defined order of objectivity would have two empirical bases to support its validity, and without there being anything to ensure the convergence of the observational assessments delivered by each.

The solution, as we have seen (cf. 1.2.1.5) consists in recognizing a theoretical character to phenomenological analysis, which would then require an "auxiliary" empirical complement, the role of which the neurosciences will need to assume. In this manner, the initially triangular configuration will have been linearized: With respect to the plane of observations, semiolinguistic theory as such rests upon a phenomenology which is itself validated by neurobiological observation. By transitivity, we will expect the semiolinguistic concepts to also be correlated to neurobiological configurations which validate them at this empirical level. From this experimental architecture, we will retain the following which is very particular: The theoretical plane (concepts, relations, and semiolinguistic functions) and the phenomenological plane gain autonomy with respect to the neurobiological empirical foundation. This has the consequence of opening the possibility for observing interactions between the order of the phenomena as such and the order of semiolinguistic functionings (regularities, constraints, transformations...), which are manifested by these phenomena and for which theory specifically accounts for. Indeed, with "classical" configurations, such interaction is out of the question. Thus, for example, a kinematics (spatio-temporal forms) will be constructed independently from the dynamic device (mass and force) supposed to account for the observable paths. Conversely, in our device, the manifested phenomena and functionings are subject to separate observations, and the possibility for recording relations of interdependence between these two planes (of objectivity and of phenomenality) is thereby given.

This observational latitude, offered by the theoretical-empirical construction established here, is neither a useless extension of the "classical" configuration, nor the effect of serendipity. In truth, it is the practical and well adapted response to an ontological complexion, in which phenomenality and objectivity interpenetrate one another, and of which it will allow, as we will see, to found the empirical truth. And it is to this original complexion that the undertaking initiated here in view of different ends will progressively lead us.

Because, having Husserlian phenomenology as an observational component, it remains necessary to have a main component adapted to it, that is, a theoretical apparatus of which the conceptions may be confronted to phenomenal data (as described by Husserl). In order to do this, and following the direction induced by our examination of Husserlian analysis, we have had recourse to Saussurean structuralism which we have established in morphodynamic terms.

It is at this stage of our demonstration's progression that the results which had no place in the elaboration of semiolinguistics as an empirical discipline arose, the reason being that it was possible to establish that the functional architecture of the Saussurean sign, which delivers the forms of semiolinguistic objectivity, has a phenomenological signification—specifically, that the various planes of internal articulation of the Saussurean sign define so many phases of consciousness within which the sign-phenomenon is elaborated.

Thus, what was prepared, even foreseen, at the end of an epistemological examination concerning the forms of empirical validation of knowledge in terms of signs and meaning, found itself to be established by means of a technical analysis of the forms of semiolinguistic structural objectivity. But this result required both an empirical corroboration as well as a detailed explanation.

In what concerns the empirical corroboration, the terrain was therefore prepared, and it was a matter (cf. Chap. 8) of showing that the main neurophysiological correlates of semiolinguistic activity validate both (i) the phenomenological inter-pretation of the morphodynamic device (amplitude of the N400 wave as indexing the traversal of the strata of verbal consciousness, serving thereby as a marker of the deployment of semiolinguistic intentionality), as well as (ii) the functional orga-nization of this same device (LAN and P600 waves as markers of a process of emergence, respectively direct and retroactive, of differential structures within the substance of content). Moreover, the interdependence of the forms of semiolin-guistic objectivity and phenomenality is attested by the various intercorrelations of the "syntactic" waves (LAN and P600) and the "phenomenological" wave (N400).

But if the overlapping of the forms of semiolinguistic phenomenality and objectivity find a sort of guarantee in the theoretical assembling inducing it, and if the elements of empirical attestation are moreover available, its comprehension remains, however, delicate. Because the idea of a phenomenality, hence of a mode of appearing, which would be subject to the forms of a conceptualization, therefore, of a manner of conceiving phenomena, is resolutely paradoxical. It is because phenomena have their own forms proper, because they proceed from specific regimes of constitution, and that it is through them that things such as regularities and constraints are manifested, these demonstrating an underlying order of objec-tivity which is then to be grasped by thought. Thus, phenomena are prior to the thoughts they entail and are therefore not subjectable to them. This is to say that conversely to *data*, phenomena do not enter the circles of self-constituency: Whereas data are amorphous factualities raised to the status of objects of obser-vation under the instruction of regulatory concepts—and when these concepts are those belonging to the theory to be tested, the data thus constructed ever only have an affirmative impact—the phenomena, for their part, are the givens of a sensibility administered by these specific forms which make up perception. Therefore, the mutual envelopment of the forms of the perceived and of the conceived, in spite of the theoretical enhancements brought by Kant's Third Critique, is quite problematical.

Taking a step away from any philosophical and epistemological investigation, it is herein from a specifically and "technically" semiolinguistic angle that this question has been addressed. By doing so, we are probably restricting the scope of the conclusions, but what we lose in terms of extension, we gain in terms of comprehension, and doubly so: We provide ourselves with an insight regarding the object/phenomenon duality at the same time as we resituate the nature and scope of semiolinguistic legality.

Because, concerning this last point, we have seen (cf. 3.2.8) that from the perspective of a transcendental phenomenology, the grammatical laws are *laws of essence* in that they pertain to the very existence of semiolinguistic phenomena. Dually, the judgments of grammaticality have an apodictic character: Their certitude bears the mark of evidence and necessity.

It is precisely such admissibility judgments (in terms of grammaticality, of semanticity…) which linguistic science solicits as empirical foundations, precisely because they constitute an intuitive and ensured basis: a basis which is *intuitive* in that, pertaining to immediate knowledge, admissibility judgments do not require to be demonstrated, and *ensured* by virtue of the nature of the laws of which they relate the respect or violation. Because the grammatical laws which semiolinguistic science has the ambition of uncovering, and in the manner of the laws of essence of Husserlian phenomenology, concern the very existence of the semiolinguistic fact, not in its phenomenal nature, but in its objective being. In this respect, for example, Berrendonner (1983, p. 22) notes that "[the] grammatical is opposed to [the] agrammatical, which is synonymous with 'non-sentence'" and that "an agrammatical sequence is nothing more than a sequence to which has been denied [...] the status of a sentence [stemming] from competence." Also, knowing that competence delimits the field of objectivity in language, an agrammatical sequence is not an object of language. This same conception can be found with Bach (1973, p. 25) who, after recalling that the objective of a linguistic theory lies in that it "characterizes and provides for all sentences of language and only them", observes that "the instruction 'all and only them' is almost a tautology, a bit like the proposition 'an adequate physical theory must account for all physical phenomena' (and not for theological phenomena, etc.)."

This double scope (phenomenal and objective) of the grammatical (and semantic) laws is precisely rendered in the morphodynamic device of the Saussurean sign in that these laws, from the standpoint of their objective content, relate constraints upon the emergence of differential structures of meaning (stabilization paths—cf. 6.3.2), an emergence to which the actualization of the sign-phenomena is suspended.

But in truth, grammatical reality is much more flexible. To the judgments which are supposedly assured regarding the deviance of a verbal assemblage with respect to the regimes (for instance, grammatical) of linguistic objectivity, we must oppose in practice the multiple acceptations of this same assemblage, which then, far from escaping the field of linguistic existence and effectiveness, is shown to attest to or to induce new meanings, new regimes of signification, and, more broadly, new ways of being in language.

The principle of this apparent paradox is well known: It is that the judgment of grammaticality, the recognition of existence *versus* non-existence in language, always proceeds from a point of view regarding the being of language; either, for the simple speaker, regarding the project (expressive, communicational…) assigned to it within a given socio-cultural context, or, for the linguist in his or her reflective attitude, regarding the *a priori* form and nature of its object. Thus, and very radically, for grammaticality, semanticity, or any other form of differential qualification of semiolinguistic data, "it seems impossible to find [...] a source which is not the theory itself" (Berrendonner 1983, p. 26). Therefore, from the moment we leave the perspective of a certain conceptualization, which under a previous perspective may have appeared to be devoid of semiolinguistic constituency, it may always (or almost always) recover its body and consistence when approached by virtue of a different semiolinguistic rationality.

Does this amount to saying that this "solid reference" (Milner 1989) that constitutes the 'possible *versus* impossible' differential in language, and of which various admissibility judgments deliver us the multiple facets, is but unstable grounds? Then, does this imply that any semiolinguistic legality is but an illusion, and that no empirically founded objectivity regarding signs and meaning is therefore accessible? This could indeed be the case, at least if semiolinguistics respected the classical divide between phenomenality and objectivity. But we have seen that in the semiolinguistic field, such duality is overcome through a reciprocal assimilation of its terms—a reciprocal assimilation which remains enigmatic indeed, but which, we will see, enables to ensure the consistency and legitimacy of semiolinguistic knowledge.

To this end, we will transpose into our field of questioning a few of the considerations which M.-P. developed regarding the perspective representation of space and of the objects which populate it. Firstly, M.-P. insistently emphasized that perspective representation is not an identical replication of natural sight: "[I]t is certain that classical perspective is not a law of perceptual behavior. [It is] one of the ways man has invented for projecting before himself the perceived world".[2] Yet, a sort of natural evidence seems to indeed inhabit the world of perspective representation, to a point of "imposing itself"[3] as a form of sensibility. But perspective representation is never but a mode of geometrization of spontaneous vision, a representation which draws from it without however replicating it. This point is essential: Perspective is neither the truth of perceived space, nor is it an arbitrary and unconnected reconstruction. It is simply a geometrical rationalization which space as it is experienced and practiced accepts as a legitimate interpretation. The rules of perspective "form an optional interpretation [of spontaneous vision], although perhaps more probable than others—not because the perceived world contradicts the laws of perspective and imposes others but rather because it does not demand any one in particular and belongs to another order than these rules".[4]

In its principle, the transcription of the world of spontaneous vision into the format of perspective is an operation which muzzles the expressive spontaneity of objects and of their positioning, an expressive spontaneity which constitutes their originary form of appearing. Thus, the conversion operated by perspective representation brings into a homogeneous space and under a common measure the multiple and mutually irreducible signifying values which weave the appearing of a

---

[2]*PW*, p. 51.

[3]"Malraux sometimes speaks as if the *senses* and sense-data had never varied throughout the centuries and as if the classical perspective was imperative so long as it referred to them" (*PW*, p. 51).

[4]*PW*, p. 51.

world of which the things solicit our gaze. Thus, for example, "In spontaneous vision, things rivaled one another for my look and, being anchored in one of them, I felt the solicitation of the others which made them coexist with the first. Thus at every moment I was swimming in the world of things and overrun by a horizon of things to see which could not possibly be seen simultaneously with what I was seeing but *by this very fact* were simultaneous with it. But in perspective I construct a representation in which each thing ceases to demand the whole visual field for itself, makes concessions to the others, and agrees to occupy no more space on the paper than the others leave it".[5] Thus, also, the free and plentiful diversity of things offering itself to be explored through time and following an order which is by no means imposed, is distributed over a single plane of simultaneous existences, one where a cluster of converging lines fully administers a gaze which is then made to be encompassing. It is also the aggressiveness of what is close and the lost character of what is afar that becomes erased, always to the benefit of a (geometric) order fully governing a thus homogeneous and coherent universe, one in which each thing has its place and receives its qualities from an unequivocal system of relations.

But, we should insist, this reconformation of the perceived world, if it is phe-nomenologically "denaturalizing", it is not, however, phenomenologically incon-sistent. By putting natural vision into relation with a geometrical format, a rupture is not made with the full reality of spontaneous vision—simply, we suspend its vital principle in order to only retain one of its possible forms, one of the ways in which it lends itself to be represented, that is, as simultaneously conceived and perceived. It is in this manner that geometrical reason preserves an authentic visual content, at least sufficient content so as to give the illusion of being a replica.

What this short examination of the links between perspective representation and spontaneous vision teaches us is that the latter is not intrinsically reducible to a specific order of determination, but that the phenomenal field (in this case, visual) in which are configured the signifying values of a world (milieu) instituted with regard to the vital exercise of a subject (who thus resides within it), this phenomenal field, therefore, beyond the practical significations which are instituted within, lends itself to various phenomenological reconfigurations which relate the conceptions it induces upon itself. In other words, the originary phenomenal field is likely to manifest (here: to produce as a specific phenomenology) the principles of order or of the regimes of functioning following which it may be conceived—principles and regimes which thus find themselves, to put it as such, to be certified by a specific phenomenological expression.

This problematic configuration is directly transposable to the semiolinguistic field, and it enables to shed light on the enigmatic interdependence of the orders of semiolinguistic phenomenality and objectivity. It is that, to borrow the views of M.-P., the originary "speaking mass" does not deliver itself in the format of a sign-phenomenon and of its possible assemblages. At first, we should recall

---

[5]Ibid., p. 52.

(cf. Chap. 7), there is a diacritical act which posits a figure and its ground, following the mode of a dialectic and dynamic gap—in the sense where the ground here designates the "rustling silence", the still mute signifying mass, attempting to find itself through speech ("[the] invisible ground which composes all possible meanings and which makes possible the genesis of meaning based on diacritical relations between differential signs"[6]) and of which the dually posited figure delivers a sort of resolution.

This speaking mass upon which any speech constitutively takes ground and which always opens onto other speech is then, in the manner of a phenomenal field, the possible locus of a finalized investment. Just as visual perception takes a perspective form when, under the horizon of a culture in which the values of order, of rationality, and of universality progressively impose themselves, spontaneous vision abandons its grip on its vital dimension to the benefit of geometric rationality, likewise, linguistic rationality carried by such or such cultural project (for example, classical language, "conceptual" and rigorous: "whatever is well conceived is clearly said") will instruct the originary speaking mass with its requirements, its principles, and its forms, so as to produce *in fine* an adjoining phenomenality.

It is thus that the judgment of grammaticality, and more generally any admissibility judgment, as a correlative of a certain intention of bringing into practical linguistic existence, finds an authentic phenomenological content which preserves it from arbitrariness, inasmuch as it confers it the characters of evidence and apodicticity.

Thus, and in other words, the rationalizing intention underlying admissibility judgments finds itself to have an empirical signification. The project of semiolinguistic legality, the perspective of an order of systematicity and of constraints pertaining to concrete signs, is translatable towards the plane of semiolinguistic empiricity by the installation of a specific phenomenality. And as perspective representation may have appeared as a direct transcription of spontaneous vision when it is but one of its possible representations, dictated by geometrizing thought, the sign-phenomena, as delivered by Husserlian phenomenology and which the morphodynamics of the Saussurean sign accounts for, are, echoing the idea of a language as a system and the idea of a language disposing of univocal and well determined objects, the empirical manifestations of an originary speaking mass of which the praxical essence is at first situated beyond the field of any conceptualization.

We have shown how this epistemic configuration is accomplished on the plane of theoretical architectures. Admissibility judgments have a double value—objective and phenomenological—in that, on the one hand, as expressions of semiolinguistic legislation, they govern the constitution of sign complexes, and, on the other hand, in that they determine the actualization of differential structures into a

---

[6]Kearney (2013, p. 189).

substance of content, that they govern the production of signifieds, and therefore also, of sign-phenomena.

We may add that the originary speaking mass is of a diacritical texture. We have previously seen that the mute background in which an expression is sought is not a primary material which exists prior to any speech—its substantial crucible, so to speak—but rather the correlative backdrop of an upsurge of speech. It thus clearly appears that the differential, topological, and dynamic forms, which the morpho-dynamics of the Saussurean sign expose, are the analogues of the geometrical forms of perspective representation. In the same way as geometry substitutes the expressivities and live tensions of experienced space with an order of homogeneity and of common measures, likewise, the differential forms of the Saussurean conception, in that they operate in a substance of content postulated as a homogeneous continuum, reduce the diacritical gaps animating speech to relations of discontinuity within a homogeneous substrate.

The differentiality which administers both the order of phenomenality and the order of objectivity thus also appears as the principle linking a semiolinguistic phenomenality having an objective value with an originarily diacritical speaking mass. Thus, a thread will have been extended between living speech and the possible forms of semiolinguistic knowledge.

References

Bach, E. (1973). *Introduction aux grammaires transformationnelles*. Paris: Armand Colin, coll. Linguistique.

Berrendonner, A. (1983). *Cours critique de grammaire générative*. Lyon: Presses Universitaires de Lyon.

Kearney, R. (2013). Écrire la chair: l'expression diacritique chez Merleau-Ponty. *Chiasmi International, 15*, 183–196.

Mesure, S. (1990). *Dilthey et la fondation des sciences historiques*. Paris: PUF.

Milner, J.-C. (1989). *Introduction à une science du langage*. Paris: Le Seuil, coll. Des Travaux.

Bibliography

Benoist, J. (1997). *Phénoménologie, sémantique, ontologie: Husserl et la tradition logique autrichienne*. Paris: PUF, coll. Epiméthée.

Benoist, J. (2001b). *L'idée de phénoménologie*. Paris: Beauchesne, coll. Le grenier à sel.

Benoist, J. (2002). *Entre acte et sens, Recherches sur la théorie phénoménologique de la signification*. Paris: Vrin, coll. Problèmes et Controverses.

Benoist, J. (2004). *De la perception comme langage au langage comme perception. Texte de l'exposé au séminaire Husserl*. Paris: ENS-CREA.

Besnier, J.-M. (2005). *Les théories de la connaissance*. (Vol. 3752). Paris: PUF, coll. Que sais-je?

Besson, M., & Kutas, M. (1997). *Manifestations électriques de l'activité de langage dans le cerveau* (S. Robert & C. Fuchs, Eds., pp. 251–271).

Bitbol, M. (2004). Néo-pragmatisme et incommensurabilité en physique. *Philosophia Scientiae, 8* (1), 203–234.

Bouquet, S. (1992). La sémiotique linguistique de Saussure. *Langages, 107,* 84–95.

Bouquet, S. (1997). *Introduction à la lecture de Saussure*. Paris: Payot, coll. Bibliothèque scientifique.

Bouveresse, R. (1978). *Karl Popper ou le Rationalisme critique*. Paris: Vrin.

Boyer, A. (1994). *Introduction à la lecture de Karl Popper*. Paris: Presses de l'École Normale Supérieure.

Brain and Language (Éditorial). (2003). *Understanding Language*, 86, 1–8.

Chalmers, A. (1987). *Qu'est-ce que la science?*. Paris: Editions La Découverte.

Chomsky, N. (1969). *Structures syntaxiques*. Paris: Le Seuil.

Chwilla, D., & Kolk, H. (2000). Mediated priming in the lexical decision task: Evidence from event-related potentials and reaction time. *Journal of Memory and Language, 42,* 314–341.

Chwilla, D., & Kolk, H. (2003). Event-related potential and reaction time evidence for inhibition between alternative meanings of ambiguous words. *Brain and Language, 86,* 167–192.

CLG/B: de Saussure, F. (1959). *Course in general linguistics* (W. Baskin, Trans.).

Cruse, D. (1986). *Lexical semantics*. Cambridge: Cambridge University Press, coll. Cambridge textbooks in linguistics.

Culioli, A. (1990). *Pour une linguistique de l'énonciation; Opérations et représentations*, T. 1. Paris: Ophrys.

de Saussure, F. (1959). *Course in general linguistics* (W. Baskin, Trans.). New York: Philosophical Library.

de Saussure, F. (1972). *Cours de linguistique générale*. Paris: Payot, coll. Bibliothèque Scientifique.

de Saussure, F. (1974). *Cours de linguistique générale* (Édition critique par R. Engler). Wiesbaden: Harrassowtitz.

de Saussure, F. (2002). *Ecrits de linguistique générale*. Paris: Gallimard, coll. Bibliothèque de philosophie.

© Springer International Publishing AG, part of Springer Nature 2018
D. Piotrowski, *Morphogenesis of the Sign*, Lecture Notes in Morphogenesis,
https://doi.org/10.1007/978-3-319-89848-3

de Saussure, F. (2005). *Edition des notes d'Emile Constantin du Troisième Cours de Linguistique Générale. Cahiers Ferdinand de Saussure* (Vol. 58).

Deleuze, G. (1987). *La philosophie critique de Kant*. Paris: PUF, coll. Le Philosophe.

Ducrot, O. (1967). La commutation en glossématique et en phonologie. *Word, 23*, 101–121.

Eco, U. (1972). *La structure absente*. Paris: Mercure de France.

Eco, U. (1988). *Sémiotique et philosophie du langage*. Paris: PUF, coll. Formes Sémiotiques.

Etard, O., & Tzourio-Mazoyer, N. (2003). *Cerveau et langage*. Paris: Hermès.

Friederici, A., et al. (2001). Syntactic parsing preferences and their on-line revisions: A spatio-temporal analysis of event-related brain potentials. *Cognitive Brain Research, 11*, 305–323.

Friederici, A., Hahne, A., et al. (1998). First-pass versus second-pass parsing processes in a Wernicke's and a Broca's aphasic: Electrophysiological evidence for a double dissociation. *Brain and Language, 62*, 311–341.

Garnero, L. (1998). *Les bases physiques de physiologiques de la Magnétoencéphalographie et de l'Electroencéphalographie*. http://www.labos.upmc.fr/center-meg/media/ecp2001/Meg11.pdf

Garnero, L., et al. (1998). Magnétoencéphalographie, Electroencéphalographie et imagerie cérébrale fonctionnelle. *Annales de l'Institut Pasteur/Actualités, 9*, 215–226.

Gilloux, M. (1989). *L'articulation syntaxe sémantique dans LFG et GPSG. Actes du séminaire* SEMANTICA: *les modèles sémantiques pour le traitement automatique du langage*. Paris.

Goethe, J.-W. (2001a). *Maximes et réflexions*. Paris: Payot et Rivages, coll. Rivages Poche.

Goethe, J.-W. (2001b). *La Métamorphose des plantes*. Paris: Triades.

Hahne, A., & Friederici, A. (2002). Differential task effects on semantic and syntactic processes as revealed by ERPs. *Cognitive Brain Research, 13*, 339–356.

Holcomb, P., & McPherson, B. (1994). Event-related brain potentials reflect semantic priming in an object decision task. *Brain and Cognition, 24*, 259–276.

Hopf, J.-M., Bader, M., et al. (2003). Is human sentence parsing serial or parallel? Evidence from event-related brain potentials. *Cognitive Brain Research, 15*, 165–177.

Husserl, E. (1969). *Recherches logiques (prolégomènes)*. Paris: PUF, coll. Epiméthée.

Husserl, E. (1985). *L'idée de la phénoménologie*. Paris: PUF, coll. Epiméthée.

Husserl, E. (1991). *Recherches logiques (Recherches 1 et 2)*. (Vol. 2, Part 1). Paris: PUF, coll. Epiméthée.

Husserl, E. (1992). *Méditations cartésiennes*. Paris: Vrin, coll. Bibliothèque des Textes Philosophiques.

Husserl, E. (1993a). *Idées directrices pour une phénoménologie*. Paris: Gallimard, coll. Tel.

Husserl, E. (1993b). *Recherches logiques (recherches 3, 4, 5)*. (Vol. 2, Part 2). Paris: PUF, coll. Epiméthée.

Husserl, E. (1995). *Leçons sur la théorie de la signification*. Paris: Vrin, coll. Bibliothèque des textes philosophiques.

Husserl, E. (2000). *Recherches logiques (Recherche 6)*. (Vol. 3). Paris: PUF, coll. Epiméthée.

Husserl, E. (2001b). *Logical investigations: Prolegomena, investigations III, IV, V & VI* (J. N. Findlay, Trans.). London & New-York: Routledge.

Indefrey, P., Hagoort, P., et al. (2001). Syntactic processing in left prefontal cortex is independent of lexical meaning. *NeuroImage, 14*, 546–555.

Jakobson, R. (1963). *Essais de linguistique générale: 1*. Paris: Editions de Minuit, coll. Arguments.

Jakobson, R. (1973). *Essais de linguistique générale : 2. Rapports internes et externes du langage*. Paris: Éditions de Minuit, coll. Arguments.

Jakobson, R. (1976). *Six leçons sur le son et le sens*. Paris: Editions de Minuit, coll. Arguments.

King, J., & Kutas, M. (1998). Neural plasticity in the dynamics of human visual word recognition. *Neuroscience Letters, 244*, 61–64.

Kleiber, G. (1999). *Problèmes de sémantique : la polysémie en questions*. Villeneuve d'Ascq: Presses Universitaires du Septentrion, coll. Sens et Structure.

Kumar, N., & Debruille, J.-B. (2004). Semantics and N400: Insights for schizophrenia. *Journal of Psychiatry and Neuroscience, 29*(2), 89–98.

Kuperberg, G., et al. (2003). Electrophysiological distinctions in processing conceptual relationships within simple sentences. *Cognitive Brain Research, 17,* 117–129.

Lazard, G. (1999). La linguistique est-elle une science? *Bulletin de la Société de linguistique de Paris, 94*(1), 67–112.

Lazard, G. (2006). *La quête des invariants interlangues. La linguistique est-elle une science?* (Vol. 23). Paris: Champion, coll. Bibliothèque de grammaire et de linguistique.

Le Bihan, S. (2006). La conception sémantique des théories scientifiques. *Matière Première, 1,* 215–249.

Levinas, E. (1967). *En découvrant l'existence avec Husserl et Heidegger.* Paris: Vrin.

Lévi-Strauss, C. (1973). *Anthropologie structurale.* (Vol. 2). Paris: Plon, coll. Agora.

Lévi-Strauss, C. (1985). *Anthropologie structurale.* (Vol. 1). Paris: Plon, coll. Agora.

Lopez, A., & Sere De Olmos, A. (1992). *Où en est la linguistique; Entretiens avec des linguistes.* Paris: Didier Erudition.

Maess, B., et al. (1997). *A high density auditory ERP study: The processing of words, pseudowords and non-words* (pp. 42–45). Max Planck Institute Annual Research Report.

Martin, R. (1992). *Pour une logique du sens.* Paris: PUF, coll. Linguistique Nouvelle.

Martin-Loeches, M., et al. (2004). Electrophysiological evidence of an early effect of sentence context in reading. *Biological Psychology, 65,* 265–280.

Merleau-Ponty, M. (1964b). *Signs* (R. C. McCleary, Trans.). Evanston: Northwestern University Press.

Merleau-Ponty, M. (1969). *La Prose du monde.* Paris: Gallimard.

Merleau-Ponty, M. (1973). *The prose of world* (J. O'Neill, Trans.). Evanston: Northwestern University Press.

Merleau-Ponty, M. (1996). *Le primat de la perception et ses conséquences philosophiques.* Paris: Verdier.

Merleau-Ponty, M. (2001). *Phénoménologie de la perception.* Paris: Gallimard, coll. Tel.

Merleau-Ponty, M. (2003a). *Le langage indirect et les voix du silence. In Signes.* Paris: Gallimard, coll. Folio-Essais.

Merleau-Ponty, M. (2003b). *Signes.* Paris: Gallimard, coll. Folio-Essais.

Merleau-Ponty, M. (2012). *Phenomenology of perception* (D. A. Landes, Trans.). London: Routledge.

Osterhout, L., et al. (1997b). Event-related brain potentials and human language. *Trends in Cognitive Sciences, 1*(6), 203–209.

Penny, W., et al. (2002). Event-related brain dynamics. *Trends in Neurosciences, 25*(8), 387–389.

Petitot, J. (1993). Phénoménologie naturalisée et morphodynamique: la fonction cognitive du synthétique a priori. *Intellectica, 17,* 79–126.

Petitot, J. (1996a). Forme. In *Encyclopaedia universalis.* Paris.

Petitot, J. (1996b). Objectivité faible et philosophie transcendantale. In *Journée d'étude sur la philosophie de Bernard D'Espagnat.* Paris: IHPST.

Petitot, J. (2005). *Morphologie et esthétique structurale: de Goethe à Lévi-Strauss.* (Vol. 502). Paris: CREA-CNRS: UMR 7656, coll. Rapports et Documents du CREA.

Pfurtscheller, G., & Lopes da Silva, F. (1999). Event-related EEG/MEG synchronization and desynchronization: Basic principles. *Clinical Neurophysiology, 110,* 1842–1857.

PhP/L: Merleau-Ponty, M. (2012). *Phenomenology of perception* (D. A. Landes, Trans.).

Piotrowski, D. (2012a). Morphodynamique du signe; III—signification phénoménologique. *Cahiers Ferdinand de Saussure,* 65, 103–123.

Piotrowski, D. (2012b). Sur la concrétude du signe: Saussure et Husserl. *Tribune Internationale des Langues Vivantes,* n° spécial Linguistique et phénoménologie du langage, P. Cadiot coord., mai 2012, pp. 30–43.

Piotrowski, D. (2012c). Le pont du signe: Structuralisme et phénoménologie. In A. Bondi (Ed.), *Percezione, semiosi e socialità del senso* (pp. 159–201). Milano: Mimesis Edizioni.

Piotrowski, D. (2013). L'opposition sémiotique/sémantique comme articulation de la conscience verbale. *Versus—Quaderni di Studi Semiotici*, 117, 27–52.

Piotrowski, D., & Visetti, Y.-M. (2014). Connaissance sémiotique et mathématisation: sémiogenèse et explicitation. *Versus—Quaderni di Studi Semiotici*, 118, 141–170.

Piotrowski, D., & Visetti, Y.-M. (2015). Expression diacritique et sémiogenèse. *Metodo—International studies in phenomenology and philosophy*, 3(1), 64–112 (Num. Spécial: Phenomenology & Semiotics).

Piotrowski, D., & Visetti, Y.-M. (2017). The game of complexity and linguistic theorization. In P. Perconti, F. La Mantia, & I. Licata (Eds.), *Language in complexity*. Dordrecht: Springer, coll. Lecture notes in morphogenesis.

Popper, K. (1978). *La logique de la découverte scientifique*. Paris: Payot.

Popper, K. (1985). *Conjectures et Réfutations: la croissance du savoir scientifique*. Paris: Payot, coll. Bibliothèque Scientifique.

Price, C., et al. (2003). Cortical localisation of the visual and auditory word form areas: A reconsideration of the evidence. *Brain and Language, 86*, 273–286.

Prieto, L. (1988). Caractéristique et dimension; essai de définition de la syntaxe. *Cahiers Ferdinand de Saussure, 42*, 25–63.

Prieto, L. (1997). La sémiologie. *Cahiers Ferdinand de Saussure, 50*, 17–20.

PW: Merleau-Ponty, M. (1973). *The prose of world* (J. O'Neill, Trans.).

Pynte, J., et al. (1996). The time-course of metaphor comprehension: An event-related potential study. *Brain and Language, 55*, 293–316.

Rastier, F. (1987). *Sémantique interprétative*. Paris: PUF, coll. Formes sémiotiques.

Rastier, F. (1991). *Sémantique et recherches cognitives*. Paris: PUF, coll. Formes sémiotiques.

RLx/F: Husserl, E. (2001). *Logical investigations [x]* (J. N. Findlay, Trans.).

Robert, S., & Fuchs, C. (Eds.). (1997). *Diversité des langues et représentations cognitives*. Paris: Ophrys, coll. L'Homme dans la langue.

Rosenthal, V., & Visetti, Y.-M. (2003). *Köhler*. Paris: Les Belles Lettres, coll. Figures du savoir.

Rudell, A. P. (1992). Rapid stream stimulation and the recognition potential. *Electroencephalography and Clinical Neurophysiology, 83*, 77–82.

Ruwet, N. (1968). *Introduction à la grammaire générative*. Paris: Plon, coll. Recherches en Sciences Humaines.

Saddy, D., Drenhaus, H., et al. (2004). Processing polarity items: Contrastive licensing costs. *Brain and Language, 90*, 495–502.

Sarti, A., & Piotrowski, D. (2015). Individuation and semiogenesis: An interplay between geometric harmonics and structural morphodynamics. In A. Sarti, F. Montanari, & F. Galofaro (Eds.), *Morphogenesis and individuation*. (pp. 49–73). Dordrecht: Springer, coll. Lecture notes in morphogenesis.

Silva-Pereyra, J., et al. (1999). N400 and lexical decisions: Automatic or controlled processing? *Clinical Neurophysiology, 110*, 813–824.

Steinhauer, K. (2002). Electrophysiological correlates of prosody and punctuation. *Brain and Language, 86*, 142–164.

Stengers, I. (1995). *L'invention des sciences modernes*. Paris: Flammarion, coll. Champs.

Swaab, T., et al. (2003). Understanding words in sentence contexts: The time course of ambiguity resolution. *Brain and Language, 86*, 326–343.

Tartter, V., et al. (2002). Novel metaphors appear anomalous at least momentarily: Evidence from N400. *Brain and Language, 80*, 488–509.

Taylor, C. (1997). *La liberté des modernes*. Paris: PUF.

Thom, R. (1990). *Apologie du logos*. Paris: Hachette.

Van Gelder, T. (Ed.). (1995). *Mind as motion: Explorations in the dynamics of cognition*. Cambridge: MIT Press.

Van Petten, C. (1995). Words and sentences: Event-related brain potential measures. *PsychoPhysiology, 32*, 511–525.

Victorri, B., & Fuchs, C. (1996). *La polysémie: construction dynamique du sens.* Paris: Hermès, coll. Langue, raisonnement, calcul.

Vion-Dury, J., Besson, M., Cermolacce, M., Schön, D., & Piotrowski, D. (2015). Neurophénoménologie du signe linguistique: Apport du modèle Phénoménologique Morphodynamique et Structuraliste (PMS) à la compréhension des mécanismes neuraux sous-tendant la donation de sens. *Intellectica, 2*(64), 123–157.

Vion-Dury, J., & Blanquet, F. (2008). *Pratique de l'EEG: Bases neurophysiologiques; Principes d'interprétation et de prescription.* Paris: Masson.

Visetti, Y.-M., & Cadiot, P. (2006). *Motifs et proverbes. Essai de sémantique proverbiale.* Paris: PUF, coll. Formes Sémiotiques.

Weiss, S., & Mueller, H. (2003). The contribution of EEG coherence to the investigation of language. *Brain and Language, 85,* 325–343.

Wicha, N., et al. (2003). Potato not Pope: Human brain potentials to gender expectation and agreement in Spanish spoken sentences. *Neuroscience Letters, 346,* 165–168.

Wildgen, W. (1982). *Catastrophe theoretic semantics. An elaboration and application of René Thom's theory.* Amsteram: Benjamin.